MEASUREMENT, REALISM AND OBJECTIVITY

AUSTRALASIAN STUDIES IN HISTORY AND PHILOSOPHY OF SCIENCE

General Editor:

R. W. HOME *University of Melbourne*

VOLUME 5

MEASUREMENT, REALISM AND OBJECTIVITY

Essays on Measurement in the Social and Physical Sciences

Edited by

JOHN FORGE

School of Science, Griffith University,
Queensland, Australia

D. REIDEL PUBLISHING COMPANY

A MEMBER OF THE KLUWER  ACADEMIC PUBLISHERS GROUP

DORDRECHT / BOSTON / LANCASTER / TOKYO

Library of Congress Cataloging-in-Publication Data

Measurement, realism, and objectivity.

(Australasian studies in history and philosophy
of science ; v. 5)
  Includes bibliographies and indexes.
  1. Physical measurements. 2. Science—Philosophy.
3. Science—history. 4. Realism. 5. Objectivity.
I. Forge, John, 1946–   . II. Series.
QC39.M396   1987   530.1'6   87–26301

ISBN-13: 978-94-010-8238-9   e-ISBN-13: 978-94-009-3919-6
DOI: 10.1007/978-94-009-3919-6

Published by D. Reidel Publishing Company,
P.O. Box 17, 3300 AA Dordrecht, Holland.

Sold and distributed in the U.S.A. and Canada
by Kluwer Academic Publishers,
101 Philip Drive, Norwell, MA 02061, U.S.A.

In all other countries, sold and distributed
by Kluwer Academic Publishers Group,
P.O. Box 322, 3300 AH Dordrecht, Holland

TABLE OF CONTENTS

FOREWORD

The institutionalization of History and Philosophy of Science as a distinct field of scholarly endeavour began comparatively early — though not always under that name — in the Australasian region. An initial lecturing appointment was made at the University of Melbourne immediately after the Second World War, in 1946, and other appointments followed as the subject underwent an expansion during the 1950s and 1960s similar to that which took place in other parts of the world. Today there are major Departments at the University of Melbourne, the University of New South Wales and the University of Wollongong, and smaller groups active in many other parts of Australia and in New Zealand.

"Australasian Studies in History and Philosophy of Science" aims to provide a distinctive publication outlet for Australian and New Zealand scholars working in the general area of history, philosophy and social studies of science. Each volume comprises a group of essays on a connected theme, edited by an Australian or a New Zealander with special expertise in that particular area. Papers address general issues, however, rather than local ones; parochial topics are avoided. Furthermore, though in each volume a majority of the contributors is from Australia or New Zealand, contributions from elsewhere are by no means ruled out. Quite the reverse, in fact — they are actively encouraged wherever appropriate to the balance of the volume in question.

R. W. HOME
General Editor
*Australasian Studies in History
and Philosophy of Science*

INTRODUCTION

The essays in this volume are concerned more or less directly with the theory and practice of measurement. As a theme or organising principle for the collection, I suggested to contributors that they might explore some of the implications of the theory and practice of measurement for the realist interpretation of science and for the claim that scientific knowledge is objective, or, conversely, to consider some of the ways the realist interpretation and the claim that scientific knowledge is objective might influence our ideas about measurement. In other words one could say that the volume focusses on the area of intersection of the topic of measurement and the topics of realism and objectivity. My aim has been not so much to achieve a consensus on a narrow range of issues but to produce a collection of papers which reflect the interests and concerns of the Australasian community of History and Philosophy of Science. While this aim is not necessarily incompatible with that of achieving consensus, it would, in view of the richness and diversity of the activities of those who belong to the Australasian H.P.S. community, be somewhat surprising if complete agreement were reached. The unity of the present volume is thus supplied more by its overall theme rather than by its substantive conclusions. In the remainder of this introduction I shall say something about each paper, not necessarily taking them in the order which they appear in the volume, and indicate some of the points of agreement between them.

Jim Flynn's "The Ontology of Intelligence" is the first of two papers about psychology, and it deals specifically with intelligence testing in relation to the construct g, or general intelligence, of the Spearman—Jensen theory. On the first page of his paper, Flynn asserts that the concept of intelligence only has explanatory power if it can be measured, and hence the question is raised as to whether IQ tests do actually measure intelligence. This leads him to a careful and detailed discussion of what it is to measure intelligence, and then to consider the conditions under which the measurable quantity or construct g can be interpreted realistically. Flynn sets out four conditions, or assertions as he calls them, under which we would be justified in taking g to be real. These conditions are quite stringent. Not only must g have a physiological substratum, it must play a causal role in the world, which, for Flynn, amounts to its explanatory significance. The conditions for the realist construal of g are stated at roughly the half-way point of the essay. The remainder is taken up with an assessment of the Spearman—Jensen theory in the light of these conditions. Flynn concludes that the theory, though not without its strengths, requires modification in several respects.

The second of the two papers on psychology is Robert McLaughlin's "Freudian Forces". McLaughlin, unlike Flynn, is not concerned with any current theory of psychology, but rather with the historical question of the tenor of Freud's metapsychology. In particular, he is interested in the status of certain hypotheses in which Freud expresses himself in terms of "force", "energy", "resistance" and other such mechanical concepts. In considering how we should interpret such expressions, McLaughlin distinguishes between several different varieties of realism, and identifies realism in contrast to instrumentalism as the variety at issue. The question, then, is whether 'Freudian forces' are to be taken as instruments of prediction, after the manner of the typical nineteenth century interpretations of force in physics, or as real entities. One might suppose that if it were possible to assign values to these entities by well-defined procedures of measurement, then this would be enough for a realist interpretation. McLaughlin denies this, and demonstrates that values can be assigned to so-called intervening variables which by definition are instruments of prediction. As with Flynn, it is the causal explanatory role played by an entity or construct which is said to be the basis of any realist construal. McLaughlin goes further than Flynn in this regard and spells out what it is for something to have a causal explanatory role by making use of Salmon's account of the causal structure of the world.

The essays of Henry Krips and John Norton are concerned with some of the implications for measurement of, respectively, quantum mechanics and general relativity. Both papers contain some technicalities, but in neither case are the conclusions inaccessible to those unfamiliar with the mathematical apparatus employed.

The upshot of Krips' paper "Quantum Measurement and Bell's Theorem" is that we must give up the principle that for any physical system S, quantity Q and time t, there always exists a value $mv(Q, S, t)$ which S would exhibit for Q if only a measurement were performed on

S at *t*. In other words, we must renounce the idea that physical systems exhibit quantities to some definite degree at all times. Krips states that this assumption has been taken as an underlying principle of classical physics. Why must it be given up? In the first section of his paper, Krips proves that the principle is incompatible with standard quantum mechanics in conjunction with the locality postulate which states that there are no faster than light causal influences. This reasoning is typical of discussions that involve Bell's theorem. The conclusion of such a discussion is typically that we are presented with a dilemma or trilemma: either standard quantum mechanics must be renounced or locality or . . . . Most philosophers of science, Krips included, do not consider giving up quantum mechanics or locality. Hence in this case it is the above-mentioned principle that must be relinquished. Krips does not think that this is a great loss since it only amounts to abandoning determinism. The classical metaphysician may not agree.

Those readers familiar with operationism, the doctrine that quantities are nothing more than sets of measuring operations, may be aware that certain remarks made by Einstein in his 1905 paper on special relativity were in part responsible for this doctrine. Bridgman, who defended this view of quantities in his *Logic of Modern Physics*, believed that Einstein's paper exemplified operationism. Norton in "Einstein, the Hole Argument and the Reality of Space" is concerned, not with the special theory of relativity, but with the general theory. In the development of this theory also, Einstein's pronouncements have some significance for our ideas about measurement. Norton quotes passages from Einstein in which he speaks of the requirements of his theory as robbing space and time of 'objective reality' and of the reference system (co-ordinate system) as signifying 'nothing real'. If so, what are we measuring when we make measurements of lengths with measuring rods and time intervals with clocks, and what do co-ordinate systems refer to? Subjective impressions, phenomenal objects or what?

Norton, however, wants to argue against the view that Einstein was in fact working out the implications for space and time of the non-realist position of philosophers like Mach and Schlick. By a careful study of Einstein's rendering of two crucial arguments, the point-coincidence argument and the hole argument, Norton concludes that Einstein denied the substantivalist account of space and time, not the realist account. The substantivalist believes that space and time exist independently of any fields, such as the gravitational field. Norton conjectures that perhaps Einstein was an antisubstantivalist all along, and that his apparently anti-realist remarks of 1913—1916 should be construed as denying the existence of space and time independently of the fields they contain. Whatever Einstein's intentions, his arguments as reconstructed by Norton are telling only against the substantivalist. Measurement, then, is not to be conceived as reporting anything about spatial properties of objects nor the duration of events that are not determined to some extent by the nature of the fields pervading space and time. The passage from Einstein quoted in section 6 of Norton's paper makes this explicit: space has no metrical properties without the gravitational field. With regard to our understanding of Einstein, it would seem that the situation here resembles that to which I referred in connection with operationism: we should look carefully at what Einstein did, not at what he said — which is in fact his own advice to his readers!

There are three papers in the present volume that deal explicitly with questions about the objectivity of statements about measured values. Ian Lowe in "Measurement and Objectivity: Some Problems in Energy Technology" takes the most radical line on the issues. He presents a number of real-life examples in which quite outrageous claims are made on the basis of supposedly objective measurements, such as those concerning the safety of certain methods of disposing of nuclear waste. Lowe rejects the suggestion that we distinguish his examples as belonging to the sphere which Weinberg calls trans-science and hence deny that they are representative of actual scientific practice. He proposes that the lack of objectivity of measurements evident in his examples drawn from the policy sciences infects the natural sciences also. But this is not, I think, a conclusion that would be accepted without qualification by Silvio Funtowicz and Jerry Ravetz, nor would it be accepted by David Oldroyd.

"Qualified Quantities: Towards an Arithmetic of Real Experience" is co-authored by Funtowicz and Ravetz. Ravetz and Funtowicz agree with Lowe in questioning the objectivity of statements that attribute values to quantities, particularly in the context of the policy sciences. They propose a schema which they refer to by the acronym NUSAP, which stands for number, unit, spread, assessment and pedigree, in terms of which to express quantitative information. However, I take it that conformity with the dictates of the schema would produce objective knowledge. Turning to the detail of the schema, the ideas of number and unit are familar; thus "10.5 grams" expresses a number and a unit. The concept of spread is a generalisation of that of error, and is normally expressed by giving an interval of numbers in which the actual value of the quantity is presumed to lie, as in "10.5 ± 0.5 grams". Assessment refers to a judgment about how likely it is that the actual value does in fact lie within the limits set out in the spread category. For instance, a judgment that it is 90% certain that the mass of the object whose mass value is recorded above does indeed lie within the specified interval could be appended to the statement as an assessment thereof.

The first four places of NUSAP are thus commonplace. Not so the fifth. This is, in the authors' words, the most qualifying category and represents the 'border with ignorance' of a statement of quantitative information. In the pedigree, the mode of production of the information is exhibited. One might think of this as making explicit the sense in which the information is theory-laden. However, pedigree is more than this. It is in fact a matrix which shows the type of theoretical structure in which the information is embedded, the type of data from which it is determined together with the degree of consensus which it engenders. It would be interesting to construct the pedigree of the quantitative information made use of in the examples given by Lowe.

Oldroyd's study "Punctuated Equilibrium Theory and Time", to give its short title, also focusses on the objectivity of the assignment of values, but does so with regard to a particular scientific debate. The measurements concern the ages of certain rocks in volcanic deposits in Kenya. These rocks yielded large amounts of fossils which, as it happened, were of considerable significance for a debate in evolutionary theory between those who, like Darwin, were gradualists and believed that species developed relatively slowly and partisans of punctuated equilibrium theory. On the latter theory, periods in which species remain stable are interspersed with rapid punctuations of speciation events. First geophysical estimates of the ages of the rocks suggested that very rapid punctuations had taken place and so seemed to support the punctuated equilibrium theory. This then is the setting of Oldroyd's paper. The detail of the paper is concerned with the controversy that erupted when the first estimates of the rock dating was challenged.

The challenge came from both other geophysicists and from paleontologists. And some of the dissenting geophysicists used different techniques from those employed in the original investigation — all these are described by Oldroyd. But if different values for the ages of these fossil-bearing rocks are attained by reseachers using various apparently reliable methods, what is the real value of the quantity and how is it to be determined? Can one, indeed, say that there is a real value? Oldroyd considers the possibility that there is no such thing as a real value in this case, a possibility that would be congenial to Lowe, but he comes to reject it. In his view, quantities of this sort do in fact have actual values, but social considerations must be acknowledged when it comes to deciding what these values are and who, so to speak, gets the answer right. The criterion for the right answer, Oldroyd thinks, is coherence between measurements made by different methods.

The last four papers in this collection are about the nature of quantities considered from the point of view of the philosophy of science. I should explain that I originally asked Brian Ellis and David Armstrong to write full length papers. Ellis is the author of a well-known monograph on measurement, *Basic Concepts of Measurement*, while Armstrong has developed a theory of universals in several influential publications. It seemed to me that the consequences of Armstrong's theory for measurement would give rise to an account of quantities somewhat different from that which I perceived in Ellis' book. Unfortunately neither Ellis nor Armstrong were able to contribute full length papers, but they did offer to comment on any contributions that dealt with their previous work. Chris Swoyer has very ably taken up Armstrong's cause, as it were, while I myself have tried to defend a position which I took to resemble that of Ellis. Hence the small 'internal symposium' on quantities. Things have not, however, turned out as I expected. It appears that Ellis has changed sides and now believes that quantities need to be indentified with certain properties and relations conceived as universals. It was with some surprise that I read his comments, but I hope they will be of interest to those who know his views as expressed in *Basic Concepts*.

Swoyer's essay "The Metaphysics of Measurement" does more than just defend a particular account of quantities. His is a fully-fledged theory of measurement, complete with axioms, formal language and model-theoretic results, all very thoroughly presented. My own paper (which I have been tempted to re-name "Beyond Ellis' Theory of Quantities") is only an attempt to shore up the view that quantities are sequences of ordered classes of physical objects. The issue between myself and the other members of the 'symposium' is whether we do indeed need universals in accounting for the measurement of quantities. If we do not really need universals (properties), then I assume that neither Armstrong nor Ellis nor Swoyer would recommend that we introduce them just for the sake of it, for no one wants to multiply entities for no good reason. Now Armstrong and Ellis both object to the extensionalist position that quantities are ordered classes, which I take it is the only viable alternative to the quantities-as-properties viewpoint, on the ground that it is unable to distinguish those orderings that are natural and hence which constitute quantities and those that are artificial. The response that natural orders are linked to one another in laws of nature, which I suggest in my paper, does not appeal to Ellis, who states that there are many terms which figure in statements of laws that do not designate quantities. As a rejoinder to this I would say that it is correspondences between ordered classes at the level of laws of nature, the ontic level, that is relevant here, not the terms appearing in numerical laws. A task which then confronts the extensionalist is to give the semantics of numerical laws with reference to laws of nature conceived as correspondences between ordered classes. I think that this can be done, and hence that natural orders can be differentiated from those that are artificial.

Swoyer has an objection to the extentionalist programme which is more difficult to deal with. It is that the closure of the concatenation operation ∘ cannot be guaranteed if quantities are ordered classes of physical objects. To satisfy the usual axioms for the measurement of an extensive quantity q, it must be assumed that there exists a composite or concatenated object equal in q to each pair of objects having q. So if a and b have mass then it must be assumed that there exists a composite $a \circ b$ of mass equal to that of the mass of a plus the mass of b. If a and b are universals, and not physical objects, then we can posit enough of them to guarantee that the composite exists in every case and hence guarantee that the axioms of extensive measurement are satisfied. Perhaps this may ultimately force the extensionalist to reconsider his or her position. I leave it to the reader to judge.

ACKNOWLEDGEMENTS

I am very grateful to the following persons for their help: the general editor of this series, Rod Home; Randall Albury, Ron Johnston, Alan Musgrave and Jack Smart of the editorial board of the series; John Earman, John Maze, Ian McDougall, John Stachel and Phillip Staines for their assistance in refereeing; Alan Walker for the index, and Mrs J. C. Kuipers of D. Reidel Publishing Company.

JOHN FORGE

JAMES R. FLYNN

THE ONTOLOGY OF INTELLIGENCE

The purpose of this paper is to clarify the existential or ontological status of intelligence and this means an analysis of both the primitive concept of intelligence found in everyday life and scientific constructs. Scientific constructs can be understood only in the context of the theories that have generated them and I have chosen to concentrate on the Spearman—Jensen theory of intelligence and the psychometric construct called g or general intelligence. The first half of the paper will argue that psychometric g has shown considerable promise as a scientific construct; the second half will show that its successes have been accompanied by significant failures and argue that the Spearman—Jensen theory must be revised and transcended.

PSYCHOMETRIC g AND ITS CRITICS

The selection of the Spearman—Jensen theory reflects the personal view that A. R. Jensen has done most to give the concept of intelligence explanatory power. The presentation of the theory will cover the following topics: the relationship between the primitive concept of intelligence and IQ tests; the derivation of g from certain performance trends on IQ tests; the critical debate about g; the ontology and explanatory power of g. Throughout I will attempt to defend Jensen against certain of his critics, on the grounds that the theory has often been unfairly attacked and its merits unacknowledged.

Intelligence and IQ Tests

Measuring intelligence is a prerequisite to giving it explanatory power which immediately poses the question of whether IQ tests measure intelligence. Early on Jensen (1972, pp. 75—77) asserted that intelligence was by definition what IQ tests measure. Block and Dworkin (1977, p. 415) responded with a critique of this kind of crude operationalism. Any attempt to define something in terms of the readings given by a measuring instrument leads to absurdities. Defining tem-

1

John Forge (Ed.), Measurement, Realism and Objectivity, 1—40.

perature in terms of what thermometers measure denies the possibility of defective thermometers or the invention of a new instrument better than any now existent. Defining intelligence as what IQ tests measure would leave us unable to say any IQ test was better than another. Actually Jensen's so-called definition consists of a single sentence and was not really meant to define anything. It was a badly phrased attempt to introduce the construct g which is derived from performance trends on IQ tests.

The need for a concept of intelligence independent of IQ tests suggest an analysis of the primitive concept found in everyday life. Jensen (1979, pp. 80—81) argues that it arises in a comparative context. A Robinson Crusoe would become aware that he remembered things and learned things even in total isolation but would realize he was quicker than others at learning things only with a companion. Thus, we have the concept of intelligence as a mental ability distinct from memory and learning and indeed speak of people as intelligent even though they may be forgetful or ignorant. Perhaps this is why the concept has such a long history; that is, from ancient India to ancient Greece, from St. Thomas to Binet and Spearman, the notion persists that some people have 'better minds' than others and that this has to do with abstract problem-solving, induction and deduction, transfer of learning from one situation to another, the perception of relationships.

The later definitions of intelligence are not much improvement on the earlier ones and therefore the obvious next step is to move from the pre-scientific to a scientific concept, a measurable variable with explanatory power. However, this does not mean that the two concepts can proceed in complete isolation from one another. The primitive concept of intelligence describes a certain role that scientific concepts attempt to play: they are supposed to measure a mental ability with great importance for the life-histories of individuals and groups; whether someone is intelligent should tell us something about their academic and occupational achievements and groups that have a high average intelligence ought to show a high standard of cultural attainment. Moreover, if psychology discovered a potent mental ability distinct from memory and learning, and that ability happened to bear little resemblance to the primitive concept, it would be important to say so. People in general should not be allowed to think that those who score well or poorly on IQ tests are 'bright' or 'dull' if what the tests measure has little to do with the ordinary meaning of those words.

The case that IQ tests measure something close to the primitive concept of intelligence attempts to match test scores with popular assessments of intelligence and popular expectations about intelligent behaviour. For example, correlations between teacher rankings of pupils for intelligence and their tested IQs average at about 0.70; the meaning of a value like 0.70 will be explained shortly but for now, it should be taken as moderately significant. More impressive, the reasons for discrepancies between teachers and tests appear to favor the tests. Teachers tend to rank girls over boys, presumably influenced by their greater docility and application, they tend to rank extroverts over introverts, and they usually ignore age. The tests allow for age differences, that is, a 10-year-old who scores as well as an 11-year-old is given a higher IQ. The notion that intelligence increases with age during childhood is certainly not counter-intuitive; parents often cite precocious behaviour as evidence that a young child is bright (Jensen, 1980a, pp. 173—174).

The strongest piece of evidence comes from Terman's famous study of high IQ subjects (Terman and Oden, 1959). In 1921, he selected a sample of 1528 children as having IQs above 140 and their life histories are a good match for popular expectations about the real world achievements of the highly intelligent. As children they read much, had wide interests, and seven out of eight were ahead of their age group at school while none had been held back. As adults an extraordinary number had earned degrees, entered professions, achieved high positions, published books and articles, and had biographical citations in leading reference works. Children with IQs above 150 sometimes show quite remarkable ability and by the age of 12 or 13 are capable of excelling in university courses. Jensen points out that low IQ also tends to tally with what popular expectations would dictate. He reports an interview with a subject whose greatest interest was baseball (Jensen, 1981, p. 65). Despite regular attendance and watching telecasts, the low-IQ subject did not know the number of players, the names of the positions, or most of the rules.

High IQ is generally a prerequisite for high achievement in mathematics, science, and other areas dependent on academic learning, but it is not a sufficient condition. Aside from the obvious need for non-intellectual traits such as motivation, persistence, good health, there is clearly a special talent for things like mathematics and music, a creative imagination that IQ tests cannot identify. And outside academic areas, people can excel in commercial dealing, social skills, acting, and design without particularly high IQs. Correlation coefficients add confirmation: they are moderately high between IQ and standardized scholastic achievement tests, or between IQ and school grades, but quite low with things like income and job performance in most occupations (Block and Dworkin, 1977, pp. 439—444). This has caused Bane and Jencks (1977, p. 326) to conclude that IQ tests measure only one rather limited variety of intelligence. Whether one considers this kind of intelligence, call it academic intelligence, of limited importance depends on what one thinks of academic pursuits. I think it important to have people who can write and read serious works of philosophy, history, literature, mathematics, and science, even if the kind of intelligence they possess has little market value and says little about their social skills.

The contention that IQ tests measure something close to the primitive concept of intelligence has surrounded that concept with a number of limitations. However, most of these have analogues in popular opinion; that is, 'IQ' could be replaced by the word 'intelligence' in the last paragraph without doing much violence to ordinary language. We all know 'bright' people, particularly people with academic ability, who are poorly motivated, hopeless with numbers, socially obtuse, practically inept, and also people accomplished in the performing arts and organizational roles who do not seem particularly intelligent. Economos (1980, p. 342) gives perhaps the best summary of the case in favour of IQ tests: it proposes "that people who score poorly on these tests will almost always find it harder, for example, to follow advanced mathematical reasoning, or quickly to extract the meaning from a scholarly paragraph, than will people who score well on them"; but it does not say that these are the only or the most desirable or the most advantageous of human abilities.

The Construct Called g

Assume we could rank representative samples of a population on a number of non-team sports, golf, bowling, archery, shooting, the ten events of the decathalon with a long-distance event, say the half-marathon, thrown in to make a total of 15. It becomes apparent that the same people tend to be good or bad at most of the sports in question. That is, the variation in the percentile scores of one person across the 15 sports tends to be less than the score variance of 15 randomly selected people on any one sport. Therefore, we speak of general athletic performance, meaning for the moment nothing more than to describe the tendency towards consistent quality of performance across sports. This is of course something that impresses us in real life: people who beat us at practically everything, no matter how hard we try, people of whom we say, he or she is just a better athlete than I am.

Assume we invented a mathematical technique for measuring the tendency towards consistent performance across these 15 events or sports and, moreover, invented an artificial event, no more complicated than real events, on which a person's level of performance was an index of their general level of performance. Indeed, it predicts general athletic performance better than what might seem to be the obvious method: taking a sample from various sports, that is, a test that included one hole of golf, one sprint, one weight event, and so forth. Having operationalized the tendency towards consistent performance, we would be justified in calling it an ability. After all, a person's performance on this artificial event would tell us something about their potential performance across sports in general, including those they had never tried. We would therefore call it a measure of general athletic ability. This would not mean that it measured something unitary in the sense of being one simple skill. It would measure a functionally interrelated set of skills no more and no less complex than those used in a particular sport, say golf or the pole vault. At present, nothing would be known about the physiological prerequisites of good performance, or the causes, or how to enhance performance.

The trend towards consistency of performance is also manifest across a variety of mental tests and this despite great diversity of content. No matter whether tests feature vocabulary, general information, verbal oddities, scrambled sentences, logical reasoning, inferential conclusions, number series, pictorial oddities, spatial analogies, figure generalization, or completion of matrices, the same people tend to be good or bad at them. The construct called g is essentially a mathematical device that measures this tendency towards consistent performance across tests. The great psychologist Charles Spearman (1904) invented factor analysis and derived the first g, although initially from things that today would hardly qualify as mental tests. Rather than presenting a technical account of factor analysis, I will try to convey the logic behind how g is calculated. The following assumes that all tests have standard scores and show a bivariate or multivariate normal distribution; and that the population mean and standard deviation have been set at 100 and 15 respectively.

Another way of describing the above tendency is to say that all IQ tests have a positive correlation with one another and therefore, the best starting point is to explain what a correlation coefficient is. A correlation coefficient measures the slope of the regression line, for example, take two tests with a positive correlation of 0.70. If we used the first test to select out an elite group with a mean IQ of 110, that group would also be an elite on the second test with a mean IQ of 107. In other words, the correlation coefficient told us that a group 10 points above average on the first test would keep 70 per cent of their advantage on the second, that is, 7 points. If the tests had a perfect correlation of 1.00, they would keep all 10 points and remain steady at an IQ of 110. If the correlation was nil or 0.00, they would regress all the way to the population average and have an IQ of 100. Clearly the correlation coefficient is a measure of the tendency of those who do well on one test to excel on another.

Now g is essentially nothing more than a super correlation coefficient that measures the slope of the regression line when you have many tests, say 15 rather than just two. The g loading for any one test of the 15 tells how much an elite on that test will regress in terms of their overall performance on all 15 tests taken collectively. If the first test had a g loading of 0.70, an elite group with a mean IQ of 110 would score 107 on their overall performance; if the second test had 0.60, a similar elite would score 106; and so forth. The value for g itself represents the correlation between a test whose loading is typical of the 15 tests and overall performance on all 15 tests collectively. Jensen (undated, p. 19) has tried to approximate the value one would get for g from a total of more than 70 mental tests, deriving an estimate of 0.65. In other words, given a collection of all mental tests in common use, the typical regression would be from a performance of 110 on a particular test to a performance of 106.5 on all tests taken together. Clearly g measures a significant tendency of those who do well on one test to excel on all mental tests.

Indeed, the scores of one person across 15 tests tend to vary less than the scores of 15 randomly selected people on any one test. Jensen's estimate of the percentage of variance accounted for by g is 42.7 percent: just as taking the square root of variance accounted for will give the correlation coefficient, so the square root of 0.427 was used to derive the value of 0.65 for *g* noted above. We can also calculate the variance *not* accounted for by *g* as follows: 1.000 — 0.427 = 0.573; and 0.573 × 225 (total variance) = 129. So variance of performance across tests is only 129 compared to variance within a particular test of 225. Since the standard deviation is the square root of variance, the across test *SD* is 11.35 compared to a within test *SD* of 15. What this means in practice is this: take a group with a mean IQ of 115 on a particular mental test, one with a typical *g* loading. This performance would put them at the 84th percentile of the whole population on that test. If they then took a battery of 15 mental tests, they would tend to score below average or below the 50th percentile on only one or two tests. An elite on a particular test shows a strong tendency to remain an elite on other tests. Therefore we speak of general mental test performance, meaning for the moment nothing more than to describe the tendency towards consistent quality of performance across all mental tests.

During the 1930s, L. Penrose and J. C. Raven invented a new test, Raven's Progressive Matrices or Ravens for short, in an effort to maximize *g* loading. They largely succeeded, that is, a group that scores above or below the population mean on Ravens will hardly regress towards the mean at all in terms of their overall performance on a diverse collection of tests. Indeed, it predicts general performance, say on a collection of 15 tests, better than what might seem to be the obvious method: making up a composite test by random selection of items from each of the 15 tests! The fact that Ravens operationalizes the tendency towards consistent performance across mental tests raises this question: can we now move from saying *g* measures general mental test *performance* to saying it measures general mental test *ability*? I would answer in the affirmative, not merely because of what Ravens predicts but because of what it operationalizes.

The mere fact something predicts general mental test performance would not in itself encourage us to say that it measures an ability. Socioeconomic status also predicts test performance and it is a measure of things like income. It is the content of Ravens that is so impressive. Each item presents a pattern in which there is a gap, followed by six alternatives each of which would fit that gap like fitting a piece into a jig-saw puzzle. The subject must choose the piece with the correct markings, those markings which alone would render the total pattern complete. The patterns can be made very complex, with few or many elements, some shaded and some unshaded, some derived from others by rotation, flipping over, or as mirror images. In other words, Ravens consists entirely of perceptual analogy items based on geometrical patterns, items that require making comparisons, seeing similarities and differences, reasoning by analogy, and perceiving a consistent pattern amid irrelevant complexity. This fact plus the fact Ravens is such an excellent predictor of general mental test performance, the two in combination, make a *prima facie* case for the following: those who possess a relatively small number of functionally interrelated mental skills have a significant advantage on a huge diversity of mental test items. Moreover, these items have little apparent functional relationship with one another; once again they range from vocabulary to number series, general information to logical reasoning, verbal comprehension to figure generalization, verbal oddities to coding digits.

The fact that people who possess a set of functionally interrelated skills have an advantage over other people on a wide range of apparently unrelated tasks is all I mean when using the word 'ability'. Similarly we speak of someone with good timing, fast reflexes, and good coordination as possessing athletic ability across a wide range of sports that apparently are unrelated in terms of tasks. Therefore I am prepared to speak of general mental test ability. Thus far, that ability refers only to good performance on mental tests and possible links to the real world, both causal and consequential, remain to be discussed.

Cattell (1963) has made a start towards explaining how the ability in question could cause good performance on the tasks in question. He noted that mental tests with heavy *g* loadings divide themselves into two very different sorts: tests that have little informational content but demand the ability to see relationships between relatively simple elements, such as Ravens, which he called tests of *fluid g*; and tests that emphasize already acquired knowledge, such as vocabulary, general information, and arithmetic, which he called tests of *crystalized g*. The hypothesis is that Ravens and these other tests usually correlate so highly because a person with the ability Ravens measures will, given normal cultural opportunities, be the sort of person who acquires a large vocabulary, wide general information, and so forth. Indeed, when subjects from different backgrounds are tested, it is sometimes found that they can score equally well on Ravens despite differential vocabularies and stores of information, that is, they have the same level of ability but have applied it to different cultural raw material.

The growth curves of fluid and crystalized g are consistent with this hypothesis. Ravens performance improves throughout childhood, holds at a stable maximum between ages 18 and 25, then begins a gradual decline which accelerates after 60. Performance on vocabulary and general information tests can increase throughout life right up to 60 or 70 years of age and decline thereafter only gradually if at all. This certainly does not contradict observations from everyday life: that basic mental ability or agility declines after youth but that accumulated knowledge and its attendent skills can increase until old age.

Controversy About g

There has been much critical debate about g and while some of this is best postponed, certain objections are so fundamental that they must be addressed at once.

First, since g is essentially a correlation coefficient calculated from performance on a collection of mental tests, it will differ from one collection to another. Even were it possible to calculate it for all mental tests in common use, this collection alters over time as tests are added or discarded. Thurstone (1940, p. 208) asked how g could have any psychological significance given that it measures performance on "an arbitrary collection of tests anyone happens to put together", and that it alters radically if one first includes only spatial tests, then only verbal ones, then only numerical tests, and so forth. Jensen (1980a, pp. 233—234) answers that based on very high correlations, the g of one set of tests is very much the same as the g of another, just so long as the sets are both large and diverse. Each should include 10 or more tests and sample a wide range of information, tasks, and materials inclusive of verbal, figural, and numerical items. He stresses that the best contemporary IQ tests, such as the Wechsler, have 10 or more subtests and that these easily possess the required diversity.

The presupposition that lends strength to this answer is that such IQ tests measure a g which is a good bet to win the prize dangled before our eyes by the primitive concept of intelligence: that they measure a mental ability which may well have explanatory power when applied to the life-histories of individuals or groups. This can only be established by testing g repeatedly in a programme of scientific research. There is no problem about ensuring that g will remain the same throughout:

Ravens can be used as the marker test of fluid g and the Wechsler tests as markers when a mix of both fluid and crystalized g is required.

However, the above objection makes a point Jensen would not dispute: the quality of g will be a function of the quality of the tests from which it is derived. If we had only tests of school learning, general performance on these would give deceptive expectations about able people only recently exposed to good education. If we had only memory tests, general performance would not differentiate black and white Americans in the same way current IQ tests do. And if, for some reason, the best of present-day IQ tests were found wanting, then we would need both new tests and a new psychometric g.

A second objection: if g is test relative, are not the tests themselves culturally relative? This is undoubtedly to some degree the case: a modern industrial society dependent on science and technology emphasizes abstract problem-solving; a pre-industrial society may emphasize the memory needed to absorb oral tradition or the information-processing skills needed to survive in the bush. But it would be much to the credit of psychology if it could measure even those mental abilities needed in industrial society. Moreover, cultural relativism offers little to those groups some psychologists believe to be genetically inferior in terms of g. American blacks want to succeed in American society and if they are genetically handicapped for that, that is what counts. It is no solace to be told that they would have a genetic advantage were they still living in the environment of their African ancestors. As Lewontin (1977, p. 81) has pointed out, the argument of specific cultural origins of IQ testing cuts both ways.

A third objection is based on factor analysis, that is, the mathematical technique by which g is calculated. Gould (1981, pp. 310 and 314) emphasizes that factor analysis of performance on a collection of tests can be used to extract either the general factor called g or a multiplicity of other factors. He argues that since mathematics gives no guide, whether one calculates g or other factors becomes a matter of personal preference or bias. However, he grants that the scientific status of various factors could be tested by real-world data, such as evidence of a link to biological entities. Unless biology has a monopoly on the vindication of g, Gould's admission that scientific considerations can give a rational guide to our preferences about factors robs his objection of its force.

Factor analysis and performance trends on IQ tests do leave us with choices but these are not mutually exclusive. There is a strong tendency towards consistency of performance across IQ tests in general, but there is of course an even stronger consistency within subgroups of similar tests, that is, within the subgroup of purely verbal tests, or numerical tests, or spatial tests. Therefore, if a collection of 15 tests divides into three distinct subgroups, one can either calculate a general factor as a measure of the positive correlation of all the tests with one another, or a number of specific factors as a measure of the correlations within each kind of test, in this case a verbal factor, a numerical factor, and a spatial factor. In order to extract maximum scientific value from performance trends on IQ tests, it seems sensible to do both. The verbal factor may be the best predictor of marks in English courses and g the best predictor of results if students decide to broaden their education. As for seeking biological or physiological correlates, both general and specific factors might prove valuable.

Returning to sports, focusing on those who do well on a specific event, like distance running, has turned up something interesting. There is a high correlation between good performance and pulse rate, particularly how long it takes the pulse to return to normal after it has reached its maximum rate during vigorous exercise. This suggests investigating the efficiency of the cardio-vascular system as a factor in endurance. On the other hand, focusing on those who do well across a variety of sports and calculating a general factor could also prove valuable. The sporting dominance of American blacks has been so striking as to prompt investigation of physiological correlates: the major one discovered thus far is a faster conduction time from nerve to muscle as measured by electrodes. That is, there may be a correlation between general good performance in sport and the speed of stimulus-response or the reflex arc. This suggests investigating certain areas of body chemistry, such as the electrolyte or salt balance, and the roles of sweating and diet. The data is tentative and the status of general factors is not high at present in exercise physiology. But the point is that no-one can know in advance what factors will prove of scientific interest. Those who wish to investigate psychometric g have every right to invest their time in so doing.

The critics of a general factor often list the things it cannot do: ranking people by g does not tell us who will understand what ails a troubled friend; ranking students by g does little to diagnose their specific academic problems; ranking applicants by g is not a proper method of selecting a police force. As McClelland (1977, p. 58) points out, criterion sampling should be used to test applicants for most jobs. It creates tests based on analysis of what police actually do and should do, the tasks they perform, the vocabulary they use to communicate with the public, the fact they should not be racially biased. But it was a mistake to ever suppose a measure of general mental ability could do all these things. The primitive concept of intelligence hardly suggests that intelligence rankings would provide a detailed diagnosis of academic problems or an adequate criterion for selecting a police force. Psychometric g need not do everything in order to have its own unique scientific value.

A fourth objection contends that the positive correlations between IQ tests and between test items may appear synthetic but they are operationally analytic. As Block and Dworkin (1977, pp. 463—464) point out, within each test, both items and subtests are discarded unless they have a positive correlation with the test as a whole; and when a new test appears, it is simply vetoed unless it has a positive correlation with those tests already accepted. If this is what creates the consistency of performance across the diverse tests and items that g measures, the general factor is an artifact of how psychologists validate IQ tests. They rig the tests to produce g and then hail it as significant.

Jensen (undated, pp. 4—5) replies that when a variety of items selected only because they have some plausible claim to test mental abilities are given to representative samples of a population, there are many positive correlations, very few negative ones, and the positive correlations are always higher. As sample size increases, the negative correlations tend to disappear, suggesting that they are largely due to measurement error. He challenges anyone to take negatively correlated items and try to construct a mental test. Jensen (1982, p. 132) notes that Thurstone spent years trying to design tests that would not correlate, tests that would measure only verbal, numerical, or spatial factors, and that he failed: when administered as a collection, these tests so intercorrelate that a second-order g emerges which accounts for twice the variance of the special factors combined. Finally, Jensen (1980a, p. 230) isolates tests which have low g loadings, that is, correlate least with other mental tests. These turn out to be things like counting dots, making dots, crossing out designated letters or numbers, tests that hardly seem to be mental tests at all. They also have low external validity, that is, show little correlation with people's eventual academic achievement or socioeconomic status.

However, Block and Dworkin (1977, pp. 444—447) argue that the correlations with scholastic and occupational success have also been built into the tests. They note that Alfred Binet, the inventor of the first really useful mental test, used teachers' judgments of intelligence as a guide to selecting test items; and that since then, test after test has been discarded because it did not yield the proper correlations which means that extant tests are simply those which have jumped the correlational hurdles. They add that educational attainment is an important avenue towards occupational success, so rigging the tests in favour of the former automatically produced a correlation with the latter. In reply, it is mistaken to think that when a certain criterion has been used to screen test items, this provides a sufficient explanation of the correlation between the overall test score and that criterion. Binet did use teachers' assessments, that is, he observed the children described as 'bright' and selected his items from among the everyday tasks they could perform. But he was soon able to predict which children would fail in school better than the teachers themselves. As we have seen, present-day IQ tests do not merely reflect teachers' assessments but improve on them: they correct for irrelevant factors like docility, extroversion, and age-advantage.

Once a test isolates the mental processes that lie behind a correlation the magnitude of that correlation can deviate from the built-in value. Ravens is a good example of this: as a test of fluid g, it measures mainly mental ability only and thus has a lower than usual correlation with academic achievement; tests of crystalized g measure mental ability plus effort and thus have higher correlations.

Jensen (1979, pp. 83—93; 1980a, pp. 229—232 and 247—248; 1980b, pp. 365—366; 1982, p. 134) defends both g and IQ tests by way of a whole range of correlations that no-one has ever built into the tests. Moving from items and tests with high g loadings to those with low g loadings, the contrast is clear: a move from problem-solving to mechanical skills, from cognitive complexity to simplicity, from manipulation of materials to simple feed-back. It seems significant that items lose their g loading if not enough time is given for reflection, or if items become so complex subjects fall back on trial and error. Everyday life provides an example of a correlation with cognitive complexity: the task of making jelly-rolls is more g loaded than the simpler one of scrambling eggs. Learned tasks in general are more g loaded when they require conscious mental effort rather than mere memorization, when they are hierarchical (later learning is dependent on earlier learning), when they must be transferred out of the original learning context, when proofs rather than theorems are learned, and so forth.

Binet did not use intelligence assessments from other cultures to design his tests. Jensen grants that other cultures may value certain skills more than we do, for example, hunters may put a higher value on speed and motor coordination than on abstract problem-solving. However, the individuals they call intelligent tend to exhibit g. The Kalahari Bushmen of Africa call some of their tribe the 'clever ones' and these tend to score better than average on performance IQ tests. In fact, g has some application even across species. Human children can be given certain of the performance tests designed for animals and this reveals that g-loaded tasks put apes ahead of monkeys, monkeys ahead of dogs, dogs ahead of chickens. Gould (1981, p. 318) expresses horror: speaking as a paleontologist, he accuses Jensen of ranking all animal species, each of which possesses its own solution to its own environmental niche, according to human standards. Surely that is the whole point: human beings rank animals using a distinctively human concept of intelligence, the primitive concept found in everyday life, and these rankings correlate with g. Moreover, from time immemorial, human beings have associated low intelligence with certain genetic abnormalities and with inbreeding or the practice of mating with close relatives. There is a negative correlation between g and inbreeding, as measured by the offspring of cousin marriages in Japan, and carriers of the *recessive* gene for PKU (cretinism) average 10 IQ points below their noncarrier full siblings. Vogel (1980, p. 358) adds that IQ tests have proved sensitive to quite subtle effects from known genetic conditions, a spatial defect from Turner's syndrome and a slight verbal weakness from the recessive PKU gene.

In sum, Jensen believes that these correlations put the link between g and the primitive concept of intelligence beyond reasonable doubt. I believe he has earned the right to back g as a potentially useful scientific construct — and to hope that the evidence will vindicate his choice.

The Ontology of g

The case for the reality of g rests on four assertions: (1) That it describes a phenomenon, namely, the tendency towards consistent performance across mental tests; (2) That it measures an ability, that is, when g is operationalized it reveals that people have an advantage across mental tests because they possess a limited set of interrelated skills; (3) That it plays a causal role in the real world, that is, has explanatory power concerning the life-histories of individuals and groups; (4) That it has a physiological substratum, that is, correlates with certain elementary cognitive tasks and with evoked electrical potentials of the cerebral cortex.

The first and second assertions have already been established and the second adds much to the first, indeed, the fact that Ravens has operationalized g is crucial. Those who can solve perceptual analogy items possess g and this forecloses the possibility that so-called general ability is no more than an average of performance on functionally unrelated mental tasks. In sport, it is discouraging that no-one has been able to operationalize a general ability factor and it is no substitute to simply average someone's scores for archery and golf and pole-vaulting, any more than it makes sense to average marks for cooking, cleaning, and gardening. However, the first and second assertions have an important limitation, namely, they both apply only to the world of mental tests.

The case for a causal role means that for the first time, aside from intimations that the venture might prove worthwhile, we will be attempting to link g to the real world. On the personal as distinct from the group level, g is supposed to have a potent effect on academic achievement and a significant though lesser effect on socioeconomic status (SES), that is, on a person's ability to qualify for certain occupations and enjoy upward social mobility. But correlations cannot in themselves establish causal links between g and these variables. The claim has often been made that IQ tests are no more than academic achievement tests in disguise and, therefore, of course they correlate with academic achievement, just as academic achievement tests correlate with one another (Green, 1974). As for the correlations between IQ and SES, Block and Dworkin (1977, pp. 411—414, 431—432, and 453—454) argue that IQ tests may be little more than measures of social privilege and attitudes that go with privilege such as positive self-image and persistence. Therefore, good performance on IQ tests may be essentially an epiphenomenon, that is, pure effect with no causal role of its own. Someone from a privileged home would both do well on IQ tests and find an easy path to high SES; but privilege rather than IQ would be the real causal factor.

Jensen cites the work of Crano (1974; also Crano, Kenny, and Campbell, 1972), Gibson (1970), and Waller (1971) concerning the relationship between IQ and attainment. He stresses that early IQ predicts later academic achievement better than early academic achievement predicts later IQ. This is based on 5495 pupils who took both kinds of tests in Grade 4 (ages 9—10) and also took them two years later in Grade 6 (ages 11—12). Jensen (1980a, pp. 241—242) concludes that we are dealing with two distinct variables and that the causal line runs from individual differences in IQ towards individual differences in academic achievement, rather than in the reverse direction. This conclusion rests on an unstated assumption: that achievement differences should come into line with ability differences over time given that ability remains relatively stable. The unstated assumption makes sense if we also assume that children tend to perform above or below their ability at earlier ages, pushed at home or neglected, and that these discrepancies should disappear later thanks to formal education and a more uniform learning environment in general. These assumptions are at least plausible: the tendency of force-fed children to lose their achievement advantage as they progress in school is common.

The apparent effects of IQ were not uniform. As Brody and Brody (1980, p. 335) point out, when the sample was divided into mainly white suburban children and mainly black inner-city children, the predictive advantage of IQ was enhanced for the former but absent, even slightly outweighed by academic achievement, for the latter. When the sample was divided into high-IQ and low-IQ children, IQ had a predictive advantage for both but the advantage was much greater for the high-IQ subjects. I believe that these so-called disparate results fall into a coherent pattern. In a suburban middle-class school, mental ability claws its way towards a commensurate academic achievement. But when a child enters an inner-city school, a serious academic deficiency feeds on itself: with every year that passes, the child falls further and further behind. The tendency for initial differences in academic achievement to widen overwhelms mental ability as a causal factor. On the other hand, the higher the level of mental ability, the more it has the power to assert itself. All of the results are at least consistent with the hypothesis that mental ability, as measured by IQ, is a real-world causal force.

Studies of social mobility show that IQ has a differential effect on the life-histories of individuals even when social privilege is held constant.

In England, when siblings reared in the same family differ significantly in IQ, those above the family average tend to move up the SES scale and those below to move down. In America, a son who has an IQ 8 or more points above or below his father's (taken at the same age) tends to attain higher or lower social status accordingly. When the difference is as much as 23 points, the tendency is overwhelming. The average IQ difference between parent and child, and also between siblings, is 12 points and more than 50 per cent of children move into a different SES category as adults.

Jensen (undated, pp. 36—40) makes an important supplementary point. Although more evidence is needed, he gives results from three samples each composed of subjects who took a collection of mental tests and were numerous enough to include many siblings. The performance of sibling and co-sibling is the closest we can get to simulating a classless society, given that class differences operate primarily between families. Jensen did a factor analysis of the tendency of sibling and co-sibling towards consistent performance across mental tests, one tending to do better the other worse. The g that emerged was virtually identical to that extracted from the performance of members of different families who, of course, represent our present society with all its class divisions. Since g remains unaltered when no classes exist, it cannot be primarily a measure of social privilege. Jensen concludes that it appears to measure an ability that ranks people within as well as between classes. I agree, although this does not commit me to the ultimate conclusion that the ability measured can be identified with the primitive concept of intelligence.

Ever since Spearman invented factor analysis, the adherents of g have yearned for a physiological substratum. Gould (1981, p. 310) cites the fact that "no concrete tie has ever been confirmed between any neurological object and a factor axis" as crucial. Jensen has taken up this challenge on two fronts.

First, Jensen (1980a, pp. 686—692) and others have discovered correlations between the performance of subjects on Ravens, used as a marker test to measure how much subjects differ for fluid g, and their reaction times (RTs). Simple reaction time measures how long it takes to remove the index finger from a 'home' button so as to press another button when a light adjacent to the latter goes on. Choice reaction time measures the time taken to release the home button when the subject is confronted with a choice, say sees four digits on a screen, then is shown one digit, and must decide whether the latter was one of the former by pressing a 'yes' or 'no' button. The highest correlations are obtained between g and choice reaction times and usually range from 0.35 to 0.45; these are of course rather low accounting for only 12% to 20% of variance, that is, for only a small percentage of individual differences in g (variance accounted for = correlation squared, for example, 0.35 × 0.35 = 0.12).

Second, as an even closer approach to a physiological substratum, Jensen (1980a, pp. 707—710; undated, pp. 31—33) cites correlations between g and various measurements of the brain's electrical potential. Two or more small electrodes are gummed to the subject's scalp and a flash of light or a 'beep' is presented at random intervals every few seconds over a session of about 10 minutes. The experimenter can view the subjects' 'brain waves' on an oscilloscope but the real readings are done by computer. The computer scans the brain's electrical responses to the stimuli over the entire session arriving at an average evoked potential (AEP) which represents the brain's characteristic response. The principal readings are AEP latency, the time between the stimulus and the brain's response, AEP amplitude, the height of the graphically recorded waves, AEP complexity, a measure of the shape of the waveform, and AEP habituation, the difference between wave height (amplitude) during a first session of exposure to stimuli and a second. Until recently, correlations were very low but Eysenck and Barnett (1985) report a value of 0.60 between Wechsler IQ and AEP complexity, Scafer (1985) 0.73 between Wechsler IQ and AEP habituation, Wechsler tests being marker tests for a mix of fluid and crystallized g.

While granting that this research is at an exploratory stage, Jensen (1980a, pp. 700—704 and 707—708; undated, pp. 33—34 and 41) believes it has great theoretical importance. He believes the RT correlations alone completely refute the notion that individual differences in g are largely the result of individual differences in learned strategies. Students may learn strategies that give them 'short-cuts' to solving mathematical problems on achievement tests, but when they excel on Ravens, they do so because of the superior speed and efficiency with which they can execute basic cognitive operations, such as how quickly things held in short-term memory can be scanned and retrieved. He predicts that an item's g loading will turn out to be a function of the number or importance of the neural processes that are involved in the item's solution. The AEP correlations also dispel the notion that g is no more than skills people learn at home or in the larger culture for, after all, the brain's electrical potential is even further from cultural influence than reaction times. He believes the AEP research may prove to be the 'Holy Grail' that links g with the brain's neural efficiency; eventually correlations between g and neural structure may be found all the way down to a level just short of the molecules or even atoms that compose the brain.

Later we will have reason to refer back to Jensen's contention that g is an ability dependent on the neural substratum of mental activity and not much influenced by learned strategies or cultural factors. For now, I have tried to give a fair summary of his views about the ontological status of g. This is important because Gould (1981, pp. 24, 151, 159—160, 239, and 317—320) launches a fierce attack on Jensen on this very point. Jensen above all has committed the sin of reification: he takes the diverse set of capabilities called intelligence and labels them with an abstraction called g; having reduced them to one concept, he then converts that concept into a thing located in the brain. Well, Jensen's views are more complicated than that. If he can show that g measures a tendency to do well on a variety of mental tasks, that it can be operationalized as an ability, that it has true causal potency, and that it has a physiological substratum, he can make the appropriate claims about its reality or ontological status. If he cannot show these things, the claims must be withdrawn. Perhaps the sin of reification is making such claims without sufficient evidence; if so, we would do well to drop the word and focus on the evidence.

Sometimes Gould seems to be simply saying that intelligence is more than academic intelligence, that even the latter includes diverse abilities, and that we overlook this diversity at our peril, for example, how could anyone give a sensible diagnosis of a case of mental retardation just by saying a person is deficient in g? These points have already been conceded. But once again, the fact we cannot measure all forms of intelligence should not prevent us from measuring academic intelligence, the fact that academic intelligence comprises many abilities does not mean they are all as scientifically interesting as g, and the fact that g cannot tell us everything does not mean it cannot tell us anything.

THE FALSIFICATION OF g

Arguing both that the merits of a theory have often been overlooked

and that it must be revised and transcended encourages a certain
amount of schizophrenia. Thanks to the above account, the Spearman—
Jensen theory may appear to have enjoyed success after success and
stand at the apex of its scientific plausibility. It is time to set the record
straight: over the last 40 years, psychologists have discovered something that the theory can neither explain nor accommodate. I refer to
the phenomenon of massive IQ gains over time, that is, the tendency
for each generation to have a higher average IQ than its predecessor.
Until recently, the theory progressed unaffected by this phenomenon,
but that was because reports of massive IQ gains were either little
known or largely ignored.

The interaction between the Spearman—Jensen theory and the
phenomenon of IQ gains over time is a classic example of the
degeneration of a theory in the face of evidence. The following is
borrowed from Lakatos with a few changes in terminology: (1) A
theory generates a series of *original* hypotheses which flow naturally
from the theory; (2) The theory itself cannot be falsified but its
hypotheses can; (3) If that happens, defenders of the theory put
forward *protective* hypotheses — these are designed to defend the
theory but may well make interesting predictions and possess scientific
respectibility; (4) If these in turn are falsified, defenders of the theory
may begin to put forward *ad hoc* hypotheses — these too defend the
theory but at the price of doing nothing else and are scientifically
bankrupt; (5) When it becomes clear that nothing better than *ad hoc*
hypotheses are forthcoming, it is time for the theory to be modified or
replaced by another.

The Theory and Its Hypotheses

The Spearman—Jensen theory has three essential parts: IQ tests
measure a mental ability called g; g bears a close resemblance to the
primitive concept of intelligence; intelligence is a mental ability with a
potent causal role, that is, it is productive of achievements of the kind
usually associated with academic ability. How closely g matches
popular notions of intelligence is not important. But if it did not refer to
a mental ability other than learning or memory that played the above
causal role, the theory would not deserve to be called a theory of
intelligence at all. The theory logically entails the following hypothesis:
if the people of a modern industrialized nation make massive gains in g

over time, they will make corresponding gains in terms of creative achievement, scientific and mathematical discovery, and technological progress.

This hypothesis implies a comparison between groups, that is, an ability comparison between the present generation of a given nation and past generations. Up to now, our account of the theory has focused on its explanatory power concerning the life-histories of individuals, but its proponents have always assumed that group differences in IQ or g are highly significant. Some group differences are quite large, ranging from 10 to 15 or even 20 points. These can be best appreciated if we take into account that IQ scores have a normal distribution with a standard deviation (SD) of 15. This means that if a particular group has a mean IQ 15 points below another, only 16 per cent of its members will exceed the average of the superior group and, in addition, it will be on the wrong end of a 17 to one ratio at high IQ levels, that is, 130 and above. A deficit of 20 points means that only 9 per cent of the inferior group will exceed the superior group's average and now, the ratio will run 57 to one against it at high IQ levels.

When Lynn (1982) discovered that Japan was 11 points above America on the performance half of the Wechsler, he suggested that this might have been a significant factor in Japan's outstandingly high rate of economic growth since World War II. Nichols (1987) focuses on the fact American blacks are 15 IQ points below American whites and argues that, thanks to that, nothing can be done about how unfavourably blacks compare to whites in terms of real-world achievement, indices such as income, occupational status, symptoms of family demoralization, and crime statistics. Eysenck (1985) sums up his survey of racial differences as follows: Mongoloid peoples have the highest mean IQ, particularly the Chinese and Japanese, and the scores then decline through Northern Europeans and their descendants, through Southern Europeans and Indians, down to Malays and Negroid groups at the bottom. He sees a close correlation between mean IQ, socio-economic status, and level of cultural achievement. Harlow and Harlow (1962, p. 34) generalize: human beings collectively have little more than the minimum intelligence needed for social progress and a mean IQ significantly below that level would mean that there could be no civilization as we know it.

Jensen (1980a, 1981, 1985) has made a detailed comparison of American blacks and American whites. IQ has the same moderate

success in predicting academic achievement, as measured by standardized tests and university grades, for blacks as it does for whites.
There is a slight tendency for blacks to underachieve as compared to
whites with the same IQ. The primary factor that differentiates black
and white test performance is a racial difference for g and the under-
representation of blacks in elite occupations makes sense in terms of g
differences between the people who normally staff various occupations.
Black and white have similar hierarchies of item difficulty on IQ tests
which means the same items are difficult or easy for both. Insofar as
black children of a given age have a different item hierarchy from
whites of the same age, it mimics a maturity difference, that is, the
performance of black 13-years olds is indistinguishable from that of
white 11-year olds. The tallies with the hypothesis that black—white g
differences account for black—white test score differences in that fluid
g increases with age up to about 18 to 20. Jensen concludes that the
15-point black IQ deficit represents a real 'mental maturity' or ability
difference between the races, rather than the effects of tests culturally
biased against blacks.

Jensen's analysis shows that in many respects, American blacks
perform as if they were a group of less able whites, a group selected out
of the white population on the basis of below-average IQ. However,
blacks differ from that pattern in other respects. When black children
are compared to their own parents, their IQs regress to their own
population mean rather than to the white mean, reflecting the fact that
they constitute a separate breeding group with its own environment.
They have much lower incomes and socioeconomic status than would
a group of whites that match them for IQ (Flynn, 1987a) and they
self-select in terms of occupational aspirations by way of ability
comparisons with one another rather than with the larger white
community (Jensen, 1980a, p. 101).

When between-group hypotheses compare Japanese and Americans
or Northern Europeans and Malays, they may not seem to be a fair
test of the Spearman—Jensen theory. When groups speak different
languages, practice radically different customs, have different histories
and incentive-systems, stand on opposite sides of the industrial revolution, and so forth, this raises questions about the ability of IQ tests to
bridge cultural distance. The proponents of the theory argue that the
right tests correctly administered can answer such questions, but they
may be ignoring common-sense limitations on its explanatory power

and a theory should not be measured against the audacity of its proponents. On the other hand, between-group hypotheses about successive generations within nations, ones long part of the technologically developed world, should constitute an ideal test of the theory: common language, common history, continuity of custom, tests long introduced and easily administered. If the theory is to generate no hypotheses about such groups, it is difficult to see that it can lay claim to any between-group explanatory power.

As we have seen, massive IQ gains from one generation to another enormously increase the proportion of the population at high IQ levels. Jensen (1980a, pp. 111—114) makes clear what real-world behaviour we have a right to expect from those at high IQ levels: above 130 they find school easy and can succeed at virtually any occupation; above 140 their adult achievements are so extraordinary they fill the pages of *American Men of Science* and *Who's Who*; above 150 they amaze their teachers with their precocity and begin to duplicate the life histories of famous geniuses who made creative contributions to our civilization. Jensen asserts that the quality of a society's culture is highly determined by the fraction of its population that is highly endowed and that they are the main source of philosophical insights, mathematical and scientific discoveries, practical inventions, masterpieces of literature and art, and so forth. Clearly massive IQ gains should bring a cultural renaissance too obvious to be overlooked.

The hypothesis that massive IQ gains will mean corresponding gains in terms of real-world achievement is implied by the Spearman—Jensen theory, but the theory has an accompanying expectation, namely, that generational IQ gains will be small. Large intelligence differences between races and between nations may seem plausible, at least to some, but huge intelligence differences between generations seem absurd: we live in intimate daily contact with our children and they do not appear that much more brilliant than ourselves and we do not remember our parents as being intellectually limited. Therefore it is not surprising that for 40 years studies reporting massive generational IQ gains were largely neglected, although admittedly many of them simply did not circulate across national boundaries. Here is a partial list of studies, each followed by the reason for its neglect: Tuddenham (1948) — U.S. Army tests said to be more akin to achievement than intelligence tests; Elley (1969) — New Zealand results and thus ignored in accord with international custom; Bouvier (1969) — Belgian Army

pamphlet published in French only; Thorndike (1973) — doubts about
the American Binet samples; Girod and Allaume (1976) — published
in French only; Mehlhorn and Mehlhorn (1981) — East German
results published in German only; Rist (1982) — Norwegian Army
pamphlet published in Norwegian only; Lynn (1982) — emphasized
comparison between Japan and America rather than Japanese gains;
Flynn (1984) — Wechsler test gains in America seemed offset by
Scholastic Aptitude Test (SAT) losses; Leeuw and Meester (1984) —
published in Dutch only; Schallberger (1985) — technical report in
German only.

However, massive IQ gains have become so universal throughout the
technologically developed world and so well documented that they can
no longer be dismissed. It was something of a turning point when Lynn,
long skeptical about the radical malleability of IQ, accepted the
phenomenon as real and actually began to add to the evidence (Lynn
and Hampson, 1986). Flynn (1987b) has analyzed post-1950 data from
14 nations: the present generation has a mean IQ 5 to 25 points above
the last; the advantage varies from nation to nation with a median of 15
points or a full standard deviation. Some of the most impressive gains
have been in Western Europe, for example the Netherlands gained 20
points from 1952 to 1982 and France some 20 to 25 points from 1949
to 1974 with 20 points being acceptable as a conservative estimate.
Therefore, the potentially creative elite of both of these nations multiplied over that period by about 57 times.

Flynn (1987b) verified that neither the Netherlands nor France had
any perception that they were in the midst of a cultural renaissance.
There is not a single reference to a dramatic increase in genius or
mathematical and scientific discovery in the present generation. No one
has remarked on the superiority of contemporary schoolchildren. The
number of patents granted for inventions has diminished in both
nations. Jensen (1987) concurs that if the real-world achievements we
associate with intelligence had escalated in accord with massive IQ
gains, the results would be too obvious to be missed: the aging survivors
of the last generation would be perceived by almost everyone as
border-line mentally retarded; Dutch university professors would be
amazed at the prevalence of genius in their classes.

The facts are these: when IQ tests rank people at a given place and
time the results make sense in terms of relative intelligence; when IQ
tests rank generations over time they give nonsense results. These facts

imply that IQ tests do not measure intelligence but rather a correlate with a weak causal link to intelligence. Imagine we could not directly measure the population of cities but had to take aerial photographs which gave a pretty good estimate of area. In 1952, ranking the major cities of New Zealand by area correlated almost perfectly with ranking them by population and in 1982, the same was true. But if anyone found that the area of cities had doubled between 1952 and 1982, they would go far astray by assuming that population had doubled. The causal link between population and its correlate is too weak, thanks to other factors that intervene such as central city decay, affluent creation of suburbs, more private transport, all of which can expand area without the help of increased population.

The implications for IQ tests as measures of intelligence is serious: their explanatory power appears to be a function of cultural homo-geneity. They are at their best when accounting for the differential real-world achievements of siblings raised within the same family; they do reasonably well with Dutch or French or Americans of the same generation sharing the same formative environment; they fail com-pletely when they attempt to bridge the cultural distance that separates generations in modern industrial societies. This last means that all between-group comparisons which attempt to bridge even greater cultural distance are suspect. I refer to Lynn's use of IQ differences to explain differential real-world achievements by the Japanese and American economies and all other cross-national and cross-racial comparisons.

It may be said that to reject Jensen's comparison of American blacks and American whites is question-begging. If blacks are so assimilated that they mimic a subgroup of the white population, the question of cultural distance does not arise and the racial IQ difference could be identified with an intelligence difference. In response, some of Jensen's evidence is less impressive after analysis of the between-generations data. Recall that the fact black 13-year olds mimic the test performance of white 11-year olds was used to suggest a real ability difference between the races. W. B. Dockrell (personal communication, 7 August 1985) has supplied data on Scottish IQ gains between 1965 and 1982. These are greatest on those Wechsler subtests that measure reasoning ability such as similarities and comprehension. The performance of yesterday's 13-year olds on these two subtests mimics the performance of the 10 or 11-year olds of today. So the generations also appear to be separated by a real ability difference and yet, unless evidence of extraordinary achievement is forthcoming, that ability difference cannot be identified with an intelligence difference. At the very least, when we have new tests that are better measures of intelligence, blacks will get a second chance to improve their performance.

The between-generations data also solves a problem I have always found disturbing. Elsewhere I have attempted to show that the black IQ deficit is probably due to environmental rather than genetic inferiority (Flynn, 1980; 1987c). It hardly seemed plausible that environmental differences between the races could cause g differences and nothing else; that is, no differences such as blacks being less familiar with certain test items than whites. And yet, black and white item hierarchies differed in terms of g loading and little else — that was why older blacks mimicked the performance of younger whites. The between-generations data proves that all of this is possible: the generations are separated almost entirely by environmental differences; and yet, the older Scottish children of yesterday mimic the younger Scottish children of today.

It is important to avoid confusion. In order to shed light on this particular black-white problem, generational IQ differences must reflect g differences only, at least in one nation. But generational IQ differences threaten the Spearman—Jensen theory just so long as they signal massive g gains unaccompanied by achievement gains. Additional gains or losses caused by enhanced or diminished familiarity with items can provide an unwelcome complication but they in no way reduce the threat.

Protective and Ad Hoc Hypotheses

The proponents of a theory under threat can either question the evidence or surround the theory with a belt of protective hypotheses or both. The proponents of the Spearman—Jensen theory of intelligence have done both.

From 1982 to 1984, the English-speaking world was aware primarily of evidence for massive IQ gains in Japan and America. Wechsler and Stanford—Binet standardization samples, selected over a period of a generation or more, had performed better and better on a variety of Wechsler and Binet tests. The samples were stratified rather than random, numbered from about one to three thousand subjects, and in most cases test content varied somewhat because of revision or the age-group for which the test was intended. Jensen (1980a, p. 570; personal communication, 3 February 1983) stressed the possibility of sample bias: he stipulated that samples should be extremely large, say comprehensive testing of draft registrants, and the tests identical. These conditions have now been met: Dutch, Norwegian, and Belgian IQ gains are based on military testing of virtually the entire population of young adult males and the tests were unaltered.

Turning to protective hypotheses, these are entirely respectable just so long as they suggest testable predictions. When Newton's gravitational theory failed to explain the orbit of Mercury, the hypothesis offered was that it was being influenced by an undiscovered planet closer to the Sun. The orbit required for the undiscovered planet was calculated, it was tentatively named Vulcan, and the French Academy awarded the Legion of Honour to M. Lescarbault for its discovery. The sighting was wishful-thinking and Newton's theory had to be transcended, but the hypothesis was potentially fruitful. The planet Neptune had been discovered in precisely this way, that is, as a posited influence on the orbit of Uranus.

When the Spearman—Jensen theory failed to explain the phenomenon of massive IQ gains, there were three obvious targets for protective hypotheses: the subjects who took the tests, the tests themselves, and the character of the enhanced performance.

Hypothesis: Massive IQ gains represent early maturation rather than gains at full maturity (Jensen, 1980a, p. 570; personal communication, 12 January 1983). If generational gains were large among young children, small among older children, and non-existent at full maturity, there would be no reason to expect enhanced adult achievement and lack of such would not weaken the link between IQ or *g* and intelligence. Full maturity refers to the age of peak raw-score performance on a particular test. It is not easy to establish such an age for tests of crystalized *g* because they emphasize acquired knowledge and additional knowledge can be acquired into old age. However, performance on tests of fluid *g* like Ravens maintains a stable maximum from ages 18 to 25, then begins a gradual decline which accelerates in old age. *Prediction*: IQ gains will decline with age among school children and will be absent by ages 18 or 19 on tests of fluid *g*.

Hypothesis: Massive IQ gains represent learned item gains rather than *g* gains (H. J. Eysenck, personal communication, 14 December

1982; Jensen, 1980a, pp. 569 and 635; Jensen, personal communication, 12 January 1983). Tests of crystalized g emphasize acquired
knowledge and therefore, really measure mental ability or g only when
all subjects have had an equal opportunity to acquire the knowledge in
question. This situation may obtain for all or most members of a
particular generation but not from one generation to another: over 30
years, general educational advances may give the current generation a
large advantage over the last on learned content items. Take Plato or
Aristotle or Archimedes: they would do badly on a modern general
information subtest or an arithmetic subtest using modern notation.
However, they should do very well indeed on tests of fluid g like
Ravens. Once they had become familiar with the directions and multiple
choice format, they would need to demonstrate only decontextualized
problem-solving ability and their performance would be a true measure
of g.

As the elaboration of this hypothesis makes clear, it protects the
Spearman—Jensen theory by arguing that massive IQ gains may represent only a larger repertoire of learned items and not massive g gains.
Without massive g gains, we would expect no escalation of intelligence
or real-world creativity hence lack of such does not count against the
theory. There are difficulties with this hypothesis: on the one hand, we
are told that the knowledge IQ tests require is so elementary and
universally accessible that the tests are fair measures of g differences
between subjects; on the other hand, we are told that this knowledge
was so poorly acquired by one generation that huge improvements were
made by the next, so that the tests are not fair measures of g differences
between generations. At any rate, this hypothesis clearly tells us what
to expect. *Prediction*: IQ gains will be found primarily on tests of
crystalized g and be small or absent on tests of fluid g such as Ravens.

Hypothesis: Massive IQ gains represent enhanced test sophistication
rather than g gains (H. J. Eysenck, personal communication, 14
December 1982; J. C. Loehlin, personal communication, 3 January
1983; J. Ray, personal communication, 7 August 1986). Once again,
the theory is protected by breaking the link between IQ and g. Test
sophistication refers to the fact that people who have never taken
formal tests or a particular kind of test are unfamiliar with format and
strategy. When first confronted with multiple choice items, they may go
through the series of suggested answers in the order presented, rather
than scanning them to see if there is an obviously correct choice.

Therefore, they are at a disadvantage compared to more sophisticated subjects in a way that has nothing to do with mental ability or g. Fortunately, test sophistication has been much studied and the gains involved carefully estimated. Jensen (1980a, pp. 590—591) concludes that even working with totally naive subjects, repeated testing with parallel forms of IQ tests gives gains that total only 5 or 6 points; moreover, increments after the first exposure to testing rapidly diminish and approach a threshold or limit. *Prediction*: IQ gains will exhibit a declining rate of gain and approach a limit of 5 or 6 points.

Flynn (1987b) shows that all of the predictions engendered by these protective hypotheses have been overwhelmingly falsified. Over the last 30 years, there have been massive gains on tests of fluid *g* in the Netherlands, Norway, France, and Australia, gains ranging from 12 to 25 points. Belgium is also impressive with 7 points in only 9 years. Lest it be thought that Ravens is the only test of fluid *g* that shows gains, the Australian gains are on Jenkins, the Belgian Shapes Test matches their Ravens gains, and recently Lynn, Hampson, and Mullineux (undated) have shown that British gains on the Cattell are actually greater than British Ravens gains. Gains on tests of crystalized *g* are sizable but lag behind tests of fluid *g*. In other words, fluid *g* gains are actually greater than the typical IQ tests would convey: the acquired knowledge content of the typical test actually disguises the magnitude of *g* gains rather than explaining them away! The Netherlands, Norway, France, and Belgium all give us subjects aged 18 or 19 which means they are fully mature in terms of fluid *g*. Five nations give data for school children of various ages and four of these show no tendency for gains to decline with age.

As the above shows, generational IQ gains do not follow the pattern of test sophistication gains. They are usually far greater than 5 or 6 points and while trends over time vary, there are nations like the Netherlands where gains have continually accelerated over 30 years rather than declining as they approach a limit. It may be suggested that perhaps some new form of test sophistication has appeared which gives average gains three or four times as great as anything we know and which feeds on itself. This is purely a verbal strategem. Test sophistication proper refers to the limited gains arising out of familiarization with test setting and format. To turn it into something radically different is merely to give cases in which people have more ability of the sort IQ tests measure a new name, perhaps 'super test sophistication'. This is like attributing the ever better performance of athletes to some new and incredibly potent form of 'competition sophistication'. The actual experience of competition produces limited gains that are well known and there is no point in applying the term to a quite different phenomenon.

, In a recent publication, Jensen (1987, pp. 379—381) rehearses some of the protective hypotheses, emphasizes that IQ gains have not meant intelligence or real-world achievement gains, and concludes as follows: "It seems much more plausible that the reported test score increase of twenty points does not reflect a corresponding change in g or its real-life correlates, but is rather the result of some artifact not yet identified." Given the context, this assertion could mean several things. Jensen was reacting primarily to the Dutch data only and perhaps all he means is that we should wait until all the evidence is in concerning the protective hypotheses stated thus far. All well and good, but the evidence is now in and the protective hypotheses have been falsified. Perhaps the reference to an "artifact not yet identified" is crucial. This would be tantamount to an expression of hope that some new as yet unformulated protective hypothesis will come along that will not be falsified, that a 'factor X' will be discovered that successfully severs the link between massive Ravens gains and massive g gains.

The proponents of the Spearman—Jensen theory can legitimately ask for a breathing space to rethink their position. However, such a period should not be too prolonged. Jensen himself (1973a, pp. 135—139 and 188—189; 1973b, pp. 351 and 413—414) has ridiculed those who have nothing better to offer than an unknown 'factor X' that yields no testable predictions, for example environmentalists at a loss for an explanation of racial IQ differences. He also points out the extraordinary difficulty of finding a factor that will explain away between-group differences and not also explain away within-group differences. For example, he believes that any plausible factor that differentiates blacks from whites would differentiate blacks from one another. Now he is in precisely the dilemma he describes: finding an artifact that will work between generations but have no effect within generations. The Spearman—Jensen theory cannot afford an artifact that severs the link between Ravens and g for individual differences within a generation, as well as severing the link between generations. Ravens is the marker test for fluid g in all their physiological research; IQ tests would be as suspect in ranking individuals for intelligence as they are in ranking groups. And yet, if 'super test sophistication' or 'acquired knowledge' varies from one generation to the next, how likely is it that there is no significant variance for the individuals who comprise a particular generation?

Jensen's assertion that generational IQ gains cannot be plausibly construed as g gains is susceptible to a third interpretation. At times, he refers to the well-established correlates between g and real-world achievements and uses the phrase "intelligence or g" as if the two were interchangeable. Perhaps Jensen would reject the following interpretation, but I will proceed because I have no doubt that it will tempt many of the proponents of the Spearman—Jensen theory.

The theory has had many successes in evidencing links between g and the real-world achievements we associate with intelligence. Nevertheless, these cannot justify contending either that its successes cancel out its failures or that when the appropriate real-world achievements are absent then g simply must be absent as well. These confusions can be dispelled by keeping in mind exactly what g is. It begins as no more than an ability operationalized by Ravens which allows people to do well across IQ tests. Every bit of its real-world explanatory status must be won through evidence and no bit carries over to another bit. Between-group differences are supposed to be a major area of explanation: Jensen (1980, p. 636) says Ravens can measure fluid g for groups of people of remotely different cultures. Its between-generations failure can no more be cancelled out by its successes than Newton's failure with Mercury could be cancelled out by successes with the other eight planets. Indeed, it was the gravitational theory's great success that made Mercury such a scandal.

As for positing that g must be absent whenever the real-world achievements of intelligence are absent, this would be an *ad hoc* hypothesis of the worst sort: an undefended boundary hypothesis. The construct g cannot be deemed present or absent for the convenience of the theory. Saying g simply must not apply to between-generations IQ gains is no better than saying gravitation simply must be different when you get to the planet closest to the sun. This hypothesis generates no prediction save the embarrassing one that under certain conditions, hitherto thought similar to normal conditions, the theory will not work. However, it may be said that analogies with physics are all very well, but is it really unreasonable to posit the absence of g? After all, we have posited the absence of intelligence gains when real-world achievement gains were missing. Why not posit the absence of g for the very same reason?

The analogy looks plausible only because, for simplicity's sake, I have spoken of the primitive concept of intelligence as if it referred to something known to exist which can be here or there. As the last section of this paper will show, that is not at all its ontological status. It is really like an invitation issued to various scientific constructs, asking them if they wish to play a certain explanatory role, the role of a mental ability that causes certain real-world achievements. When they use Ravens, the proponents of the Spearman—Jensen theory accept the invitation: they use Ravens as the marker test for fluid g to give g a chance to play the explanatory role. When Ravens is there, that is *prima facie* evidence that g is there and if the appropriate real-world achievements are absent, that is *prima facie* evidence that g has failed. When an actor reads for a part and fails to perform properly, it is absurd to cite his or her failure as evidence the actor did not show up for the audition. You can of course claim it was someone else in disguise: that is what the theory's protective hypotheses tried to do — to say that learned items or test sophistication showed up disguised as g.

However, they failed and for now the limitation applied to IQ tests must be applied to g: it does not measure intelligence but rather a correlate with a weak causal link to intelligence; or put differently, the ability of g to play the role of intelligence is a function of cultural homogeneity.

Future of the Theory

Assuming that no protective hypothesis comes to the aid of the Spearman—Jensen theory, it must be revised or abandoned or transcended. However, no reasonably successful theory is abandoned or transcended until a better theory comes along, and for the time being our attitude to the Spearman—Jensen theory should be as follows: use IQ tests within clearly homogenous cultural settings; abandon group comparisons across cultural distance; use IQ tests for clinical purposes where clinical psychologists have found them of diagnostic value; be wary of taking their cutting lines for mental retardation at all seriously (Flynn, 1985). Their use in schools and as university entrance examinations cannot be discussed in a sentence or two. However, excellent performance on academic achievement tests should never be discounted because of lower IQ scores; academic achievement validates IQ tests and not the reverse. The kind of criterion sampling already described should replace IQ tests in selecting people for virtually all jobs and for armed forces training programmes as well.

Until a better correlate of intelligence than g is derived, the search for correlations with physiological data can continue but in a very tentative spirit. The weaknesses of g could be either inflating or deflating the physiological correlations and their existence certainly does *not* show that g is some kind of rock impervious to the usual cultural influences. Jensen is simply wrong on this point. Huge g gains from one generation to another show that it is highly sensitive to environmental factors and some of these may be cultural factors such as learned strategies of problem-solving picked up at school, or at home, or elsewhere.

As for changing the theory, the least radical change would be revision by way of better IQ tests. The present construct of g has failed and if you want a better g, you must improve the tests that engender g. I have no special expertise here, but I find some of the suggestions of Sternberg (1985) exciting. Clearly one criterion of a better test is that aside from matching the explanatory power of current tests, it must not give huge g differences where no real-world achievement differences exist. The prospect of a fundamental rethinking of mental tests should be regarded as exciting rather than a cause for gloom. Who knows what better tests and a better g might bring: a real understanding of group differences; higher physiological correlations? Who knows how many lines of research have been crippled by the defects of current g?

On the other hand, the failures of the Spearman—Jensen theory could necessitate a conceptual revolution: Einstein did not merely alter Newton's predictions in selected cases; he revolutionized the foundations of gravitation theory and our concepts of space, time, and light. If plugging a better g into the Spearman—Jensen theory does not work, its whole conceptual system may have to be replaced on a theoretical level. When a theory is replaced, this may take a more or less radical form, that is, it may either be abandoned or transcended. Like Velikovsky, it may be completely discredited because for every fact or supposed fact it explains, it generates several falsified hypotheses. Or like Newton, it may be transcended by a new theory: the conditions under which it works may be explained, its failures explained, and its predictions displaced only in certain cases where the stipulated conditions do not hold. I very much suspect that the Spearman—Jensen theory will not be abandoned but will find a place within a new theoretical structure.

The conceptual revolution that may be necessary can best be envisaged by positing what might at first appear to be a mere protective hypothesis: that massive g gains did bring massive intelligence gains but that the real-world achievements associated with intelligence were suppressed by other trends. Take the prediction the generational data falsifies: if the people of a modern industrialized nation make massive g gains over time, they will make corresponding gains in terms of creative achievement, scientific and mathematical discovery, and technological progress. This assumes, at least in the cultural context named, that intelligence is so incredibly potent a factor that dramatic effects should be visible no matter what else is going on. Perhaps academic intelligence has been diverted away from academic pursuits, perhaps factors like motivation have collapsed, perhaps educational institutions have so lost their way as to squander enhanced intelligence, perhaps a new character type means people do not express enhanced intelligence in the extroverted wit and sparkle which would make its presence evident. In other words, this kind of protective hypothesis would force us to face up to something everyone knows but all of us dislike because it complicates our task: the theory of intelligence can make no real progress until scientific constructs of intelligence are put in the context of larger theories of personal and social development.

The Spearman—Jensen theory has a tendency to treat g as if its explanatory potential could be assessed in virtual isolation from other factors. A decade ago Block and Dworkin (1977, pp. 416—419) demanded a more comprehensive theory to explain why IQ tests sometimes work; this demand can no longer be ignored now that we need to know why IQ tests sometimes fail. It is instructive to note how economics has reacted when its theories have faced similar failures.

For example, Keynes (1936) developed a general economic theory in which the disposable income available to households was the principal determinant of consumer demand. Within-generation data suggested that the percentage of income spent on consumption decreased as disposable income increased, that is, within each generation, the affluent logically enough spend a smaller proportion of their income on consumption than the poor. This led to a between-generations prediction that as people in general became more affluent, there would be a disastrous fall-off of consumer demand. In fact, the next higher-income generation spent the same percentage on consumption as the previous lower-income generation. Between-generations income differences did not produce the same effects as within-generation income differences. The mistake had been to assess the explanatory power of income in virtual isolation and once it was put in the context of a larger complex of variables, the failure was explained. A psychological variable was crucial: when people over time begin to believe that an increase in their incomes is permanent, they begin to spend a higher proportion on consumption and the average percentage is no less than when incomes were lower. The moral, of course, it that psychology may explain its between-generation failures when intelligence is related to other factors in a larger theoretical framework.

My own guess is that the paradoxes of massive IQ gains over time will be solved neither by better tests and a better g alone, nor by a more comprehensive theory of personal and social development alone. Social trends and a new character type may well have suppressed some of the real-world achievements associated with the primitive concept of intelligence, but it just does not seem plausible that the effects of such huge intelligence gains could be completely nullified. Therefore, tests of fluid g must be at fault when they suggest that the intelligence gains have been so large. Probably the Spearman—Jensen theory will have to be both revised and transcended.

The Primitive Concept of Intelligence

We end as we began with the primitive concept of intelligence. This concept should be construed not as something to be located or measured, but rather as a piece of advice to psychologists about research strategy. It says in effect: when you formulate a theory to explain the life-histories of individuals or groups, your theory will lose explanatory power unless it includes a mental ability or abilities distinct from memory and learning.

When Kepler realized that the orbits of the planets deviated from 'natural' or circular motion, he seized upon the primitive concept of celestial influence. He was a Pythagorean Sun-worshipper and the Sun was so big and so close it seemed that it just must have some effect on the planets that raced around it. The primitive concept of celestial influence said in effect: when you formulate a theory to explain the motions of the planets, your theory will lose explanatory power unless it includes the notion of influence by heavenly bodies that are large and proximate. The primitive concept had no specificity or explanatory power of its own. It was up to astronomers to supply that by putting forward theories and scientific constructs which would yield predictions that matched the phenomena in question. There have been a variety of such: Kepler speculated that the Sun acted as a magnet; Descartes that it rotated setting up a vortex or whirlpool that swung the planets around; Newton that its mass exerted a gravitational pull inversely as the distance squared; Einstein that its mass influences the shape of the space through which the planets move.

The latter constructs have had great explanatory power and have proved that the advice of the primitive concept of celestial influence was sound: our theories would be impotent if they omitted the facts of the Sun's size and proximity. But it would have been absurd to attempt to sharpen and measure the primitive concept. The thing to do was take its advice and find a scientific construct that had explanatory power within the context of a comprehensive theory that interrelated all the factors affecting celestial motion. The advice might have been bad, in which case all such theories would have failed. Then a new primitive concept would have come forward, perhaps one of internal propulsion: that planets have an internal mechanism or guidance system that propels them along their orbits, witness the force emanating from within in the form of volcanoes and earthquakes. Astronomers would sink shafts deep into the earth rather than merely gaze at the heavens.

The primitive concept of intelligence gives rise to a heuristic for guiding theory construction and has no ontological status beyond that. It gives advice to social scientists, that is, it warns us that we omit an 'intelligence factor' from our theories of human behaviour at our peril. There is nothing to be gained from trying to sharpen or measure the primitive concept of intelligence. The thing to do is follow its advice and formulate a scientific construct of a mental ability or abilities other than memory and learning that has explanatory power. Jensen has pursued the right path in worrying little about the 'nature' of intelligence and attempting to exploit the potentialities of g: it is g that must earn real-world ontological status by explaining the life-histories or achievements of individuals and groups. A lot of time is being wasted asking people about their concepts of intelligence and examining ordinary language. Or better, this research has a purpose but it is purely one of communication rather than advancing explanation. Knowing how people in general use the words intelligence and its derivatives like 'bright' and 'dull' will indicate how far these differ from our scientific constructs and whether we can use such words in describing our results without being irresponsible. It is of course irresponsible to use them where there is any reason to believe our scientific constructs have failed, such as in explaining between-group differences. Even if the link with ordinary language usually exists, it is then broken.

Lack of interest in the primitive concept of intelligence does not entail the dilemmas of crude operationalism. Intelligence is not what IQ tests measure; and therefore, there is no problem about how we can say that present IQ tests are not perfect measures of intelligence. Intelligence is a description of an explanatory role IQ tests aspire to play and therefore, IQ tests must be revised when they do not play that role well — when IQ differences occur and predicted real-world achievement differences do not. When you formulate a theory to explain the life-histories of individuals and groups, your theory will lose explanatory power unless it includes a mental ability or abilities distinct from memory and learning. That cannot be entirely bad advice despite the selective failures of g and the Spearman—Jensen theory.

The failures of the theory are not Jensen's failures. As a working scientist, he has rarely put a foot wrong and he is certainly not guilty of the philosophical mistakes of which he has been accused. IQ tests simply have not given him a good enough g and psychology has not advanced far enough to integrate scientific constructs of intelligence into a more complex and comprehensive theory of personal and social development.

University of Otago
Dunedin, New Zealand

BIBLIOGRAPHY

Bane, M. J., and Jencks, C.: 1977, 'Five Myths About Your IQ', in N. J. Block and G. Dworkin (eds.), *The IQ Controversy: Critical Readings*, Quartet Books, London, pp. 325—338.
Block, N. J., and Dworkin, G.: 1977, 'IQ, Heritability, and Inequality', in N. J. Block and G Dworkin (eds.), *The IQ Controversy: Critical Readings*, Quartet Books, London, pp. 410—540.
Bouvier, U.: 1969, *Evolution des Cotes à Quelques Tests* (Evolution of Scores from Several Tests), Belgian Armed Forces, Center for Research into Human Traits, Brussels.

Brody, N. and Brody, E. P.: 1980, 'Differential Construct Validity', *Behavioral and Brain Sciences* **3**, 335—336.

Cattell, R. B.: 1963, 'Theory of Fluid and Crystalized Intelligence: A Critical Experiment', *Journal of Educational Psychology* **54**, 1—22.

Crano, W. D.: 1974, 'Cognitive Analyses of the Effects of Socioeconomic Status and Initial Intellectual Endowment on Patterns of Cognitive Development and Academic Achievement', in D. R. Green (ed.), *The Aptitude-Achievement Distinction*, McGraw-Hill, New York, pp. 223—253.

Crano, W. D., Kenny, D. A., and Campbell, D. T.: 1972, 'Does Intelligence Cause Achievement? A Cross-Legged Panel Analysis', *Journal of Educational Psychology* **63**, 258—275.

Economos, J.: 1980, 'Bias Cuts Deeper Than Scores', *Behavioral and Brain Sciences* **3**, 342—343.

Elley, W. B.: 1969, 'Changes in Mental Ability in New Zealand Schoolchildren', *New Zealand Journal of Educational Studies* **4**, 140—155.

Eysenck, H. J.: 1985, 'The Nature of Cognitive Differences Between Blacks and Whites', *Behavioral and Brain Sciences* **8**, 229.

Eysenck, H. J., and Barnett, P.: 1985, 'Psychophysiology and Measurement of Intelligence', in C. R. Reynolds and V. Wilson (eds.), *Methodological and Statistical Advances in the Study of Individual Differences*, Plenum, New York, pp. 1—49.

Flynn, J. R.: 1980, *Race, IQ, and Jensen*, Routledge, London.

Flynn, J. R.: 1984, 'The Mean IQ of Americans: Massive Gains 1932 to 1978', *Psychological Bulletin* **95**, 29—51.

Flynn, J. R.: 1985, 'Wechsler Intelligence Tests: Do We Really Have a Criterion of Mental Retardation?', *American Journal of Mental Deficiency* **90**, 236—244.

Flynn, J. R.: 1987a, 'Flynn Replies to Nichols', in S. Modgil and C. Modgil (eds.), *Arthur Jensen: Consensus and Controversy*, Falmer Press, Lewes, Sussex, pp. 234—235.

Flynn, J. R.: 1987b, 'Massive IQ Gains in 14 Nations: What IQ Tests Really Measure', *Psychological Bulletin* **101**, pp. 171—191.

Flynn, J. R.: 1987c, 'Race and IQ: Jensen's Case Refuted', in S. Modgil and C. Modgil (eds.), *Arthur Jensen: Consensus and Controversy*, Falmer Press, Lewes, Sussex, pp. 221—232.

Gibson, J. B.: 1970, 'Biological Aspects of a High Socio-Economic Group: I. IQ, Education and Social Mobility', *Journal of Biosocial Science* **2**, 1—16.

Girod, M., and Allaume, G.: 1976, 'L'Évolution du Niveau Intellectuel de la Population Francaise pendant le Dernier Quart de Siècle' (The Evolution of the Intellectual Level of the French Population during the Last Quarter Century), *International Review of Applied Psychology* **25**, 121—123.

Gould, S. J.: 1981, *The Mismeasure of Man*, Norton, New York.

Green, D. R. (ed.): 1974, *The Aptitude-Achievement Distinction*, McGraw-Hill, New York.

Harlow, H. F., and Harlow, M. K.: 1962, 'The Mind of Man', in *Year-Book of Science and Technology*, McGraw-Hill, New York, pp. 31—39.

Jensen, A. R.: 1972, *Genetics and Education*, Methuen, London.

Jensen, A. R.: 1973a, *Educability and Group Differences*, Methuen, London.
Jensen, A. R.: 1973b, *Educational Differences*, Methuen, London.
Jensen, A. R.: 1979, 'The Nature of Intelligence and Its Relation to Learning', *Journal of Research and Development in Education* 12, 79—95.
Jensen, A. R.: 1980a, *Bias in Mental Testing*, Methuen, London.
Jensen, A. R.: 1980b, 'Author's Response: Bias in Mental Testing', *Behavioral and Brain Sciences* 3, 359—368.
Jensen, A. R.: 1981, *Straight Talk About Mental Tests*, The Free Press, New York.
Jensen, A. R.: 1982, 'The Debunking of Scientific Fossils and Straw Persons', *Contemporary Education Review* 1, 121—135.
Jensen, A. R.: 1985, 'The Nature of the Black-White Difference on Various Psychometric Tests: Spearman's Hypothesis', *Behavioral and Brain Sciences* 8, 193—219.
Jensen, A. R.: Undated (but circa 1985), 'The g Beyond Factor Analysis', unpublished manuscript, courtesy of A. R. Jensen, University of California, Berkeley, California.
Jensen, A. R.: 1987, 'Differential Psychology: Towards Consensus', in S. Modgil and C. Modgil (eds.), *Arthur Jensen: Consensus and Controversy*, Falmer Press, Lewes, Sussex, pp. 353—399.
Keynes, J. M.: 1936, *The General Theory of Employment, Interest and Money*, Macmillan, London.
Leeuw, J. de, & Meester, A. C.: 1984, 'Over het Intelligente-Onderzoek bij de Militaire Keuringen vanaf 1925 tot Heden' (Intelligence — as Tested at Selections for the Military Service from 1925 to the Present), *Mens en Maatschappij* 59, 5—26.
Lewontin, R. C.: 1977, 'Race and Intelligence', in N. J. Block and G. Dworkin (eds.), *The IQ Controversy: Critical Readings*, Quartet Books, London, pp. 78—92.
Lynn, R.: 1982, 'IQ in Japan and the United States Shows a Growing Disparity', *Nature* 297, 222—223.
Lynn, R., & Hampson, S.: 1986, 'The Rise of National Intelligence: Evidence from Britain, Japan and the U.S.A.', *Personality and Individual Differences* 7, 23—32.
Lynn, R., Hampson, S. L., and Mullineux, J. C.: undated (but circa 1986), 'The Rise of Fluid Intelligence in Great Britain, 1935—1985', unpublished manuscript, courtesy of R. Lynn, University of Ulster, Coleraine, Londonderry, Northern Ireland.
McClelland, D. C.: 1977, 'Testing for Competence Rather Than for "Intelligence" ', in N. J. Block and G. Dworkin (eds.), *The IQ Controversy: Critical Readings*, Quartet Books, London, pp. 45—73.
Mehlhorn, G., and Mehlhorn, H.-G.: 1981, 'Intelligenz-Tests und Leistung' (Intelligence-Tests and Achievement), *Wissenschaft und Fortschritt* 31 (9), 346—351.
Nichols, R. C.: 1987, 'Racial Differences in Intelligence', in S. Modgil and C. Modgil (eds.), *Arthur Jensen: Consensus and Controversy*, Falmer Press, Lewes, Sussex, pp. 213—220.
Rist, T.: 1982, *Det Intellektuelle Prestasjonsnivaet I Befolkningen Sett I Lys av den Samfunns-Messige Utviklinga* (The Level of the Intellectual Performance of the Population Seen in the Light of Developments in the Community), Norwegian Armed Forces Psychology Service, Oslo.
Scafer, E. W. P.: 1985, 'Neural Adaptibility: A Biological Determinant of g Factor Intelligence', *Behavioral and Brain Sciences* 8, 240—241.

Schallberger, U.: 1985, *HAWIK und HAWIK-R: Ein Empirischer Vergleich* (HAWIK and HAWIK-R: An Empirical Comparison) (Tech. Rep.), Psychologisches Institut der Universität, Zurich.

Spearman, C.: 1904, ' "General Intelligence": Objectively Determined and Measured', *American Journal of Psychology* **15**, 201—292.

Sternberg, R. J.: 1985, *Beyond IQ: A Triarchic Theory of Human Intelligence*, Cambridge University Press, Cambridge, England.

Terman, L. M., and Oden, M.: 1959, *The Gifted Group at Mid-Life*, Stanford University Press, Stanford, California.

Thorndike, R. L.: 1973, *Stanford-Binet Intelligence Scale 1972 Norms Tables*, Houghton Mifflin, Boston.

Thurstone, L. L.: 1940, 'Current Issues in Factor Analysis', *Psychological Bulletin* **37**, 189—236.

Tuddenham, R. D.: 1948, 'Soldier Intelligence in World Wars I and II', *American Psychologist* **3**, 54—56.

Vogel, F.: 1980, 'Genetic Influences on IQ', *Behavioral and Brain Sciences* **3**, 358.

Waller, J. H.: 1971, 'Achievement and Social Mobility: Relationships Among IQ Score, Education, and Occupation in Two Generations', *Social Biology* **18**, 252—259.

HENRY KRIPS

QUANTUM MEASUREMENT AND BELL'S THEOREM

INTRODUCTION

The concept of measurement favoured by Einstein, Bohm, Putnam, *et al.* in the context of Quantum Theory (QT) is a realist concept. It encapsulates the realist idea that in measurement we are about the task of reporting an independently existing objective reality. Moreover for Einstein, Bohm, Putnam *et al.* this objective reality is to be characterised classically in terms of physical quantities possessing particular values.[1] Thus an "ideal measurement" is characterised as a process in which a measured value for some physical quantity is produced and that measured value reflects the value possessed by the physical quantity.

Moreover it is traditionally taken as part of this classical realist concept of measurement that we can make sense of a counterfactually construed "measured value" — a value which Q would be measured to have were it measured. And for ideal measurements not only must actual measurement accurately report the possessed values but so also must hypothetical measurements, i.e. the counterfactually construed measured value for Q in a system S at time t even were Q not actually measured in S at t must equal the value possessed by Q in S at t. In short the value possessed by Q in S at t is traditionally seen as responsible for the disposition for S at t to produce a particular response in an ideal Q-measurement apparatus, *viz.* the response of that apparatus registering the particular possessed value.

Formally we can write the metaphysical assumptions just enunciated in the form of two principles. The first principle is the principle of possessed values:

(Poss)   For any system S and time t during the life-time of S and any physical quantity Q for the system S, Q possesses a unique value.

We let '$Q(S, t)$' abbreviate 'the value which Q has in S at t'.

The second principle is the principle of faithful measurement:

41

John Forge (Ed.), Measurement, Realism and Objectivity, 41–58.

(FM)   For any Q, S, t if $Q(S, t)$ exists so does mv(Q, S, t), the counterfactually construed measured value of Q for S at t, and mv(Q, S, t) = $Q(S, t)$.

The measurement referred to in (FM) is implicitly taken to be ideal.

The result which I shall prove here is that these two traditional metaphysical principles together with the "locality principle", which says that there are no faster than light transmissions of causal influences, cannot all be imposed upon standard QT. Indeed an even stronger result can be obtained, *viz.* that standard QT together with locality are not consistent with the following consequence of (FM) and (Poss):

(CD)   For any Q, S and t during the life-time of S, mv(Q, S, t) exists.

('(CD)' here stands for 'Counterfactual Definiteness'.) Thus the traditional realist notion that measurement results are the displays of some "real world" dispositions — dispositions which inhere in the real world irrespective of whether measurements take place or not — must be given up if we are to preserve locality and QT.

It may be pointed out that this result is not all that surprising even from a classical point of view. After all even from a classical point of view the notion that there are ideal *qua* perfectly accurate measurements can be questioned. Thus verificationists may point out that for continuous-valued physical quantities it makes no sense to talk of a perfect measurement since none can be achieved in practice. At best, they may argue, it only makes sense to talk of measurements which are "as perfect as one likes", or (at worst) perfect to within a certain finite degree of accuracy. But it will turn out that the proof here still works if we restrict ourselves to discrete-value (indeed two-valued) physical quantities, and even if we only suppose approximately faithful measurements, i.e. if we replace the final clause in (FM) by the weaker:

$$\text{mv}(Q, S, t) \quad \text{approximately} = Q(S, t)$$

(where the degree of approximation allowable here is quite large). So the verificationists' strategy for coming to terms with the above result from within a classical point of view does not succeed.

Moreover the verificationists' remarks here simply miss the point that it is (CD), and not just (FM) plus (Poss), which is shown to be inconsistent with QT; and (CD) (unlike (FM)) makes no assumption that there are ideal (or even "semi-ideal") measurements which reflect (or nearly reflect) the values possessed by measured quantities. Indeed (CD) merely makes the traditional assumption that measured values are the displays of certain counterfactually construed dispositions which the measured systems have. Thus even if a classically minded verificationist is happy to give up (FM) the above result will present him with a contradiction between locality, QT and one of his traditional principles, *viz.* (CD).

The proof which I shall give here is adapted from the sequence of increasingly sophisticated "No hidden variables proofs" which can be found in the literature — starting with von Neumann, then Wigner, Bell, and finally Stapp and Eberhard.[2] In point of fact the proof which I shall give does not directly show an inconsistency between QT, locality, (FM) and (Poss). It presents us instead with a trilemma: if we are to retain standard QT we must *either* assume an (indefinitely improbable) coincidence *or* allow non-locality *or* give up (CD) (and hence (FM) or (Poss)). In the conclusion to this paper I shall discuss some of the options presented by this proof; and argue that it is indeed giving up the traditional notion of measurement — as enshrined in (CD) — which is the most satisfactory option. But I shall also argue, following Redhead [1983] and Hellman [1982], that this option need not unduly disturb classical sensibilities.

Finally in this introduction I note that the only restriction we will need to make on "ideal measurements" in the course of the following proof, other than (CD), is that they satisfy the restrictions placed on ideal measurements by QT, *viz.* the Born Statistical interpretation.[3] Thus all the verificationist's objections are simply bypassed.

SECTION I — THE PROOF

Consider a collection of N electron pairs at time t when each pair is in a special state called 'the singlet state', and denoted 'f'. The details of this singlet state will not concern us here; we simply note that it is a version of the famous Einstein—Podolski—Rosen state (or at least the modification of it put forward by Bohm and Aharonov).

Let L_n and R_n be the left hand and right hand members of the nth pair; and suppose L_n and R_n are well separated from each other at t and from the members of any other pair, for all n. Let 'x' denote the physical quantity which is the component of twice the "spin" of an electron along the direction **x**. This physical quantity is characterised by having only two possible values $+1$ and -1. We shall only need to consider four such physical quantities, **a**, **a'**, **b**, **b'**, where these are the spin components along the **a**, **a'**, **b**, **b'** directions respectively.

Now let 'a_n' and 'a'_n' be rigid designators for those numbers which in *this* world happen to be the respective counterfactually construed measured values for **a** and **a'** were **a**, **a'** respectively measured on L_n at t. I.e. in this world, but not all possible worlds, we have $a_n = \text{mv}(\mathbf{a}, L_n, t)$, and similarly $b_n = \text{mv}(\mathbf{b}, R_n, t)$. Introduce b_n and b'_n similarly so that these are equal to $\text{mv}(\mathbf{b}, R_n, t)$ and $\text{mv}(\mathbf{b'}, R_n, t)$ respectively in this world. Note that the existence of the $4N$ counterfactually construed measured values $\text{mv}(\mathbf{a}, L_n, t)$, $\text{mv}(\mathbf{a'}, L_n, t)$, $\text{mv}(\mathbf{b}, R_n, t)$, $\text{mv}(\mathbf{b'}, R_n, t)$ for various n is a non-trivial consequence of (FM) and (Poss), or of just (CD) for that matter; but since the various a_n, b_n, a'_n, b'_n are all numbers, their existence is trivial, although their equality with the respective measured values is not trivial of course.

We now ask the following question: what is the value which **a** would be measured to have were *both* **a** measured on L_n at t and **b** measured on R_n at t? To answer this question we invoke the following principle:

(Loc)$_1^*$ The outcome of measurement of **a** on L_n at t would be the same were a **b** measurement performed on R_n at t in addition to the **a** measurement on L_n at t.

This principle tells us immediately that the answer to the previous question is 'a_n'. Moreover the principle (Loc)$_1^*$ itself seems to be a straightforward locality requirement for the following reason. Because L_n at t and R_n at t are spatially well-separated (by as great a distance as we please) no light signal can travel from R_n at t to reach L_n before the result of the **a** measurement started on L_n at t is fixed. Hence the locality principle tells us that **b** being measured on R_n at t can have no effect on the outcome of measurement on L_n at t (otherwise such an effect would be transmitted faster than light — contrary to the requirement of locality). And this in turn implies that the measured value of **a** on L_n at t would be the same whether or not **b** were measured on R_n at t — which implies (Loc)$_1^*$.

But the preceding argument to show that (Loc)$_1^*$ is a locality measurement is invalid. The flaw in it is clear once we realise that from '**b** measurement has no effect on the outcome of the **a** measurement' it does not follow that the mv of **a** were **a** and **b** both measured is the same as if just **a** were measured. In particular it can happen that the former is true and the latter false just by virtue of there being no definite counterfactual mv for **a** were **a** and **b** both measured.[4] Having said this however, in the next section I shall show that $(\text{Loc})_1^*$ is (with one qualification) after all a locality requirement; it's just that the previous argument for showing this is too simple. For now however I simply take for granted that $(\text{Loc})_1^*$ is a locality requirement, and leave the discussion of this to the next section.

We next ask: What is the value which **a** would be measured to have for L_n at t were both **a** measured on *all* the left hand electrons at t *and* **b** measured on *all* the right hand electrons at t? The answer is again 'a_n'; this time because (as I shall show, with one qualification, in the next section) the large separation of the various electron pairs from each other together with locality imply an obvious extension of $(\text{Loc})_1^*$ to:

$(\text{Loc})_2^*$ The outcome of measurement of **a** on L_n at t would be the same were **a**, **b** measurements performed on *all* the pairs at t as it would be were **a** measured on L_n at t.

Moreover this same answer can be shown to hold whatever n.

Now abbreviate '**a**, **b** are measured on L_n, R_n at t for all n' by 'M_{ab}' and '**a** is measured to have value a_n on L_n at t' by 'A_n' and '**b** is measured to have value b_n on R_n at t' by 'B_n'. Then what we have just shown is that, assuming locality

$$(M_{ab} \rightarrow A_n) \quad \text{and} \quad (M_{ab} \rightarrow B_n), \quad \text{for all } n.$$

Hence, using the modal inference rule of addition:

$$p \rightarrow q, \ p \rightarrow r \quad \therefore \quad p \rightarrow q \cdot r$$

we see that

(i) $M_{ab} \rightarrow (A_n \cdot B_n \quad \text{for all } n)$

We also assume that whether QT holds is independent of what is measured on all the pairs at t. On a Lewis analysis of (causal) independence this implies that

(ii) $M_{ab} \rightarrow \text{QT} \quad \text{and} \quad \sim M_{ab} \rightarrow \text{QT}.$[5]

Similarly we assume that the states of the various pairs at t are independent of what is measured on all the pairs of t. (In particular there is no pre-measurement interaction which restricts the measurable states. Note that this is not contradicted by the "projection postulate", which only says that the state *after* the measurement has finished is modified from the original state unless the initial state is an eigenstate of the measured quantity.) If we let 'F' stand for 'The states of all the pairs at *t* are *f*' this means that

(iii) $M_{ab} \rightarrow F$ and $\sim M_{ab} \rightarrow F.$

These two independence conditions (ii) and (iii) can both be seen as aspects of what we may call 'the principle of Freedom of Measurement', since they say that what is measured is not restricted by the laws of nature or the state of the measured system. Applying the rule of addition to (i), (ii) and (iii), we see that

(iv) $M_{ab} \rightarrow ((A_n \cdot B_n$ for all $n) \cdot QT \cdot F)$

Now we assume that it is indeed possible to measure **a**, **b** on *all* the L_n, R_n at *t*; more briefly we assume Poss M_{ab} (where 'Poss' is the possibility operator). From this point on the proof is somewhat easier to follow if we use Lewis's possible world semantics for counterfactuals. Doing this it follows from (iv) and Poss M_{ab} that there is some possible world — let it be w_1 — at which QT and F and $A_n \cdot B_n$ all hold for all *n*. (We can let w_1 be the nearest possible world to the actual at which A_n, B_n all hold, for all *n*, if there is such a nearest possible world.) But because A_n and B_n hold at w_1 for all *n* it follows that the relative frequency with which measured values of **a**, **b** at the world w_1 take the values *x*, *y* over the *N* pairs is equal to the relative frequency with which the a_n, b_n (which are the numerical values in the actual world of the counterfactually construed measured values of **a**, **b** respectively) are equal to *x*, *y* respectively over the *N* pairs. (This is because A_n and B_n hold at w_1 for all *n*.) Formally we can write this equality as:

(v) $RF_1(\mathbf{a}, \mathbf{b} \text{ meas } x, y) = RF(a, b = x, y)$

(where the subscript '1' on the left hand expression indicates that it is a relative frequency for quantities defined at w_1).

The previous argument can then be repeated with **a′**, **b** instead of **a**, **b**, to demonstrate that there is a world w_2 at which **a′**, **b** are measured on all the pairs at *t*, each pair is in state *f* at *t*, and QT holds, *and*

(vi) $RF_2(\mathbf{a'}, \mathbf{b} \text{ meas } x, y) = RF(a', b = x, y).$

Note that w_2 must be a different world from w_1 since the physical quantities **a** and **a'** are incommensurable if, as we suppose, the corresponding directions **a** and **a'** differ. And similarly, we derive that there are worlds w_3 and w_4 at which **a**, **b'** and **a'**, **b'** respectively are measured on all the pairs at t, each pair is in state f at t, QT holds, and

(vii) $RF_3(\mathbf{a}, \mathbf{b'} \text{ meas } x, y) = RF(a, b' = x, y)$

(viii) $RF_4(\mathbf{a'}, \mathbf{b'} \text{ meas } x, y) = RF(a', b' = x, y)$

Now consider P_1 (**a**, **b** meas x, y for L_n, R_n), which is the probability at w_1 that a measurement on L_n, R_n at t of **a**, **b** yields the measured values x, y respectively. The latter probability is given by QT, since QT holds at w_1; and since the nth pair at w_1 is in the state f at t, the probability is equal to $|(f, f_{\mathbf{a},x}\mathbf{x}f_{\mathbf{b},y})|^2$, where $f_{\mathbf{a},x}$ and $f_{\mathbf{b},y}$ are the eigenvectors of **a**, **b** for eigenvalues x, y on the Hilbert spaces for L_n, R_n respectively. (f, of course, is in the direct product of these spaces.)

For our purposes we do not need to know exactly what this last expression in brackets means, except to note that (as one would expect) the probability is the same for all n. Moreover it is easily shown, from QT, that since each pair is in a pure state at t (represented by f) then the various outcomes of measurements of **a**, **b** on the various pairs at t are statistically independent, i.e. the finite set of measurements at t (taken in any order) forms a Bernoullian sequence of trials. As such the law of large numbers tells us that

$$RF_1(\mathbf{a}, \mathbf{b} \text{ meas } x, y) \approx |(f, f_{\mathbf{a},x}\mathbf{x}f_{\mathbf{b},y})|^2,$$

where the approximation here is in the usual statistical sense. (I.e. the probability that the left hand quantity differs from the right hand quantity by any finite amount can be made as small as we like by letting N be large enough.)

Hence, by (v), we see that

(v)' $RF(a, b = x, y) \approx |(f, f_{\mathbf{a},x}\mathbf{x}f_{\mathbf{b},y})|^2.$

And we similarly derive:

(vi)' $RF(a, b' = x, y) \approx |(f, f_{a,x}\mathbf{x}f_{b'y})|^2$

(vii)' $RF(a', b = x, y) \approx |(f, f_{\mathbf{a}',x}\mathbf{x}f_{\mathbf{b},y})|^2$

(viii)' $RF(a', b' = x, y) \approx |(f, f_{\mathbf{a}',x}\mathbf{x}f_{b'y})|^2.$

(Note that these same key equalities could have been derived without the Lewis possible world machinery.) From (iv) and Poss M_{ab} it follows that

(ix) $\text{Poss}((A_n \cdot B_n \text{ for all } n) \cdot QT \cdot F)$,

because of the inference rule:

$$p \to q, \text{ Poss } p \therefore \text{ Poss } q.$$

But $A_n \cdot B_n$ for all n trivially implies that $RF(\mathbf{a}, \mathbf{b} \text{ meas } x, y) = RF(a, b = x, y)$. Hence (ix) implies:

(x) $\text{Poss}((RF(\mathbf{a}, \mathbf{b} \text{ meas } x, y)$
$= RF(a, b = x, y) \cdot (A_n \cdot B_n \text{ for all } n) \cdot QT \cdot F)$

since if p implies r then Poss $p \cdot q$ implies Poss $p \cdot r \cdot q$. But, as we have seen, $(A_n \cdot B_n$ for all $n \cdot QT \cdot F)$ together imply that $RF(\mathbf{a}, \mathbf{b} \text{ meas } x, y) \approx |(f, f_{\mathbf{a}, x} \mathbf{x} f_{\mathbf{b}, y})|^2$. Hence, from (x),

$$\text{Poss }(RF(a, b = x, y) \approx |(f, f_{\mathbf{a}, x} \mathbf{x} f_{\mathbf{b}, y})|^2)$$

And (v)$'$ immediately follows since both the left hand and right hand side of this last equality are rigid designators for numbers rather than world-dependent numerical functions. And (vi)$'$—(viii)$'$ can be derived similarly.)

Now with the help of these equalities (v)$'$—(viii)$'$ we can derive the required contradiction. Define

$$E(\mathbf{a}, \mathbf{b}) = \sum_n a_n \cdot b_n / N.$$

By grouping together all those terms of form $a_n \cdot b_n$ in this last sum for which a_n, b_n take the same values x, y we see that

$$E(\mathbf{a}, \mathbf{b}) = \sum_{x, y} x \cdot y \cdot (\text{number of } n \text{ such that } a_n, b_n = x, y)/N$$

$$= \sum_{x, y} x \cdot y \cdot RF_0(\mathbf{a}, \mathbf{b} \ x, y)$$

$$\approx \sum_{x, y} x \cdot y \cdot |(f, f_{\mathbf{a}, x} \mathbf{x} f_{\mathbf{b}, y})|^2.$$

But the right hand side of this last equality is well known in QT as the "expectation value" of $\mathbf{a} \times \mathbf{b}$ in state f and is traditionally denoted

'$\langle \mathbf{a} \times \mathbf{b} \rangle$' (the '$f$' dependence in this last expression is left implicit). Thus we have shown

(xi) $E(\mathbf{a}, \mathbf{b}) \approx \langle \mathbf{a} \times \mathbf{b} \rangle.$

And similarly we derive:

(xii) $E(\mathbf{a}, \mathbf{b}') \approx \langle \mathbf{a} \times \mathbf{b}' \rangle, E(\mathbf{a}', \mathbf{b}) \approx \langle \mathbf{a}' \times \mathbf{b} \rangle, E(\mathbf{a}', \mathbf{b}') \approx \langle \mathbf{a}' \times \mathbf{b}' \rangle.$

We now introduce a straightforward theorem of QT, *viz.* that for certain $\mathbf{a}, \mathbf{a}', \mathbf{b}, \mathbf{b}'$

(xiii) $|\langle \mathbf{a} \times \mathbf{b} \rangle + \langle \mathbf{a} \times \mathbf{b}' \rangle + \langle \mathbf{a}' \times \mathbf{b} \rangle - \langle \mathbf{a}' \times \mathbf{b}' \rangle| > 2.$

Indeed QT tells us that this expression has a value which is as much as 5/2 for certain $\mathbf{a}, \mathbf{b}, \mathbf{a}', \mathbf{b}'$, irrespective of the size of N.

But consider:

$$|E(\mathbf{a}, \mathbf{b}) + E(\mathbf{a}, \mathbf{b}') + E(\mathbf{a}', \mathbf{b}) - E(\mathbf{a}', \mathbf{b}')|.$$

This is just $\left| \sum_n (a_n b_n + a_n b'_n + a'_n b_n - a'_n b'_n)/N \right|$, which is

$$\leqslant \sum_n \left| (a_n b_n + a_n b'_n + a'_n b_n - a'_n b'_n) \right|/N$$

$$= \sum_n \left| a_n(b_n + b'_n) + a'_n(b_n - b'_n) \right|/N.$$

Since b_n and b'_n are both $+1$ or -1 there are two cases: either $b_n = b'_n$ or $b_n = -b'_n$. In the first case $|a_n(b_n + b'_n) + a'_n(b_n - b'_n)|$ is equal to $|a_n 2 b_n|$, which, since $|a_n| = |b_n| = 1$ is just 2. Similarly in the second case the expression takes the value 2. And hence

(xiv) $|E(\mathbf{a}, \mathbf{b}) + E(\mathbf{a}, \mathbf{b}) + E(\mathbf{a}', \mathbf{b}) - E(\mathbf{a}', \mathbf{b}')| \leqslant 2$

(which is just Bell's inequality).

But (xiv) can be made to contradict (xiii) together with (xi) and (xii), by simply making N large enough so that the degree of approximation in (xi) and (xii) is small enough. Or, more correctly, we should say that the probability of there being a contradiction can be made arbitrarily close to 1 by making N large enough. Note that we can take the limit here and let N go to infinity, since the Hilbert space of a countable number of systems each with a finite dimensional Hilbert space is still separable, and hence comes within the province of QT. The Strong Law of large numbers then enables us to take the probability as equal to 1 in the limit.[6]

Thus we have shown that the traditional metaphysical assumption (CD) (*and a fortiori* (Poss) together with (FM)), together with the relevant locality assumption, contradict QT with a probability which can be made as near as we like to 1 (and in the limit, as N goes to infinity, is equal to 1). Moreover we see that this same result can still be derived if we consider non-ideal measurements which merely satisfy the Born Statistical interpretation approximately.[7] In other words, short of admitting a coincidence which has a probability as small as we like, we must give up either (CD) *or* locality if we are to preserve the laws of QT.

SECTION II – INTERPRETATION

There is some unfinished business left over from the previous section, *viz.* showing that the various principles of form $(Loc)_2^*$ used in the proof are indeed locality requirements. These various principles can be brought together into the following form:

(Loc)*   For all **a** (and **b** respectively) and for all n, if the value which **a** (**b** respectively) would be measured to have were it measured on L_n (R_n respectively) at t exists then so does the value which **a** (**b** respectively) would be measured to have for L_n (R_n respectively) at t were **a**, **b** measured on *all* the pairs at t, and these two measured values are equal.

('a' and 'b' have here been redefined as variables with 'a', 'a''' and **b**, 'b''' respectively as their values, and n has been taken as a variable with range 1 to N). More succinctly we can write (Loc)* in terms of the abbreviations of the previous section as:

(Loc)*   For all   **a, b,** n if $a_n \to A_n$ then $a_n \cdot M_{ab} \to A_n$.

where 'a' is now a variable with the four directions **a, a', b, b'** as its values, and 'a_n' abbreviates either 'a is measured on L_n at t' or 'a is measured on R_n at t' (whichever is appropriate). Thus (Loc)* failure involves the truth of some '$a_n \to A_n$' together with the falsity of the corresponding '$a_n \cdot M_{ab} \to A_n$', for some **a, b,** n. (Note that given (CD) all the '$a_n \to A_n$' will be true.)

But the falsity of '$a_n \cdot M_{ab} \rightarrow A_n$', even where '$a_n \rightarrow A_n$' is true, may simply arise because $mv_{ab}(a_n)$, the value which a would be measured to have for R_n at t were a, b measured on all the L_n, R_n at t, does not exist, e.g. if there are several different "equally possible" measured values for a under the conditions of a, b being measured on all the L_n, R_n at t. And *prima facie* it would seem that this need not involve any objectionable sort of non-locality, *qua* the faster than light transmission of causal influences. Indeed as we shall see shortly, the non-existence of $mv_{ab}(a_n)$ may only involve an assumption of *indeterminism*. However for now let us bracket off this possibility — of the non-existence of $mv_{ab}(a_n)$. If we do this then the falsity of (Loc)* does indeed imply that, for some a, b, n,

(xv) $a_n \rightarrow A_n$ and $a_n \cdot M_{ab} \rightarrow {\sim} A_n$.[8]

And this together with Poss M_{ab} can then be shown (see next note) to imply both that

(xvi) $a_n \cdot {\sim} M_{ab} \rightarrow A_n$ and $a_n \cdot M_{ab} \rightarrow {\sim} A_n$

and that the counterfactuals in (xvi) are "non-trivial" in the sense that their respective antecedents are possible.[9]

But it is easy to see that (xvi) involves a form of non-locality provided we adopt a traditional model for causal relevance as given by:

(CR)   The event described by p is causally relevant to the event described by q if p is a necessary part of a sufficient condition for q, i.e. there is an event described by r for which non-trivially

$$r \cdot p \rightarrow q   \text{and}   r \cdot {\sim} p \rightarrow {\sim} q.[10]$$

I.e. from (CR) and (xvi) it follows that ${\sim} M_{ab}$ is causally relevant to A_n, i.e. that whether or not measurements take place on all the electrons at t is causally relevant to the outcome of the a measurement at t on L_n/R_N (whichever is appropriate). And given the large spatial separation of the various electrons this does involve an objectionable form of non-locality.

Although it must immediately be added here that this failure of locality only has a metaphysical significance. The change in measured value of a for L_n/R_n at t caused by the measurement of a, b on the other electron pairs at t is not one which is ever instantiated but simply remains as a possibility. This is because it also follows from (xv) that ${\sim} M_{ab}$.[11] And hence the particular a, b for which (xv) holds are not in fact ever measured. They are always among the "other possible" directions for measurement. (This result squares well with Redhead's result that "Bell telephones" cannot be constructed [1986]).

In sum then we see that given just the one condition, *viz.* that the $mv_{ab}(a_n)$ exists, (Loc)* is indeed a locality requirement, i.e. its failure does involve a locality failure. And hence we see that if (as I shall now do for argument's sake) we discount the possibility of an indefinitely improbable coincidence and accept the laws of QT then the proof of the previous section presents us with the following options:

> Either we must admit non-localities (of an objectionable kind) *or* allow the non-existence of counterfactual measured values of the form $mv_{ab}(a_n)$ *or* allow the non-existence of counterfactuals of the form $mv(\mathbf{a}, L_n, t)$ (or $mv(\mathbf{a}, R_n, t)$ whichever is appropriate).

The third of these three possibilities is just the failure of (CD), whereas, given (CD), the first two possibilities are ways for (Loc)* to fail. But note that neither of the last two possibilities involve objectionable forms of "non-localities".

I shall now argue that it is one of these last two options, in particular the last one (of giving up (CD)), which even for a *classical* metaphysicist is the most natural and least demanding option. To show this let us ask what is involved in giving up (CD). It means that we must give up the view that there is always some counterfactually definite measured value. And this in turn means that we lose the explanation of measurement results as the displays of underlying dispositions, dispositions which are related to corresponding possessed values via (FM).

But is this such a great loss? Other perfectly satisfactory explanations of measurement results are possible, even within the classical metaphysical scheme, depending on how broadly we draw the category 'classical'. For example suppose we allow classical metaphysics to include *indeterministic* processes, processes for which there are several possible outcomes none of which is determined by the history of the process. We can explicate such indeterminism within a possible world framework as saying that there is no one possible outcome which occurs in all nearest possible "reruns" of the process. (Although on a Lewis model for possible worlds this analysis will only work for strictly counterfactual — i.e. non-actual — processes, since according to Lewis the nearest possible reruns of a process will be the actual process itself if that process takes place in the actual world.) Or, using the more general framework of counterfactuals, we can say that if we let 'P' stand for a historical description of the process and 'O_i' stand for the various possible results, $i = 1, 2, \ldots$, then P is indeterministic if '$P \rightarrow O_i$' is false for each i. If we allow such indeterministic processes then we immediately see how (CD) can be rejected consistently with explaining the results of individual measurements. We simply postulate that the measurement processes (of QT) are indeterministic, so that counter-factually construed measured values (*qua* results of measurement) may not exist; and then explain the distribution of results of individual mea-surements as a manifestation of this indeterminism. (*Cf.* the Salmon—Jeffreys statistical model for explanation [1971].) Both Redhead [1983] and Hellman [1982] make this point.

Moreover this sort of indeterminism is, I claim, perfectly understand-able even within a classical metaphysics. After all, as Popper has pointed out, it is a fallacy to see classical physics as deterministic. (Although 'deterministic' is used by Popper in the somewhat different sense that the equations of motion allow the states of systems to be fixed by past states and suitable "boundary conditions".[12]) In particular Popper points out that it is only a small number of (admittedly rather central) examples of classical processes which are deterministic in this last sense; and as soon as we consider classical equations of any complexity (e.g. with velocity-dependent forces) then determinism in this sense is seen to fail. Although whether we then have "indetermin-ism" in the *counterfactual* sense introduced above is of course another matter. Nevertheless the point remains that determinism in Popper's sense fails within classical physics, and this, I claim, is at least enough to cut the ground out from under a presumption in favour of determinism in the counterfactual sense in classical metaphysics.

It is however important to note (and here I go beyond Redhead and Hellman) that this postulate of indeterministic measurements can be understood in either a radical or a conservative way. The radical way is to see (FM) as totally failing: the measured values simply emerge as they would from a random number generator, with no connection to an independently existing domain of possessed values (if indeed such exists). There is also however a conservative way of giving up (CD) and postulating indeterminism, *viz.* to preserve both (Poss) and what we will call '(Local FM)':

(Local FM) Were a measurement performed then it would be the case
          that the result of measurement agrees with the possessed
          value.

Note that in (Local FM) the possessed and measured values are
referred to *within* the scope of the counterfactual, by contrast with the
earlier (FM). This (Local FM) says that were a measurement to take
place then, however the possessed and measured values may vary
across the various equally nearest reruns of the measurement, on any
one rerun the measured value which emerges is always equal to the
particular possessed value for that rerun. And so construed (Local FM)
does of course leave it perfectly open for the measurement process to
be indeterministic but nevertheless faithful. Indeed it will be both
indeterministic and faithful precisely if and only if the possessed values
which are faithfully reflected in the measurement process are themselves indeterministic (i.e. vary across the nearest reruns). And it is
precisely this last condition which I take to characterise a "conservative" account of the indeterministic nature of the measurement process.
Note that on this account (unlike the Copenhagen account) it is what is
measured rather than the process of measurement itself which generates
the indeterminism. I.e. the measurements are seen as measurements on
an already (i.e. independently) indeterministic reality, not as generators
of that indeterminism.

It is this conservative option which I take to be the one most strongly
indicated among those left open by the results of section I. It is an
option which has the virtues of allowing the classical metaphysicist to
maximally preserve his principles (e.g. some idea of faithful measurement) without the need to give up locality. This may make it seem that I
am claiming the results of section I do not have *any* significant consequences even for classical metaphysics. But that is far from the case.
The results of section I, I claim, point to the need for indeterministic
processes even at a fundamental level, and more particularly to the
indeterministic nature of the measurement processes of QT (albeit not
necessarily of the measurement processes themselves).

SECTION III

What we have shown (in agreement with both Redhead and Hellman) is
that Stapp and Eberhard type proofs pose no threat to either realism or

causal principles such as locality, provided we countenance indeterminism. What we have also shown (going beyond Redhead and Hellman) is that if we try to preserve the existence of certain counterfactual mvs (and in particular (CD)) then we are committed to a form of nonlocality which is objectionable, *qua* involving faster than light transmission of causal influences (albeit not allowing the construction of a "Bell telephone"). This last conclusion then puts extra pressure in favour of the indeterministic option of rejecting (CD).

I shall conclude by suggesting how the conservative option of indeterminism plus (Poss) and (Local FM), which I canvassed at the end of the last section, can be made more plausible. In particular I shall give some reasons in favour of the supposition which I canvassed as part of that option that the values possessed by some physical quantities are indeterministic in the sense that they differ across different nearest possible reruns of the experimental set-up stipulated. In particular I shall argue for the indeterminism of the values taken by **a** for L_n at t given that $L_n + R_n$ at t are in the singlet state f.

The singlet state of $L_n + R_n$ at t is, as I have indicated, essentially an Einstein—Podolski—Rosen state. In particular this means that the density operator for L_n at t is

$$\tfrac{1}{2}\mathbf{P}(f_{\mathbf{a},1}) + \tfrac{1}{2}\mathbf{P}(f_{\mathbf{a},-1})$$

($f_{\mathbf{a},1}$ here is the eigenvector of the operator representing **a** for eigenvalue $+1$). This diagonal form for the density operator of L_n at t means that from the point of view of observations on L_n alone at t there are no "interference effects" between the possible values for **a**. And I suggest that we take this last fact as warrant for saying that **a** really does possess a particular value, indeed that it has value 1 with probability $\tfrac{1}{2}$ and value -1 with probability $\tfrac{1}{2}$. Moreover these probabilities (of **a** having particular values) are *intrinsic*, i.e. not reducible to 1 or 0 by increasing the information we have about the value of **a** for L_n at t without also changing the state of the *total* experimental set up consisting of the L_n and R_n for all n.[13] And I further suggest that it is because there is this non-trivial intrinsic probability distribution over the possible values of **a** for L_n at t that we are entitled to assert that the value of **a** for L_n at t is *indeterministic*, in other words:

(P)   Non-trivial intrinsic statistical distribution over possible results entails indeterminism.

The justification of this last principle takes us well beyond what we can discuss here, but I think the idea is clear enough: If reruns of an experiment produce a statistical spread over various possible results, and this ·spread is *not reducible* (in the above sense), then it does indeed seem wrong to single out *one* of the possible results as really "nearest". (This is particularly so if, as in the case here considered, the probabilities are all equal.)

These last suggestions must of course contend with the following criticism. Although there are no interference effects between possible values of **a** for L_n at t from the point of view of observations on L_n alone at t, there are (in the case considered here) interference effects between the possible values of **a** for L_n at t from the point of view of observations on $L_n + R_n$ at t.[14] And this latter fact takes away any initial plausibility which the view that **a** possesses a particular value for L_n at t may have had (had in virtue of the absence of interference effects between various possible values of **a** for L_n at t from the point of view of L_n *alone*).

I suggest taking a tough line against this last criticism; and simply insisting that it is only those observations on a system for which a physical quantity is represented by a *non-degenerate* operator which are relevant to deciding whether that physical quantity does possess a value (and with what probabilities). The results of observations on more inclusive systems are simply *stipulated* as irrelevant. Hence since the physical quantity **a** for L_n at t is represented by a non-degenerate operator in L_n but *not* in $L_n + R_n$, the required result follows. This stipulation will seem arbitrary and *ad hoc* when considered just in the present context. But it is a stipulation which successfully resolves various of the well-known paradoxes of QT.[15] And when this is taken into account it is seen not as *ad hoc* but rather as an independently supported component of an overall interpretative framework for QT.

University of Melbourne, Victoria.

NOTES

[1] For a discussion of this realist idea in the context of QT see Hooker [1972] for example.
[2] See the general discussion of these proofs in Clauser and Shimony [1978].
[3] If S at t is in state f at t and the discrete-valued physical quantity Q is represented

by the non-degenerate operator $\mathbf{Q}$ on the Hilbert space for S then the probability of Q being measured to have value Q_i on S at t is $|(f_i, f)|^2$ where f_i is the unique eigenvector of Q for eigenvalue q_i.

[4] Clearly a Lewis counterfactual, rather than a Stalnaker counterfactual is assumed here — see Lewis [1973]. I.e. for Stalnaker conditionals, it is either the case that $p \rightarrow q$ or $p \rightarrow \sim q$ for any p, q and hence were a measurement performed which had some result (as ideal measurements must) then there would be some one measured value which it would have to have.

[5] See Lewis [1975]. Lewis only defines causal relevance, but we take causal independance as the obvious contrary of this.

[6] See the discussion of the laws of large numbers in Feller [1950].

[7] *A fortiori* we see that this proof shows (with probability as near to 1 as we like) that (Poss), together with a weakening of (FM) which only requires measurements to be approximately faithful, contradict QT.

[8] Because of the valid inference: $\sim (p \rightarrow q), (p \rightarrow q$ or $p \rightarrow \sim q) \therefore p \rightarrow \sim q$.

[9] The step from (xv) to (xvi) is justified by firstly pointing out that the counterfactuals in (xv) are non-trivial, i.e. their antecedents are possible, just by virtue of Poss $\mathbf{M_{ab}}$ (which clearly implies both Poss $a_n \cdot \mathbf{M_{ab}}$ and Poss a_n since $\mathbf{M_{ab}}$ implies a_n), and by using the following theorem:

If $p \rightarrow q$ *non-trivially and* $p \cdot r \rightarrow \sim q$ *non-trivially then* $p \cdot \sim r \rightarrow q$ *non-trivially*

I demonstrate this theorem using a Lewis semantics for convenience:

Proof: Assume $p \rightarrow q$ non-trivially and $p \cdot r \rightarrow \sim q$ non-trivially. On a Lewis analysis for counterfactuals this means that there is a sphere of worlds S_1 containing p-worlds so that within S_1 all p-worlds are q-worlds; and there is a sphere S_2 containing $p \cdot r$ worlds so that all $p \cdot r$ worlds within S_2 are $\sim q$ worlds. But clearly this is only possible if none of the worlds in S_1 are also r-worlds; and hence all p-worlds within S_1 are $\sim r$ worlds. Hence all $p \cdot \sim r$ worlds within S_1 are q worlds. Moreover there must be some such $p \cdot \sim r$ worlds within S_1. Hence $p \cdot \sim r \rightarrow q$ non-trivially.

[10] This analysis *qua* analysis of 'p is a cause of q' rather than of causal relevance, can be found in McKie [1975] for example. I follow Lewis however in taking the relation of being a cause of as transitive; and hence (CR) is more suitably seen as a definition of causal relevance with the corresponding notion of causation then being derived from (CR) in the usual way — as discussed by Lewis on p. 187 [1975] for example. The McKie model of causal relevance here is weaker than the Lewis model used in section I above. Note that the non-triviality clause in (CR) is needed or else (CR) would have the ridiculous consequence that any p is causally relevant to any q by simply letting the r in (CR) be $p \cdot q$, or $-p \cdot -q$ for that matter.

[11] We show this by *Reductio*. Suppose $\mathbf{M_{ab}}$ holds in the actual world. Then so does a_n, and hence the nearest possible world to the actual at which a_n holds is unicuely the actual world itself. Hence A_n holds at the actual world so that '$a_n \cdot \mathbf{M_{ab}} \rightarrow \sim A_n$' is false at the actual world — in contradiction with (xv).

[12] See Popper's discussion in his [1982].

[13] In particular if we try to increase information about the value of $\mathbf{a}$ for L_n at t by

measuring it then we change the state of $L_n + R_n$ at t by reducing it from the Einstein—Podolsky—Rosen form:

$$(f_{a,1} \times f_{a,-1} + f_{a,1} \times f_{a,-1})/\sqrt{2}$$

to a mixture with density operator:

$$\tfrac{1}{2}\mathbf{P}(f_{a,1} \times f_{a,-1}) + \tfrac{1}{2}\mathbf{P}(f_{a,-1} \times f_{a,1}).$$

[14] In particular if we observe physical quantities for $L_n + R_n$ which are not factorisable between L_n and R_n *and* are not compatible with **a**.

[15] See the discussion in Krips, *The Metaphysics of Quantum Theory* (OUP, forthcoming). In that book I take a somewhat different line to the one here. It is the existence of the intrinsic probability distribution over various possible outcomes which directly explains the non-existence of the counterfactual measured values, rather than the explanation being mediated by an assumption of indeterminism. This enables me to bypass the difficult issue of whether determinism is compatible with objective chance — contrary to (P) above.

BIBLIOGRAPHY

Clauser, J. and Shimony, A., *Reports on Progress in Physics* **41**, 1881 (1978).
Feller, W., *An Introduction to Probability Theory and its Applications* (Addison Wesley, 1950).
Hellman, G., *Synthese* **53**, 461 (1982).
Hooker, C., in *Pittsburgh Studies in the Philosophy of Science* **5**, ed. R. Colodny (Pitt. U.P., 1972).
Lewis, D., in *Causation and Conditionals*, ed. E. Sosa (OUP, 1975).
Lewis, D., *Counterfactuals* (Blackwell, 1973).
McKie, J., in *Causation and Conditionals*, ed. E. Sosa (OUP, 1975).
Popper, K., *The Open Universe* (Hutchinson, 1982).
Redhead, M., in *Space, Time and Causality*, ed. R. Swinburne (Reidel, 1983).
Redhead, M., *Relativity and Quantum Mechanics — Conflict or Peaceful Coexistence*, to appear in *New York Academy of Sciences*, 1986.
Salmon, W., *Statistical Explanation and Statistical Relevance* (Pitt. U.P., 1971).

S. O. FUNTOWICZ AND J. R. RAVETZ

QUALIFIED QUANTITIES: TOWARDS AN
ARITHMETIC OF REAL EXPERIENCE

1. INTRODUCTION

This study has been motivated by two problems at widely separated places in the methodology of the natural sciences. One is the crisis in the philosophy of science, caused by the continuing failure of all programmes to identify a logical structure which could explain the previously succesful practice of natural science (Shapere 1986). The other is the failure of the traditional methods of laboratory science to encompass problems of risks and the environment in the policy process. Few people are aware of both problems and their possible connections. Here we will indicate their common root and, while not attempting a "solution" cast as some formalism, we will exhibit a practical device whereby quantitative statements can be made in a clear and effective way.

The two problems actually come together, implicitly at least, on those issues where in one way or another the traditional methods of science have revealed their inadequacy. In the debates on environmental and occupational hazards, which are bound to increase greatly before they ever abate, popular conceptions of science tend to change drastically from naive trust to embittered cynicism. Having been told in school, in the media, and by all the accredited experts, that science (in legitimate hands) can and will solve all our technical problems, citizens may then have a very different sort of experience, frequently involving procrastination, prevarication or even concealment and deception at the hands of the very experts employed to protect them against their hazards. All scientific expertise then tends to become used as a debating tool, at the level of courtroom psychiatry. In debates on large scale problems, such as engineering projects constituting "major hazards" or major environmental intrusions, or in the speculative technologies of nuclear armaments, the dividing line between science, nonsense and fantasy becomes very difficult to discern. The traditional methodologies of scientific research offer insufficient protection against the corruption of reason that modern conditions encourage, even in our dealings with the world of Nature.

59

John Forge (Ed.), Measurement, Realism and Objectivity, 59—88.

Our contribution is a new notational system (Funtowicz and Ravetz 1986) for the expression of quantitative information, one which provides places for each of the judgements describing the different sorts of uncertainty with which every quantitative statement is qualified. We call it NUSAP, an abbreviation for the categories *Numeral, Unit, Spread, Assessment* and *Pedigree*. The last three convey inexactness, unreliability and "border with ignorance", respectively. Familiar analogues exist for the first two of these, *spread* and *assessment*, in variance or experimental error for the first, and confidence limits or systematic error for the second. The last one, *pedigree*, does not have a precedent in ordinary scientific practice or statistical technique; we define it as an evaluative history of the process whereby the quantity was produced. By means of that history, we characterise the state of the art of the production of the quantity. This exhibits the inherent limitations of the knowledge that can be achieved thereby, and in that sense demarcates the border with ignorance in that case. The first two places, *numeral* and *unit* are close enough to their traditional analogues to need no explanation as yet. Within each place, or box, appropriate notations, depending on the applications, may be employed.

The usefulness of a tool like NUSAP for application to what we may call "policy-related research" or "public-use statistics" is not too difficult to imagine. If highly uncertain quantitative information were required to be written with all its qualifying places explicit, we could more quickly identify pseudo-precise or scientifically meaningless quantitative statements. In this respect the NUSAP notational scheme could function as an instrument of quality-control, in an area where it is both urgently necessary and extremely difficult.

On the side of epistemology, the contribution cannot be so direct; but we hope that it will provide a basis for transcending the 17th-century metaphysics in which geometrical reasoning was to supplant human judgement as the route to real knowledge. Instead of erecting some general, all-encompassing, polar-opposite alternative to our dominant "reductionist" science, be it in the form of a "holistic", "romantic", "idealist" or "voluntarist" philosophy, we can in a practical way exhibit the essential *complementarity* of the more quantifying with the more qualifying aspects of any quantitative statement. Human judgements are then seen, not as inhabiting some separate realm from exact mathematical statements, bearing a relation which is either hostile, mysterious, or non-existent; but rather as a natural and essential complement to the more impersonal and abstract assertions embodied in a numerical expression. When this insight, made familiar in everyday experience, is available for philosophical reflection, then we may be in a position to go beyond Galileo's (1632) classic pronouncement that the conclusions of natural science are true and necessary and that "l'arbitrio humano" has nothing to do with them. Thus NUSAP may make a practical contribution to the recently developed tendency in the philosophy of science, which gives some recognition to the informal aspects of scientific argument and rationality (Putnam 1981, Jiang 1985).

2. THE PROBLEM: UNCERTAIN QUANTITATIVE INFORMATION REPRESENTED BY A "MAGIC NUMBER" FORM

The problems associated with the provision and communication of quantitative information for policy-making in economic and social affairs are well known. It might be thought that the difficulties of producing "usable knowledge" (Lindblom and Cohen 1979) in these fields are caused mainly by the inherent limitations of definition and measurement of their relevant aggregated statistical indicators. But it is increasingly recognised that in policy-making for technology and for the natural environment, similar difficulties arise. Planning for investment in technological and industrial developments is characterized by frequent uncertainty and occasionally by irremediable ignorance (Collingridge 1980).

The matter now takes on some urgency, in view of the growing proportion of scientific effort that is devoted to the understanding and control of the environmental and health consequences of technology and industry. Increasing space, both in the media and in research journals, is occupied by such topics as radioactive pollution, acid rain, agricultural chemicals and pharmaceutical products. A variety of research fields are called on to provide quantitative technical information which, it is hoped, will contribute to the resolution, or at least to the definition, of these practical problems.

Such issues are the subject matter for the policy-related sciences, whose function is to provide this new sort of usable knowledge. Because of the complexity and frequent urgency of some of these issues, the research communities do not always possess the knowledge and skills required for immediate effective solutions. Even experienced advisors may find it difficult to convey to policy-makers an accurate reflection of the scope and limits of the results that can be achieved under these constraints. Solving the problems of representing and evaluating technical information in these contexts, and also of identifying meaningless quantitative expressions, thus becomes of great importance for the proper accomplishment of public policy in these areas.

Policy analysts have long been aware of this problem, and have searched for means of expressing strongly uncertain information. Thus,

One of the thorniest problems facing the policy analyst is posed by the situation where, for a significant segment of his study, there is unsatisfactory information. The deficiency can be with respect to data — incomplete or faulty — or more seriously with respect to the model or theory — again either incomplete or insufficiently verified. This situation is probably the norm rather than a rare occurrence. (Dalkey 1969).

In spite of these manifest inadequacies in the available information, the policy-maker must frequently make some sort of decision without delay. The temptation for her/his advisors is to provide her/him with a single number, perhaps even embellished with precise confidence limits of the classic statistical form. When such numbers are brought into the public arena, debates may combine the ferocity of sectarian politics with the hyper-sophistication of scholastic disputations. The scientific inputs then have the paradoxical property of promising objectivity and certainty by their form, but producing only greater contention by their substance (Nelkin 1979).

Indeed, there is now an increasing tendency for public debate to focus more on the various uncertainties surrounding the numbers than on the policy relevant quantities themselves. This has happened most notably in the cases of "the greenhouse effect" and acid rain. Such debates on the uncertainties will always be inherently more difficult to control and comprehend than those at the policy level. They unavoidably involve all aspects of the issue, from policy to methodology and even to state-of-the-art expert practice in the relevant scientific fields.

In all the fields of formalized decision analysis (e.g. Risk Analysis, Multi-Attribute Utility Theory, Operational Research, Decision Research, "Hard" Systems Theory), practitioners are now searching for means of expressing subjective factors. This endeavour frequently confuses very different aspects of technical information, such as social value-commitments, group interests, and personal judgements, as well as qualifying attributes of quantities. Innovations in statistics have not proved adequate to resolve such confusions. Under these circumstances, there is a real possibility that risk-analysis practitioners and those they advise will despair of objectivity, and in the resolution of policy issues will oscillate between emotional interpersonal contacts and ruthless power politics. Some even argue as if "pollution is in the nose of the beholder", and reduce all environmental debates to a conflict between sensible and sectarian lifestyles (Douglas and Wildavsky 1982). We believe that the core of objectivity in policy decisions must be analysed and exhibited afresh, so that consistent and fair procedures in decision-making can be defended and further articulated.

Thus, the traditional assumption of the robustness and certainty of all quantitative information has become unrealistic and counter-productive. The various sorts of uncertainty, including inexactness, unreliability and ignorance, must be capable of representation. The task was well described by W. D. Ruckelshaus (1984), when Administrator at the U.S. Environmental Protection Agency:

First, we must insist on risk calculations being expressed as distributions of estimates and not as magic numbers that can be manipulated without regard to what they really mean. We must try to display more realistic estimates of risk to show a range of probabilities. To help to do this we need tools for quantifying and ordering sources of uncertainty and for putting them in perspective.

The above reference to "magic numbers" is not merely rhetorical. Our culture invests a quality of real truth in numbers, analogous to the way in which other cultures believe in the magical powers of names. The classic statement is by Lord Kelvin,

I often say that when you can measure what you are speaking about, and express it in numbers, you know something about it; but when you cannot measure it, when you cannot express it in numbers, your knowledge is of a meagre and unsatisfactory kind. (Mackay 1977)

A quantitative form of assertion is not merely considered necessary for a subject to be scientific; it is also generally believed to be sufficient. Thus the problems discussed here are not only related to the inherent uncertainties of the subject matter (as for example in risks and environmental pollutants); they originate in an inappropriate conception of the power and meaning of numbers in relation to the natural and social worlds. By their form, numbers convey precision; an "uncertain quantity" seems as much a contradiction in terms as an "incorrect fact". But this image must be corrected and enriched if we are to grow out of the reliance on magic numbers; only in that way can we hope to provide usable knowledge for policy-decisions, including those for science, technology and the environment.

3. NUMERICAL LANGUAGE: PATHOLOGIES AND PITFALLS

The new requirements on quantitative information for policy-making have revealed inadequacies in the traditional numerical means of representation and in the implicit beliefs underlying them. But we should not think that a natural and faultless inherited numerical system is suddenly being stretched beyond its limits of applicability. Reflection on the history and existing uses of numerical systems shows that they contain many pathologies and pitfalls. These derive from the traditional basic conception of numbers as designed for counting collections of discrete objects. For measurement of continuous magnitudes, the traditional tool was geometry, with an "analogue" rather than "digital" approach. The combination of counting and measuring in practice involves estimation, for which no notational systems were developed until quite recently. But the uncritical use of numbers, with their con-notation of discreteness and hence of absolute precision, still causes blunders and confusions at all levels of practice.

Such imperfections are not advertised by teachers adhering to a "pure mathematics" pedagogical tradition. The subject of "estimation" had indeed flourished in 19th-century "practical arithmetic". But the influence of modern academic research in mathematics, culminating in the "new mathematics", encourage the teaching of the elite skills of manipulating abstract structures to schoolchildren. These did not complement traditional skills, but effectively alienated even arithmetic from practical experience (Kline 1974). (Such abstraction, perhaps based on disapproval of rote-learned practical craft-skills, had its analogue in the "global method" for teaching reading while ignoring the alphabet. For several decades children emerged from the best schools unable either to count or to spell!) Such fashions in abstraction enable us to extend the insight of Gödel's famous theorem (1931) to the present context. As Kline (1972) expressed it,

Gödel showed that the consistency of a system embracing the usual logic and number theory cannot be established if one limits himself to such concepts and methods as can formally be represented in the system of number theory.

Here we are dealing with understanding rather than proof. In the "new math", a more logical and complicated formalistic language for arithmetic was achieved at the expense of the loss of comprehension of the rich and contradictory world of the practical experience of quantity.

The confusions of arithmetic could be safely ignored so long as tacit craft-skills were adequate for coping with the ordinary problems of application. The tasks of programming computers for calculations, where nothing can be left tacit, has forced some awareness of the practical problems of managing the uncertainties in all quantitative information. There is already a flourishing literature on "numerical analysis" at all levels; but as yet no coherent and effective exposition of the management of the different sorts of uncertainty is available. Hence the ordinary practice of calculation is still afflicted with paradoxes and blunders, the sort that "every schoolboy" should know, but doesn't.

For our first example, we may consider the representations of fractional parts of unity. We may say

$$1/4 = 0.25 \quad \text{but} \quad 1/4 \text{ inch} \neq 0.25 \text{ inch}$$

In the first case we are dealing with "pure arithmetic", and the equality results from a simple calculation. But in the second case we are dealing with measurements, in this case inches; our objects are not "points on the real line" but "intervals of estimation", characterized by a "tolerance". Each representation has its own implied "tolerance" (or, as we shall call it *spread*), and so 1/4 inch and 0.25 inch mean quite different things. In the former case, the next lower unit of magnitude (implying the interval of inexactness) is likely to be 1/16 inch, while in the latter it is 0.01 inch, smaller by a factor of 6. Drawings of specifications in the different units have different implied "tolerances", and thus mean very different things in practice. Managing such anomalies may be quite trivial to those involved in such work, but this is achieved by the adoption of implicit conventions for interpretation, whose understanding may be restricted to a particular specialist group. (The traditional tables of decimal equivalents of common fractions, with entries such as 1/16 inch = 0.0625 inch, are examples of the deep confusion in this practical matter.)

A mention of "tolerance" (inexactness, error or *spread*) will usually provoke the response that all that is handled by the "significant digits" (s.d.) convention. But to describe exactly what that is, turns out to be far from trivial. Indeed, the rule for preservation of s.d.s in a calculation is not at all straightforward. For example, if we wish the circumference of a circle whose radius is 1.2 cm, then we round off π to 3.1; but if the radius is 9.2 cm, then π should be 3.14, since the "proportional error" in the second radius measure is of the order of 1%, while that in 3.1 is some 3%, inappropriately large. Thus the choice of the number of s.d.s to include in a numerical expression will depend on the calculation at hand, and the rules for choice will not be trivial.

The above examples may seem to relate only to unsophisticated practice. But the following subtle blunder has been observed even in high-level tables of statistics. Suppose (for simplicity and clarity) we have a population of just seven elements divided into three groups of 2, 2 and 3 respectively. A common tabular display would be

N	%
2	29
2	29
3	42
7	100

There seems nothing wrong here, until we observe that 3/7 is strictly 42.8%; which should be rounded-up to 43%, just as 2/7 or 28.6% was rounded-up to 29%. But then the sum would be 101%; and how often do we see percentages summed to a figure other than 100%? Paradoxically, we may say that a 100% sum is most likely to be the result of fiddling the separate percentages! It inflects an incomprehension of the arithmetic of rounding-off, and is a more amusing example of educated confusion about quantities.

A particular unfortunate consequence of such blunders is that they impart an air of incompetence (however vaguely this may be articulated) to the reports in which they are manifested. Although explicit rules for the criticism of pseudo-precision are not widely diffused, many who use statistics are aware of the principle enunciated by the great mathematician Gauss:

lack of mathematical culture is revealed nowhere so conspicuously as in meaningless precision in numerical calculation (Ravetz 1971)

One simple way to avoid such blunders is to recognize units of aggregation in countings, and the possibility of "swamping" one quantity by another. Paradoxically this phenomenon is more difficult to recognize because of a fertile ambiguity in the quasi-digit 0. This can function either as a "counter" or as a "filler". Thus when we write "10", we understand "zero in the unit place" as a digit distinct from the neighbouring digits 9, and 1 in 11. but when we write "1000", this usually refers to a count of 1 on a unit of a "thousand", analogous to "dozen" or "score".

By these examples we can see that imperfect quantitative information can be managed with greater or lesser skill. Its inherent uncertainties may be hidden, leading to confusion and blunders, and to doubts of the competence of the authors. Or the uncertainties may be clearly exhibited, improving the credit of the authors and providing more useful inputs to decision-making. One can never decrease the inherent uncertainties, or enhance the inherent quality, of any given information by such means; but as we have seen it is possible to transform the information into a more effective tool for decision-making.

4. NOTATION, LANGUAGE AND THE CONCEPTS OF SCIENCE

The examples of the previous section show how numbers may sometimes convey confusion rather than clarity; and a diligent search by any reader will reveal many instances where blunders in the manipulation or interpretation of quantities occur in all fields, and at all levels of expert practice. If we accept this phenomenon as real, we should reflect both on how it has come to be, and also why it has not been noted and analysed before now. We believe that such incompetence cannot be ascribed merely to "bad teaching", when it persists in practice long after the end of formal schooling. Rather, we would say that such defects in practice, particularly because they are unnoticed, are indicators of unresolved contradictions quite deep within the "paradigm" (Kuhn 1962) that defines that practice.

The paradigm in question is the metaphysical commitment to a certain sort of world of Nature (and by extension, humanity), and to the central role of a certain sort of mathematics in the structure of that world and in our knowing it. This is the world of the seventeenth-century scientific revolution, where reality consists of the quantitative "primary qualities", and where by appropriate methods we are to gain knowledge of those qualities, with no limit in principle to its extent and comprehensiveness.

We should be clear that this world-view, although one that (like any other) imposes a structure on reality as experienced, is far from being "arbitrary" in the sense that an isolated individual can simply choose whether to adhere to it, or perhaps to switch to some other brand. It permeates not merely our conception of the role of mathematics in knowledge, but also what sort of scientific knowledge can and should be obtained. It was explicitly claimed by such 17th-century prophets as Galileo and Descartes, and implicitly accepted ever since, that this approach to knowledge is not merely quantitatively exact, but also uniquely assured of truth in its results. Other approaches to knowing, ranging from the humanistic, through the imaginative, to the inner-oriented, have all been rejected with varying degrees of severity at different critical points in the development of the scientific philosophy. Now, some three and a half centuries later, the crisis in the philosophy of science, paralleled by the crisis in the policy sciences, becomes one of confidence. Numerical expressions, representing quantities derived by accredited scientists, cannot be guaranteed to protect us against vague, ambiguous, misleading or even vacuous assertions of a scientific appearance. Where then do we find the rock of certainty on which our scientific knowledge is supposed to be based?

Since that world is our paradigm, by its nature not to be questioned or even noticed in ordinary practice, its flaws will be revealed only occasionally; and they can then be dismissed as anomalies, or as mere anecdotes. Those who would enhance awareness of the problems of the dominant paradigm must then show how a previously unquestioned practice has defects (as we have just done). Or we may show how it reveals other significant features when examined critically. For example, we may consider the language used to describe the results of measurements in the world of experience. These are traditionally said to be afflicted by "error", implying that a perfect experiment would yield a scientifically true value with absolute mathematical precision (This is reminiscent of the naming as "irrational" by the Pythagoreans of certain magnitudes that broke their rules, as $\sqrt{2}$). Even in sophisticated statistical theory, the crucial terms have a subjective cast, as in "confidence" or "fiducial" (in our work we describe the analogous properties of information as "reliability", relating to human practice to be sure, but to experience rather than to opinion). And among scientists of many sorts, the old ideal of objective, quantitative certainty has been dominant. Thus from such dissimilar figures as Einstein and Rutherford, we have the dicta, "God does not play dice", and "If your experiment needs statistics you ought to have done a better experiment" (Mackay 1977).

Being a genuine crisis, this one does not manifest itself merely at these two dissimilar areas of experience: abstract philosophical reflection and craft arithmetical practice. The present century has been the dissolution of many certainties in the mathematical conception of science. The revolutions in physics, particularly quantum mechanics, were explicitly philosophical in part; and similarly was the "foundations crisis" in mathematics, leading through Gödel's theorems to a radical loss of certainty (Kline 1980).

This erosion of the previously unchallenged epistemological foundations of a scientific world-view has thus proceeded on many fronts. It has been accompanied by an erosion of the moral certainties of science, ever since the industrialization and militarization of scientific research became recognised. As yet there has been no effective presentation of an alternative paradigm, in the Kuhnian sense of a deep scientific revolution. The critical analyses raised in the 'sixties (and echoed in Feyerabend's (1975) works) could not have a practical outcome in the absence of a wholesale transformation of society and consciousness. One modest philosophical alternative was suggested in the 1920s by Niels Bohr (Holton 1973), in his famous attempt to resolve the "dualities" of early quantum physics by means of the essentially Chinese notion of complementarity. This remained a personal, almost idiosyncratic attempt at coherence, for the physicists were able to do quite nicely in making discoveries and inventions of unprecedented power, in spite of totally incoherent basic conceptual structures. Recent attempts to interpret physics in broadly "oriental" ways have so far remained curiosities of popularised science (Capra 1975, Zukav 1979).

In the NUSAP system we put complementarity to work. Becoming familiar with the notational scheme, through use, entails acceptance of the idea that a bare statement of quantity, in the absence of its qualifying judgements, is scientifically meaningless. To paraphrase the classic formula of the logical positivists of the Vienna Circle (Ayer 1936), the meaning of a quantitative statement is contained in its mode of qualification as much as in its quantifying part. In this respect the NUSAP system makes a contribution towards an alternative approach to the philosophy of Nature.

We may ask, can notations really be so influential as we claim? The history of mathematics shows how they can encapsulate new ideas in such a way as to transform practice. This happened twice in the seventeenth century, first with Descartes' unified conception of algebra, geometry and their relationship, expressed through the symbols a, b, c,

..., x, y, z. Then Leibniz, with dx and $\int$, tamed the infinite in this new "analysis". At a less exalted level, the "arabic numerals" democratised arithmetic in early modern Europe; previously calculation had been the preserve of those who had mastered the abacus, as supplemented by a variety of special tricks. Even when symbols are not designed for calculation, but only for effective representation, they can have a deep influence on a practice and how it is understood; the history of chemical nomenclature and symbolism provides many examples of this (Crosland 1962).

We conceived and developed the NUSAP notational scheme in full awareness of the complex interaction between tools (of which notations are an example), explicit concepts, world-views, and social practice. It is designed as an instrument of analysis and criticism, in an area of practice where such activities have been generally considered to be either unproblematic or even quite unnecessary. To the extent that there has been a mystique of quantities, and that this has been supportive of a mystique of exclusive scientific expertise, the NUSAP system also has functions in the societal aspects of scientific practice. There too, it can enrich the inevitable debates on quantities that enter into policy issues, avoiding the extremes of naïvety and cynicism from which participants now have little protection. In that connection too, it can make its contribution to the development of appropriate new conceptions of science.

5. PRINCIPLES OF THE NUSAP NOTATIONAL SCHEME

The NUSAP notational scheme is a system whereby the various sorts of uncertainty contained in all quantitative information may be expressed concisely and also consistently with existing partial notations. It is designed to be applied to any expression given in the form of numbers or more generalised notations. By its means, nuances of meaning in quantitative statements can be conveyed clearly and economically; and various aspects of the quality of the quantitative information may also be expressed. Users need master only the very simplest skills, and the underlying ideas are familiar to all those with experience of successful practice in any quantitative discipline or craft.

Should it come into standard use, there will develop a more competent general level of criticism of quantitative assertions, both among experts and the interested public. Just as quality-control is now recognised as an essential component of industrial production, meriting emphasis and appropriate organizational structures, so we can expect that with the adoption of the NUSAP system, quality-control of quantitative statements will eventually become standard practice.

NUSAP was designed with several criteria in mind. In addition to the ordinary properties of a good notation (simplicity, naturalness, flexibility, etc), it enables the distinction between meaningless and meaningful quantitative statements. Further, it protects against the misleading use of quantitative information by preventing the isolation of the "quantifying" part of an expression from its "qualifying" part. All this is accomplished because the notational system can distinguish among three sorts of uncertainty which characterize every quantitative expression. These are: inexactness of measurement and of representation; unreliability of methods, models and theories; and the border between knowledge and ignorance revealed in the history of the quantity.

The NUSAP notational scheme is a "system" because it is not simply a collection of fixed notations. Rather, it is a set of determinate *categories*, each of which can be filled by particular notations appropriate to the occasion. The names of the five categories (or boxes, or places in a string) make up the acronym NUSAP. Considering the expression as proceeding from left to right, we start with those which are more familiar, the quantifying part of the expression; and conclude with those less familiar, forming the qualifying part of the expression. With such complementary aspects of the expression conveyed in a convenient and standard form, some of the classic dilemmas of subjectivity and objectivity in science can be resolved in ordinary practice.

Considered as a formal structure, NUSAP is more than a convenient array of symbols conveying uncertainties in technical information. It is a "notational scheme" which provides a general framework so that an unlimited variety of particular notations may be employed unambiguously. It is a string of five positions corresponding to the categories of *numeral, unit, spread, assessment* and *pedigree*. By means of this place-value representation, each category can be expressed simply, without need for its explicit identification (this is a "scheme" of notations as the most abstract level). For each category, there are many possible sets available for conveying particular desired meanings (thus in *unit* we may have Imperial, CGS, MKS or S. I. units). Any particular array of such sets, we call a "notation". Given such a notation, any particular case of representation will be an "instance" of the notation.

Such distinctions enable great flexibility and power in the expression of quantitative information. In this respect it is analogous to the notational system of "Arabic" numerals, where the meaning of a digit depends on its place, thereby enabling a small set of digits to be used for the representation of any possible integer. By means of this flexibility we can escape from the "vicious circle" of digital representations, whereby even those notations used to qualify an expression are themselves afflicted by pseudo-precision (as "95% confidence limit").

The first category, in the left-to-right order, is *numeral*. We use this term rather than "number" as a reminder of the flexibility of the system. The place can be filled by a whole number, a decimal expansion, a fraction, or even a representation of an interval, or a qualitative index. Next is *unit*, which can be a compound entry, consisting of *standard* and *multiplier*. This can be important for the representation of aggregated quantities, as "$k", or perhaps "$10^{12}". The middle category is *spread*, generalizing the traditional concept of error. Although this is normally expressed in arithmetical form (perhaps by $\pm$, % or *fn*, for "to within a factor of *n*") there is a strong qualitative element about it. For *spread* cannot (except perhaps when given by a calculated statistical measure) be given precisely; it is always an estimate, whose own *spread* is not a meaningful or useful concept. There is a way of qualifying the *spread* entry; it can be done by *assessment*, the fourth category in the NUSAP system. This may be seen most familiarly as a generalisation of the confidence limits used in statistical practice. Assessment can be relevant in contexts where the problem does not admit of the calculation of confidence limits; and a great variety of notations can be deployed here, ranging from standard percentages, to a simple ordinal scale, such as "high, medium, low". The means of arriving at an *assessment* rating are equally various; it may be calculated statistically; it may be obtained by arithmetical operations from a conventional coding of the last category *pedigree*; or it may be the result of a personal judgement.

Hitherto the categories have analogues in existing practice, ordinary or statistical; and it is natural to consider the NUSAP notational scheme as an extension and ordering of existing notations. But with the *pedigree* category, a novelty is introduced. By *pedigree* we understand an evaluative history of the production of the quantity being conveyed by the notation. Histories do not normally appear as part of notations; and for this category we have developed abbreviated schemes of analysis and representation. So far there are two, one for "research information" and the other for "public-use statistics". In this paper we shall only introduce the *pedigree* for "research information".

We said before that the contents of the *numeral* box need not be ordinary numbers. Thus, if a quantity is known only to within an "order of magnitude", then an appropriate instance of *numeral* would be E6:. We remark that an instance 1:E6 denotes a determinate quantity, a million, very different from the "order of a million" conveyed by E6:. (Representations in NUSAP have the boxes in the string separated by a colon; in reading them, we express the colon by "on". For example, an instance 1:E6 — where E6 is in the units box — reads "1 on E6").

If one quantity is known only as an interval which lacks any preferred point of likelihood or of symmetry, then this should be the entry in the numeral place. Thus we could have (a, b): for an ordinary interval, $(\geq a)$: for an open-ended one. In the *numeral* place we may also find expressions of yet more general mathematical structures as numbers of a finite set representing an ordinal scale (as in much of social research), or numbers representing indices with a purely artefactual arithmetic. An extreme example of an ordinal scale with a qualitative notation of *numeral*, which is of direct practical use, is that for Geiger counter readings, as "click", "chatter" and "buzz".

By *unit* we understand the base of the physical and mathematical operations represented in the *numeral* position. We distinguish two components of the *unit*. There is the *standard*, the common or generally used unit of the relevant operations; and the *multiplier*, relating to the standard to the particular unit involved in the expression. Thus we frequently see "£342M", where the unit "£M" (with "£" as *standard* and "M" as *multiplier*) is the actual basis of the calculations reported, as distinct from the "£.p" of strict accountancy practice. The meaning of the pair *standard-multiplier* may of course vary with context; thus "kg" is now a *fundamental* unit in the S.I. system, in which strictly speaking, "1g" should be written as "1mkg". These two quantifying categories enable a refined description of topologies and scales of measurement.

Good practice in notation includes the indication of the *spread* of a quantity (which may also be called error or imprecision). For this the significant digits convention is common, as well as such statistical measures as standard deviation. In the case of highly inexact quantities, the *spread* may be conveyed by "to within a factor of *n*".

We can illustrate the application of NUSAP on some simple examples, where existing representations are inadequate. Suppose that we start with "five million", and we add some smaller quantity. If it is *very* small, as, say, 180, then the sum is normally understood still to be five million, since the latter quantity is not significant in the context. Writing the sum formally, we have, $5\,000\,000 + 180 = 5\,000\,000$. In this sum, the last three zeros are interpreted as fillers rather than true digits; and so we use an artefactual arithmetic, adopting implicit conventions for the neglect of certain digits, just as in rounded-off calculations. But if the second addendum is 180 000 it is not clear from the uninterpreted sum $5\,000\,000 + 180\,000$ just where the counter digits end and the filler digits begin. Only from the context can we know whether to apply a natural or artefactual arithmetic. A notation like 5×10^6 may help, but even that is not conclusive.

Another useful example from ordinary practice is counting in dozens; this shows more clearly the influence of the process of production of the datum, since in this base there is no ambiguity between counter and filler digits. Thus "eggs" will have, as a typical instance, $4\frac{1}{2}$:doz-eggs: rather than 54:eggs. This example exhibits the phenomenon of pseudo-precision of a *numeral* in digits, when the process has consisted of counts by dozens and half-dozen.

In the NUSAP notational scheme, we can express "five million" in the alternative forms, $5:10^6$, 5:M or 5:E6. Here it is explicit that the *unit* is millions. Although some ambiguity remains, it can be resolved by the entry in the *spread* position. But it is quite clear that 5:M + 180 = 5:M is the correct sum, unless there is an explicit note to the contrary in the *spread* position. There is no need for an artefactual arithmetic, with all its ambiguities.

NUSAP can also convey some shades of meaning that may be important in particular contexts. Thus "five million" may be better represented as 50:E5 or $\frac{1}{2}$:E7, denoting different sorts of operations in the different aggregated units. We note that the use of fractions in the *numeral* position enables us to express the meaning of a rough cutting of an aggregated unit; thus a "third of a million" is represented better as $\frac{1}{3}$:M rather than 0.33×10^6. It can be considered an advantage of a notation, that a user can represent, and even calculate with, an instance which expresses a perfectly clear statement of a quantity that previously needed a verbal form.

When representing measurements, we must distinguish between the *multiplier* and the *standard* which make up the *unit*. For an example, 5×10^3:g, represents a count of 5000 grams; and this expression implies that the measuring operations were performed in the old CGS system. Turning to $5:10^3$g, we are still in CGS, now operating in "kilo" grams, of which there are 5. If we now write 5:kg, this is the expression of a count of 5 in the MKS system, or S.I. units, were kg is fundamental. Another example of the same sort exhibits a new feature; 5:g is clearly in CGS, while $5:10^{-3}$kg tells us that we have S.I. with a scaling in thousandths of a kg. We note that here the *multiplier* represents the scaling of the measuring instrument.

For an example of the *spread* category we return to aggregated counting, with the above mentioned ambiguous case of 180000 added to "five million". It may be that the larger quantity here has such inexactness that even a tenth of it is insignificant. This could happen if it is part of a sum with much larger quantities, as 32:E6: and 155:E6:. Then the *spread* would be understood to be as large as E6, the *unit*, and therefore the 180000 or 0.18:E6 would be meaningless. In this way, the notation represents the practical situation of the swamping of a much smaller quantity in a sum; to be completely explicit we may express this as follows, 5:E6:E6. The *spread* E6, indicates that no interpolation within the scaling has been done; equivalently, every quantity in this sum has an inexactness interval which is E6 in length.

By the use of this notation, the meaninglessness of a quantitative expression can be clearly exhibited. For example, where both *unit* and *spread* are E6, the quantity 180000 would be expressed as 0.18:E6:E6. The 0.18 would be insignificant and the expression is vacuous. By contrast, if the 180000 is being added to "five million", and the *spread* is understood to be 0.1E6, then 0.18:E6:0.1E6 would be naturally rounded up to 0.2:E6:0.1E6; and this is a proper quantitative expression. The sums might read as follows. First, 32:E6:E6 + 155:E6:E6 + 5:E6:E6 + 0.18:E6:E6 = 192:E6:E6 where the 0.18 is suppressed, as being meaningless in this context. If on the other hand, our summands are, say, 3:E6, 7:E6 and 5:E6, then since these are small integers, it is likely (unless indicated otherwise) that the *spread* is less than E6, perhaps 0.1E6. In this case we may write, $3:E6:0.1E6 + 7:E6:0.1E6 + 5:E6:0.1E6 + 0.18:E6:0.1E6 = 15.2:E6:0.1E6$, where we have rounded up 0.18 to 0.2.

The notation enables us to identify pseudo-precision in measurements, even when this is forced by an accepted scaling. Thus in the S.I., where "cm" are officially suppressed, measurements which were formerly done in inches, with *spread* of $\pm\frac{1}{2}$ inch, are now frequently expressed in "mm" to the nearest ten. Thus "five feet" will be rendered as 1 520mm. In the NUSAP system, this would be properly represented as $152:10$mm. In this way we retain the *standard* required by the S.I. system, but modify by the *multiplier* 10, to express the practical scale of operation, equivalent to the illegal "cm". A somewhat less rigorous representation makes use of the *spread* category; we can keep the spurious last digit required by the S.I., but show that in practice it is not a counter. Thus we would write $1520:$mm$:10$, reminding the user that there is an effective "spread" in the number as recorded.

Strongly inexact quantities are sometimes expressed "to within a factor of n", as "5×10^6 to within a factor of 10". The convention indicates multiplicative intervals above and below the given quantity; thus the given quantity here may lie between 0.5×10^6 and 50×10^6. In the notation we write, $5:E6:f10$ meaning $0.5:E6 < 5:E6:f10 < 50:E6$. By means of such notations it is possible to convey quantities of the sort characterized by "the first law of astrophysics": $1 = 10$. We can also express inexactness given in proportional terms; for example "5×10^6 with a proportional error of 15%" is represented as $5:E6:15\%$ or as $5:E6:[15$ in $E2]$.

In the policy context, fractions less than the unity, expressed as percentages, are frequently used to indicate the division of some aggregate. The inexactness of such estimates is extremely difficult to represent in a compact notation, and a misleading impression of precision is all too often conveyed. Thus "40%" may mean "less than half but more than one-third" or perhaps "less than half but more than one-quarter". These inexact estimates may be represented as $\frac{1}{3}:1:<\frac{1}{2}$ and $\frac{1}{4}:1:<\frac{1}{2}$ respectively. Another way of expressing such estimates involves using the variable x. If there is some *unit U*, we may have $x:U:\frac{1}{3} < \frac{1}{2}$ or $x:U:\frac{1}{4} < \frac{1}{2}$. By this means, one can express quite fine distinctions among inexact estimates of fractions, avoiding the pseudo-precision of a two-digit percentage. The use of the variable x in the *numeral* place enables us to express clearly that the means the production of the quantity do not provide us with information for distinguishing among numerical values. The class to which all the relevant values belong is represented in *spread*. We can refer to this as an "indifference class", in the sense that no one numerical value can legitimately be taken as a representative of the class in preference to any other. In symbols, we write the general case as $x: U: S$.

The *assessment* category expresses the reliability of the information, generalizing not only the confidence limits of classic statistics, but also those of Bayesian statistics, interpreted as "degree of belief" (Keynes 1921) or "betting odds" (Savage 1954). Such formally defined measures are properly applicable only in special cases, and are not free of conceptual problems of their own. The *assessment* category is not to be formalised in the logical sense; but is designed to convey judgements of reliability in a convenient form. Where statistical notations are familiar and appropriate, they may be freely used. Otherwise, a more qualitative notation, such as, for example, an ordinal scale, should be adopted. Thus, we may have the set, (*Total, High, Medium, Low, None*), perhaps codified as (4, 3, 2, 1, 0) to convey this kind of judgement. As in the cases of *numeral, unit* and *spread*, a great variety of notations are available for *assessment*.

A familiar case from scientific research is that of a number which historically belongs to a sequence of experimentally-derived results describing the "same" physical quantity. It is well known that elements of such a sequence may well jump about by amounts far exceeding the *spread* of any of them; this is described as systematic error as distinct from random error. A reader of technical literature may estimate a numerical entry for the *assessment* place, by an examination of the published versions of such a variable "physical constant". With *spread* representing average, a sample case might read $4.32: \mu U: \pm 0.17: \pm 0.3:$.

In traditional statistical practice, the *assessment* (or confidence limits) is closely associated with the *spread* (or variance). Unfortunately, this association tends to conceal the radical difference between the two categories, and to inhibit the understanding of either. When we generalise *assessment* from the simplest notion of reliability, the independence of the two categories becomes apparent. For example, consider a statistical distribution where we are interested in estimating the 95th upper percentile, or the top 5%. The entry in the *numeral* place is then qualified by the expression "%95" in the *assessment* place, the order being inverted deliberately to distinguish this from the more traditional "95%" confidence limit. In such a case, the *spread* will depend on the number of trials or of simulations of the same process. So, if we are comparing the results of two different experiments involving different numbers of trials or simulations, (as for instance obtaining the top 5% of a distribution of experimental coin-toss results), we can have *spreads* varying with the size of the sample while the *assessment* entry is always "%95".

Another illustrative example is of a case where the *spread* box is empty, but where a definite (though qualitative) *assessment* is appropriate. This can happen in a "back-of-envelope" calculation, where the basic *unit* is expressed through a *numeral* entry of a small integer number. In such cases, *spread* is meaningless; but the calculation can be qualified by, say, "*Upper Limit*" (or "*U*") in the *assessment* box. This is not an ordinary sort of reliability as calculated in traditional statistical practice; but it provides the user of the information with an appropriate interpretation for reliable use in practice.

The flexibility of the system is further enhanced by the use of combinations of entries in boxes to convey nuances of meaning. A particularly direct case of this is in a trade-off between "strengths" of entries in the *spread* and *assessment* places. This also generalises statistical practice; so that we may describe a distribution more tightly by its range over the 25%—75% percentiles or more broadly over the 10%—90% percentiles. This translates directly into a lower *spread* with lower *assessment*, or higher in both categories. In NUSAP, this can be expressed by μ (the mean in the N place), S_Q and S_D (the interquartile and interdecile ranges in S). The notations would read, $\mu:U:S_Q:50\%:$ and $\mu:U:S_D:80\%:$. For an example, we imagine a distribution with a mean of 46, $S_Q = 12$, and $S_D = 20$. Alternative representations would be $46::12:50\%$ and $46::20:80\%$; the percentages in the *assessment* box relate to the amount of the total distribution represented in the *spread* place.)

When uncertain quantities are directly involved in a policy process, the flexibility of the system can be very useful indeed. An illuminating example is cited by Mosteller (1977): estimation of the number of American men who emigrated during the Vietnam war. These ranged from 2.5k through 30k to 70k or even 100k, though the higher figures were less reliable. If the absolute number is not critical for policy purposes, then a convenient NUSAP expression would be $\geqslant 3:E4::Good:$. With a one-sided interval in *numeral*, it is appropriate to leave *spread* empty. If the lower bound on the estimate is very sensitive for policy, the *numeral* entry could be reduced; and the expression reads $\geqslant 2\frac{1}{2}$:E4:+20%:High:. In this way, a policy-maker is told that s/he is very unlikely to go wrong in acting on the basis of an estimate in the range $2\frac{1}{2}$ to 3 on a *unit* E4.

These examples show how the system can be used to provide alternative communications, each valid in its own right, for a single statistical result. Each version focusses attention on a different aspect of the distribution, corresponding to different needs of users.

6. PEDIGREE FOR RESEARCH INFORMATION

In the NUSAP notational scheme, the most qualifying category, located in the far right position, is *pedigree*. This expresses the most extreme of the various sorts of uncertainty conveyed by the notation: its border with ignorance. The previously discussed categories can be seen as a preparation for the introduction of this one. Thus, *spread*, expressing the inexactness of quantities, served as a reminder that a quantitative expression is not "clear and distinct". Even if there is some realm of ideal mathematical entities (as lines without breadth), represented in necessarily true mathematical statements (as $\exp(\pi i) + 1 = 0$), the world of empirical objects and their measurements always involves "more or less", or "tolerances", about quantities possessing a fringe of vagueness. In that sense, the specification of an object in respect of its quantitative attributes implicates the rest of the world, things other than our particular object of attention, as it shades into them.

This can be seen clearly by reflection on the normal practice of indicating *spread* or the misnamed "error". When we write 4.32 ± 0.05, that extra term must surely be other than perfectly precise. How is its imprecision to be conveyed? Is 0.05 drastically different from 0.04 or from 0.06? In normal practice, we simply record an *assessment* of confidence, which is a very different kind of judgement. We do not ordinarily attempt a "spread of the spread", for many practical reasons; and also because if we were to iterate once, then why not twice or more? Hence we satisfy ourselves with an informal, tacit convention on the formal, misleadingly precise representation. Once aware of this we see how the simplest and most common of conventions for expression of the lack of perfect exactness in quantities leads us into paradoxes of infinite-regress. The border between the measured thing and its environment, or between our knowledge and our ignorance, can never be specified precisely.

Thus our quantitative knowledge can never be fully exact or perfect, even in itself. When considered in the context of its usefulness, further qualification is necessary. Even a simple assertion carries an implicit claim to be true; and therefore also to be completely reliable in use under appropriate conditions. But every statement of fact needs some sort of *assessment*, since it is impossible to achieve perfect reliability any more than perfect truth. As we have seen, technical statements involving probability and statistics include notations for the expression of their confidence limits, which can be interpreted as the odds against a "failure in use" of the information. (This interpretation is closer to practice, and also less paradoxical, than that of "confidence in its truth").

Of the three sorts of uncertainty expressed in NUSAP, ignorance is the most novel and complex, and also the most difficult to convey explicitly. In ordinary scientific practice, ignorance of a special sort is vital to the enterprise: the interesting problems which can be stated, but whose solubility is not assured. In this sense, science deals with controllable ignorance; successful science involves, in the classic formula, "the art of the soluble". Not all ignorance comes in such convenient packages; in contemporary science/technology policy, the most important problems are frequently those of "trans-science" (Weinberg 1972): problems which can be stated, whose solution can be conceived, but which are unfeasible in practice because of scale or costs. Such trans-science problems may involve ignorance that is quite important in the policy realm, as when decisions must be taken before there is any prospect of the relevant information being produced.

In the *pedigree* category, we do not characterize information (or ignorance) in technical detail. Rather we exhibit the mode of production of the quantitative information being represented, through an evaluative history. This defines the border with ignorance, through a display of what more powerful means were *not* deployed in the production of the information. Thus if we report a "computation model" as the theoretical structure for the information, that implies that there was no "theoretically based model" available, and still less "tested theories", involved in the work. Thus in each phase we are comparing existing results with conceivable alternatives of greater strength. As research fields develop through practice, early pioneering efforts may be superseded by stronger work in such a fashion as this. Hence we may imagine the choice of modes in a *pedigree* matrix as indicating the border between what is currently feasible and accepted as known, and that which is unfeasible and unknown.

In this respect a *pedigree* code is analogous to the statement of a proved theorem in mathematics. Such a statement includes more than the result; equally important are the conditions under which it holds. As to other possible conditions, there is ignorance; and the statement of a theorem constitutes an implicit challenge to explore that ignorance. Although quantitative information is not "true" in the same sense as a mathematical result, there is this analogous border between knowledge and ignorance in the specification of its production.

We may describe the three qualifying categories of NUSAP in terms of the various contexts to which they apply. In practice, they operate in interaction, so that no one is truly prior. By abstracting somewhat we may speak of contexts of production of information, of its communication and of its use. These correspond to the categories of *pedigree, spread* and *assessment* respectively. In production, the border with ignorance is shown by the limitations of each chosen mode in the *pedigree* matrix. In communication, the "unknown" is that into which the stated quantity blends by means of the (non-iterated) *spread* term. In use, the implied testing by future experience, revealing possible ignorance, is conveyed by the reliability rating of *assessment*. The order in which we have discussed these categories is not the same as that in NUSAP; in the scheme we adhere more closely to existing usages, where a notation starts with the quantifying part and proceeds towards the more qualifying.

For the evaluative history of the quantity as recorded in the *pedigree* matrix, we analyse the process into four phases. These indicate, by their various modes, the strength of the different constituents of quantitative information resulting from a research process. We have theoretical, empirical and social phases, the latter being split into two in order to encompass all the sorts of evaluation that we may want to provide. In order, the phases are: *Theoretical Structures, Data-Input, Peer-Acceptance* and *Colleague Consensus*. The *pedigree* matrix is displayed as follows (with corresponding numerical codes and abbreviations):

	Theoretical Structures	Data-Input	Peer-Acceptance	Colleague Consensus
4	Established Theory (TH)	Experimental Data (Exp)	Total (Tot)	All but cranks (All)
3	Theoretically based Model (Th.bM)	Historic/ Field Data (H/F)	High (Hi)	All but rebels (All-)
2	Computation Model (Mod)	Calculated Data (Calc)	Medium (Med)	Competing Schools (Sch)
1	Statistical Processing (St)	Educated Guesses (Ed.G)	Low (Lo)	Embryonic Field (Emb)
0	Definitions (Def)	Uneducated Guesses (Gues)	None (Non)	No Opinion (No-O)

Discussing the separate phases in order, we have first *Theoretical Structures*. Following the traditional scientific methodology, we accept that the strongest mode here is *Established Theory*. The general term "established" includes such modalities as: tested and corroborated; or theoretically articulated and coherent with other accepted theories. Thus Einstein's General Theory of Relativity was in this sense already "established" when it was tested by the famous astronomical experiment of 1919. When the theoretical component lacks such strength, and is perhaps rudimentary or speculative, then its constructs must be considered as in a "model", but one which is theoretically based; we have then the mode *Theoretically based Models*. Although still involved in explanation, such a model makes no effective claim to verisimilitude with respect to reality. In this latter respect it is similar to a *Computation Model* which is some sort of representation of the elements of a mathematical system by which outputs are calculated from inputs. In such a case, there is no serious theoretical articulation of its constructs; the function is purely that of prediction. Such a mode is particularly common in the mathematical behavioural sciences; a well known example is IQ. This mode, *Computation Model*, characterizes the use of high-speed computers for simulations where real experiments are difficult or expensive.

Important research can exist where neither articulated constructs nor elaborated calculations is present; this is the case in classic inductive science. Then, with techniques varying from simple comparisons (formalising J. S. Mill's Canons of Induction) through to very sophisticated statistical transformations, we have *Statistical Processing*. Such forms of

Theoretical Structure can provide no explanation and only limited prediction; but used in exploratory phases of research, they can yield interesting hypotheses for study. Epidemiological work of all sorts, leading to identification of likely causes of known ill effects, is a good example of this mode. Finally, we have those situations where data which is gathered and analysed is structured only by working *Definitions* that are operationalized through standard routines. This will be the case with field-data, frequently destined for public-use statistics. A *pedigree* for public-use statistics has been developed by the authors but it is not discussed in this paper.

The normative ordering among these modes is clear; the higher generally includes the lower as part of their contents. But this does not imply judgements on craftmanship, effectiveness, or on the quality of the investigators or of a field. We do not share in the traditional judgement that all science should be like physics. However, if (in its present state of development) a field can produce only relatively weak results (as gauged by the modes of this scale), that should be an occasion neither for shame nor for concealment.

The other phase deriving from traditional scientific methodology is called *Data-Input*. We use this name rather than "empirical", to include certain inputs (quite common in policy-related research), whose relation to controlled experience may be tenuous or even nonexistent. Starting again with the classical and strongest mode, we have *Experimental Data*. Not so strong, our next entry is *Historic/Field Data*; data of this sort is "accidental" in the sense of being taken as it occurs, and lacking tight controls in its production and/or strict reproducibility. *Historic Data* is that which was accumulated in the past, out of the control of the present study; *Field Data* is produced by large-scale procedures of collection and analysis.

Historic/Field Data has at least the strength of a relatively straight-forward structure, so that its possible errors and deficiencies can be identified. But sometimes *Data Inputs* are derived from a great variety of empirical sources, and are processed and synthetised by different means, not all standardised or reproducible. The numbers are then themselves "hypothetical", depending on untested assumptions and procedures. Even to estimate the *spread* and *assessment* in such cases may be quite difficult. Hence we assign *Calculated Data* to a weaker point in the scale even than the *Historic/Field Data* mode.

Traditionally, the last mode discussed would have been considered the weakest in a scientific study. But with the emergence of policy problems calling for data inputs regardless of their empirical strength, formalised techniques were created whereby opinion could be disciplined so as to provide a reasonable facsimile of facts. Such were subjective probabilites, Bayesian statistics, and other ways of eliciting quantitative estimates from experts. These we call *Educated Guesses.* Sometimes even such a mode is absent; *guesses* can be simply uneducated, and yet accepted as data, hypotheses or even facts, whichever seems plausible. In this respect *Data Inputs* in modern times have come a long way from the relative certainties of the classical methodological framework for science.

The social aspects of the *pedigree* are here given in two phases: *Peer-Acceptance* relates to the particular information under evaluation; and *Colleague Consensus* describes that aspect of the field in relation to the particular problem area. These are the phases to which users (and those who advise them) could turn first, for preliminary evaluations of possible effectiveness of the technical information. Thus if there is weak *Colleague Consensus* and a research field is seriously divided (with *Competing Schools* or perhaps only *Embryonic*) then there will be no security in any piece of quantitative information (Funtowicz and Ravetz 1984). Even the sampling of expert opinions, to obtain *Educated Guesses,* can lead to a bimodal distribution or worse; from this the policy maker learns the important lesson that scientific ignorance still dominates the problem. Stronger *Colleague Consensus,* as with *All but rebels* or *All but cranks* may well be time-bound. Since, as T. H. Huxley said: "It is the customary fate of new theories to begin as heresies and to end as superstitions" (Mackay 1977), who is a "rebel" or even a "crank" depends on circumstances. There is a real distinction between the two cases; rebels have some standing among their colleagues, whereas cranks have none.

At the other extreme from scientific orthodoxy, we have the mode *No Opinion*, where there is simply no cognitive framework or social network in which the proffered information can make any sense when it appears. This may be from its apparent lack of substance or of interest, or both.

Once we have an appreciation of the context in which peers can receive and evaluate a piece of information, it is useful to characterize that process. The modes of *Peer-Acceptance* range in linear order from *Total* to *None*. It is important to realise that the significance of any given degree of *Peer Acceptance* depends critically on the state of *Colleague Consensus*. Thus if there is a strong general *consensus* and weak *acceptance*, the information must be judged as of low quality of craftmanship (given trust in the general competence of the field). But if *consensus* is as weak as *acceptance*, even such an adverse judgement is not proper; and ignorance rules again. The degree to which consensus can be weak, even in "matured" scientific fields, is generally underestimated quite seriously by outsiders. Hence low acceptance is liable to be interpreted in a misleading fashion, as a well-founded adverse judgement on the technical information and by extension on its author as well. We have split the "social" phase into these two parts, partly to avoid such errors as this.

We now discuss various instances of quantitative information that were important in the development of science, and which illustrate significant features of our *pedigree* category.

Not all quantitative information is appreciated on its first publication; the classic example is Mendel's simple arithmetic ratios between frequencies of different sorts of hybrid peas. For the first thirty years after its publication, the *pedigree* was, as seen retrospectively by historians: (Th. bM, H/F, Non, No-O) or (3, 3, 0, 0). Of course, any contemporary who might have scanned Mendel's paper would not have been so complimentary on the cognitive side. A (reconstructed) *pedigree* code for that period would be (St, Calc, Non, No-O) or (1, 2, 0, 0). The *Calculated* mode conveys the suspicion that the simple ratios were the result of a coincidence or of "massaged" data. In the earlier twentieth century, the rediscovery of Mendel changed the *pedigree* to (Th. bM, H/F, Tot, All) or (3, 3, 4, 4). With the further development of genetics, the ratios themselves are strengthened to have a *pedigree* (Th, Exp, Tot, All) or (4, 4, 4, 4). But greater sophistication in statistics and its application to experimental design, led to a scrutiny of the aggregated numbers by R. A. Fisher, who found them "too good to be true"; and so the modern historians' judgement of Mendel's own work in his own time now has *pedigree* (Th. bM, Calc, Non, No-O) or (3, 2, 0, 0) (Olby 1966).

A sort of inverse example was provided by T. S. Kuhn (1961) in his seminal essay on measurement in science. This was an experimental value for a constant of crucial importance in the caloric theory of gases: the ratio of the two sorts of specific heat. The setting for the production of this number was quite dramatic: the Laplacian Theory of Gases could explain the experimentally known velocity of sound in air, if (and only if) the constant in question had a certain predicted value. The *Academie des Sciences* devoted its annual essay award competition to this topic in 1819; and the desired value was duly obtained by Delaroche and Berard, whose work won the prize. All was perfect; and here we have a *pedigree* (Th, Exp, Tot, All-) or (4, 4, 4, 3), the only reservation being among the nascent scientific/political opposition to the Laplace school. Unfortunately, the result was simply incorrect; and its background theory became discredited for many reasons. A retrospective *pedigree* for the result, a decade on, could be (Th. bM, Calc, Non, All-) or (3, 2, 0, 3); here the *Colleague Consensus* embraces the victorious anti-Laplacian party, comprising nearly all save the lonely disciple Poisson (Fox 1974).

There two examples of the rise and fall of *pedigree* ratings for quantitative information provide a warning that the evaluation of scientific results is a matter of judgement, which can change drastically. What is effectively scientific knowledge at any one time is very much liable to subsequent revision by the wisdom of hindsight. The reliability of quantitative information in practice does not require it to be continuously confirmed and corroborated. In this respect it can be like theories which, in spite of being superseded or perhaps refuted are still reliable in particular contexts of use ("caloric" being one good example; Newtonian mechanics is another). The *pedigree* coding, by analysing the different phases of the history of the production of the relevant quantity, can assist in the description of such changes, and perhaps thereby also contribute to the resolution of the philosophical problems of such "fallible knowledge".

Use of the *pedigree* code will also help to clarify relations between providers and users of technical information. Frequently there is a clash of interests and perceptions. Users want unambiguous and certain facts (of the sort science traditionally promises) as inputs to their decisions. This would relieve them of the burdens of evaluations of inputs and of responsibility for decisions that turn out to be "wrong". Typically, the world still expects science to define a "safe limit" for toxicants of all sorts. The scientists, however, are keenly aware of the imperfections of their offerings in such contexts (unless they are partisan experts in a dispute). Their interest is to hedge their statements with disclaimers and alternatives. With the *pedigree* evaluation, there is a means whereby the most radical uncertainties can be clearly expressed, and then form part of a reasoned discussion of the reliability of the available quantitative information.

7. CONCLUSION

We have indicated how NUSAP can contribute to the resolution of the two urgent problems in the methodology of natural science. In epistemology the problem is effectively transformed away from the need for a logical structure, independent of human judgement, whereby uncertainty and ignorance can be conquered. With the notational scheme, these complementary aspects of our knowledge are exhibited in a coherent form. Thus the experience of successful practice in the quantitative sciences is codified; and the management of the uncertainties becomes a definable task.

In those areas of policy-related research where severe uncertainty prevails, NUSAP provides a standardised means for communications. Debates on the necessarily imperfect and contentious quantities that are invoked will then have a structure and a discipline. The acceptance of NUSAP will also enhance clarity of understanding among those who provide quantitative information and contribute to the improvement of quality control. In such ways, it will increase familiarity with "uncertain quantities" among all who use them; and in that way enable a shift in "scientific common sense", so that a more mature understanding of the scope and limits of science may be achieved.

University of Leeds, U.K.

BIBLIOGRAPHY

Ayer, A. J.: 1936, *Language, Truth and Logic*, Gollancz, London.
Capra, F.: 1975, *The Tao of Physics*, Wildwood House, London.
Collingridge, D.: 1980, *The Social Control of Technology*, Francis Pinter, London.
Crosland, N. P.: 1962, *Historical Studies in the Language of Chemistry*, Heinemann, London.
Dalkey, N.: 1969, 'An Experimental Study of Group Opinion. The Delphi Method,' *Futures* 1 (5), 408—426.
Douglas, M. and D. Wildarvski: 1982, *Risk and Culture*, University of California.
Feyerabend, P. K.: 1975, *Against Method*, New Left Books, London.
Fox, R.: 1974,'The Rise and Fall of Laplacian Physics', *Historical Studies in the Physical Sciences* 4, 89—136.

Funtowicz, S. O. and J. R. Ravetz: 1984, 'Uncertainties and Ignorance in Policy Analysis', *Risk Analysis* **4** (3) 219—220.

Funtowicz, S. O. and J. R. Ravetz: 1986, 'Policy Related Research: A Notational Scheme for the Expression of Quantitative Technical Information', *J. Opt. Res. Soc.*, **37** (3), 1—5.

Galilei, G.: 1632, *Dialogue Concerning the Two Chief World Systems*, University of California (1953), 53.

Gödel, K.: 1931, *On Formally Undecidable Propositions*, Basic Books (1962), New York.

Holton, G.: 1973, *Thematic Origins of Scientific Thought, Kepler to Einstein*, Harvard U.P., 115—161.

Jiang, J.: 1985, 'Scientific Rationality, Formal or Informal', *Brit. J. Phil. Sci* **36** (4), 409—423.

Keynes, J. M.: 1921, *A Treatise on Probability*, St. Martin's Press (1952), New York.

Kline, M.: 1972, *Mathematical Thought from Ancient to Modern Times*, Oxford U.P., New York, 1206.

Kline, M.: 1974, *Why Johnny Can't Add: The Failure of the New Math . . .*, Vintage Books, New York.

Kline, M.: 1980, *Mathematics: The Loss of Certainty*, Oxford U.P., New York.

Kuhn, T. S.: 1961, 'The Function of Measurement in Modern Physical Science', *Isis* **LII**, 161—193.

Kuhn, T. S.: 1962, *The Structure of Scientific Revolutions*, University of Chicago.

Lindbolm, C. E. and D. K. Cohen: 1979, *Usable Knowledge: Social Science and Social Problem Solving*, Yale U.P., New Haven.

Mackay, A. L.: 1977, *The Harvest of a Quiet Eye*, The Institute of Physics, Bristol and London.

Mosteller, F.: 1977, 'Assessing Unknown Numbers: Order of Magnitude Estimation'. In *Statistics and Public Policy*, Fairley, W. B. and F. Mosteller (eds.), Addison-Wesley, 163—184.

Nelkin, D. (ed.): 1979, *Controversy: Politics of Technical Decisions*, Sage Publications, London.

Olby, R. C.: 1966, *Origins of Mendelism*, Constable, London, 116/182—185.

Putnam, H.: 1981, 'The Impact of Science on Modern Conceptions of Rationality', *Synthese* **46**, 359—382.

Ravetz, J. R.: 1971, *Scientific Knowledge and its Social Problems*, Oxford U.P., 158. See also Chapter 10 for 'Quality Control in Science'.

Ruckelshaus, W. D.: 1984, 'Risk in a Free Society;, *Risk Analysis* **4**, 157—162.

Savage, L. J.: 1954, *The Foundations of Statistics*, J. Wiley, New York.

Shapere, D.: 1986, 'External and Internal factors in the Development of Science', *Science and Technology Studies* **4** (1), 1—9.

Weinberg, A. M.: 1972, 'Science and Trans-Science', *Minerva* **10**, 209—222.

Zukav, G.: 1979, *Dancing Wu Li Masters: An overview of the new physics*, Morrow, New York.

DAVID R. OLDROYD

PUNCTUATED EQUILIBRIUM THEORY AND TIME: A CASE STUDY IN PROBLEMS OF COHERENCE IN THE MEASUREMENT OF GEOLOGICAL TIME (THE 'KBS' TUFF CONTROVERSY AND THE DATING OF ROCKS IN THE TURKANA BASIN, KENYA)

INTRODUCTION

The present paper was initially stimulated by attendance at a lecture in Montreal in March 1985 by a young English geologist, Peter G. Williamson, on the results of his researches on Pleistocene shells at a site in Kenya, east of Lake Turkana (formerly Lake Rudolf). Dr. Williamson, who currently holds an appointment at the Agassiz Museum of Comparative Zoology at Harvard University, is a keen exponent of the theory of 'punctuated equilibrium' in evolutionary biology. In his Montreal lecture, he claimed that his observations in Kenya provided compelling empirical evidence that shell forms could remain largely unchanged for very considerable periods, but could also change rapidly on occasions in a series of 'punctuations'.

Such claims are obviously dependent on an accurate dating of the sediments in which the shells are contained, and it occurred to me that it would be an interesting exercise to enquire exactly what measurements were employed in the determination of the ages of the strata, on which the estimates of the rates of evolutionary change were based. I little thought at the time what a remarkable story would be revealed. But in looking at this story, we find an intriguing example of how consensus about scientific 'facts' is established within the agonistic field of the scientific community. We find interaction between field observations, laboratory data, theories, hypotheses and speculations, human personalities, interests and ambitions: a fascinating dialectic between what the 'facts say' and what the researchers want them to say.

In the initial report of his investigations in *Nature*, Williamson (1981) wrote (p. 448): "the mollusc sequence can be studied within a well-documented chronostratigraphical and palaeoenvironmental context". This statement was literally correct (there being no shortage of

89

John Forge (Ed.), Measurement, Realism and Objectivity, 89—152.
© 1987 *by D. Reidel Publishing Company. All Rights Reserved.*

documents). But it swept under the carpet a decade or more of intense debate, which I shall seek to analyze in the present paper. My purpose is not to forge a weapon, either to attack or to defend the theory of punctuated equilibria. Rather, I am interested in characterizing how, in a specific case, geological time was measured; and in showing the multiplicity of 'dimensions' — physical, biological, social — that were involved in achieving a consensus as to the age of even one rock stratum. Whether, in the final analysis, Williamson's conclusions, and the theory on which they are based, are valid is a question on which science has not yet reached its final verdict, and it is one on which I do not wish to influence the jury.

In what follows, I first make some general remarks about the theory of punctuated equilibrium. I then look at the geological background to the claims made in Williamson's paper. I give a brief exposition of the principles involved in the several methods used for the estimation of numerical ages of rocks, and the kind of measurement that were involved in each case. I also describe the arguments deployed by the several protagonists as these appeared in the published papers.[1] We find that there was intense controversy over the age of one particular rock formation and examination of this controversy yields useful information bearing on the social structure of science. It also has epistemological implications which will receive consideration. Moreover, we find that the correct dating of the rock had a direct bearing on the estimates given by Williamson for the time required for a 'punctuation', so· we may see how our geological case history is intimately connected with problems in contemporary evolutionary biology. In the concluding section, I examine the developments that have occurred since 1981 in relation to the rocks of Lake Turkana, which were the subject of such contention, and I refer briefly to the present on-going controversies and their implications for the theory of punctuated equilibrium. But this can only be done in a limited fashion, and anything I say is bound to be overtaken by events. It is not, as I say, my intention to intervene in current controversies.

THE PUNCTUATED EQUILIBRIUM THEORY

The term 'punctuated equilibrium' was coined by two American palaeontologists, Niles Eldredge and Stephen Jay Gould, in a much-cited paper published in 1972 (Eldredge and Gould, 1972). They envisaged species remaining unchanged for long periods, and then evolving quite rapidly from time to time: rapid change 'punctuating' periods of biological stasis in the geological record. Such rapid changes might occur if a small 'peripheral isolate' of a species had undergone rapid evolutionary change in response to sudden environmental changes. If, then, these new conditions were to extend into the main habitat, the new forms would be likely to rapidly supersede the parent forms in that region, with the result that the stratigraphical column of the main habitat would display apparently sudden breaks in the fossil forms. Such a model does not seem particularly controversial, but it flew in the face of the 'gradualistic' doctrine that we customarily associate with Darwinian theory; and it raised questions as to *how* the apparently rapid evolutionary change might occur in the regions of the 'peripheral isolates' where the evolutionary 'punctuations' supposedly occurred. Since the publication of the celebrated 'tree diagram' of Darwin's *Origin of Species* (Darwin, 1859, following p. 116), biologists had grown accustomed to thinking in terms of evolutionary change ocurring in a gradualistic manner, with species forms slowly diverging from each other and from their ancestral types; 'phyletic gradualism', not 'evolution by jerks'.

Whether the Eldredge/Gould theory was, in truth, a radically new theory, or one that might be comfortably accommodated by neo-Darwinism, has been the subject of debate. The role of geographical isolation was admitted as essential by Darwin in the later editions of the *Origin* (though he had largely done without it in the First Edition).[2] The importance of 'peripheral isolates' in the evolutionary process had been emphasized by Ernst Mayr (1963, pp. 496, 544 and *passim*). Though he subsequently modified his views somewhat, the eminent evolutionist G. G. Simpson, back in 1944, had considered the broad implications of the palaeontological record, as known at that time, and had concluded that it suggested the occurrences of what he termed 'quantum evolution' (Simpson, 1944, pp. 206—217). Moreover, some recent commentators such as Sir Andrew Huxley, F.R.S. (Huxley, 1982), or the eminent evolutionist J. Maynard Smith (1981), have doubted whether the Eldredge/Gould theory is really so very different from orthodox theory.

However, as mentioned, Gould and Eldredge have seen their doctrine as being of radical importance and substantially different from earlier views. In a later statement of their theory (Gould and Eldredge, 1977), they argued that the palaeontological studies customarily offered as empirical exemplifications of phyletic gradualism[3] do not bear close scrutiny, and if properly construed are compatible with the punctuated equilibrium model. So the empirical basis of orthodox theory was thrown in question. Moreover, Gould and Eldredge situated their theory within a "general philosophy of change" which was stated to be dialectical in character, complaining that the customary preference for gradualism in Western science was in line with prevailing ideology. They urged pluralism in "guiding philosophies". They didn't state exactly that theirs was a new Marxist palaeontology, and it is not entirely clear to me why their theory should be regarded as dialectical or in some other way 'left-wing'; but clearly this is how it has been received in some quarters, which doubtless accounts for some of the heat that has attended the punctuated equilibrium debate.[4] Gould and Eldredge themselves see their programme as involving closer studies of fossil sequences — over extended regions, not confined to local sections or single cores. Such work should be accompanied by studies of genetic mechanisms that allow stasis punctuated by occasional periods of rapid change. The investigations may be illuminated by studies of ontogeny and phylogeny. For Gould believes that macro-evolutionary change may arise through small changes occurring early in embryonic development, accumulating during growth to generate large-scale changes in adulthood (Gould, 1977; 1983, p. 160). Also, thinking in genetic terms, he suggests that some genes may function as 'masters' over others in a gene hierarchy; in which case mutation of a 'master' gene may entail large-scale changes and be a motor of evolution (Gould, 1984, p. 196). Philosophically speaking, Gould and Eldredge see their theory as more 'holistic' than standard doctrine, and more in keeping with a dialectical world-view.

In point of fact, however, the punctuated equilibrium theory, while directing biologists and palaeontologists to re-examine their data in a manner that has proved fruitful, with some rather unexpected discoveries (Stanley, 1981), has not in itself yielded much by way of new evolutionary mechanisms. The chief new suggestion is that of 'species selection',[5] a concept propounded by Steven Stanley (1975, 1979). Here, the *whole species* is considered as a unit of selection, rather than the customary individual organism. It may, at first sight, be difficult to envisage how this might occur. But consider the following example, cited by Elliot Sober (1984, p. 366). Suppose speciation occurs in a species that tends to form isolated communities more readily than does a closely related species which interbreeds freely through a large homogenous population. Then the patterns of evolution will not be the same in the two cases, even if the total number of organisms is identical; so one 'strategy' or the other may be favoured. In thus focussing attention on *species* as the units of selection, rather than organisms, the theory is indeed antireductionist in tendency. It should be noted, however, that not all commentators allow that species selection would, in fact, be a special characteristic of puntuationism. For example, Mark Ridley (1985) claims that the same effect would result from the two 'strategies' envisaged by Sober, regardless of whether speciation occurred in jerks or according to the more usual bifurcations envisaged by neo-Darwinists.

Nevertheless, in a recent statement of his views, Eldredge (1986) has again emphasized the importance for punctuationists of the species selection hypothesis. Darwin himself thought that boundaries between species are in some measure arbitrary, for if one species gradually evolves into another over time one cannot say exactly where one species begins and another ends. On this view, therefore, species are not strictly 'real' entities. But if one species or another can be said to survive in the struggle for existence — if species constitute one possible level of selection — then one must take a different view of the ontology of species. Eldredge (1986) also suggests that after some environmental catastrophe whole species could well be in competition with one another as survivors struggle to re-establish themselves in a given area. Thus we see a kind of link back to long-relinquished catastrophist doctrines; and possibly also with pre-Darwinian 'essentialist' notions of species. (The punctuationists do not — to my knowledge — claim such intellectual roots, nor am I suggesting that they exist as such, historically speaking.) Be all this as it may, it is clear that the punctuationist hypothesis is a matter of considerable theoretical interest in modern evolutionary biology and the adequacy of the empirical evidence that it can command in its favour is certainly of the highest concern. This evidence has, therefore, attracted considerable attention, and, as may be expected, has also been the subject of much controversy.

Both Gould and Eldredge arrived at the notion of punctuated equilibrium in the course of their empirical studies as graduate students. Gould (1969) worked with Pleistocene and Recent snail shells in Bermuda; Eldredge (1972) with Devonian trilobites in the American Mid-West. In a recent semi-popular work, Eldredge (1985) has given an account of the circumstances that led to his embracing a theory of morphological stasis amongst his trilobite population, punctuated by occasional periods of rapid change. In the early part of his research, he had difficulty in finding *any* significant evidence for change in the fossils. Eventually, he discerned differences in the number of columns of lenses in the creatures' compound eyes. The fossils in some regions were discovered to have 17 columns; others 18. Eldredge then discovered changes through time also: the lower rocks had '18-column' types, which then suddenly changed to the '17-column' version. Was this sudden change due to a gap in the stratigraphical record, or was it due to a sudden evolutionary 'jerk'? The question was resolved in Eldredge's mind when he found a particular area (in a quarry to the east of where the apparent 'jerk' in fossil form occurred) in which the two forms co-existed. The conclusion was that this represented a 'peripheral isolate', evolving rapidly in response to changing environmental conditions. So the 18-column type died out in the west, but survived in the east by evolving to a 17-column form. This then supposedly spread to the western region; so when the fossil sequence is examined there it offers the appearance of sudden, discontinuous change. In fact, however, Eldredge maintained, the change occurred elsewhere (to the east) with rapid, though not instantaneous, speciation.

One can envisage several alternative shapes for the tree of life, depending on different possible modes of speciation, such as are shown in Figure 1.[6]

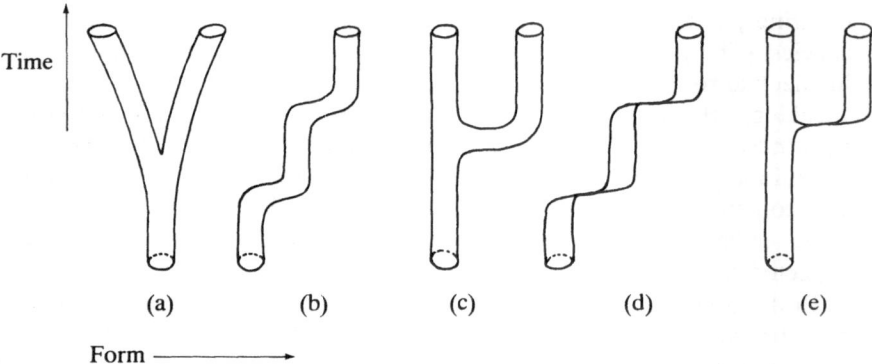

Fig. 1. Plot of time against form. Population size ∝ thickness of bar.

Scheme (a) is the model of classical phyletic gradualism. Scheme (e) is the one favoured by punctuated equilibrium theorists. But since it involves a rapid shift on the part of a small population of organisms it is natural that the actual changes in kind would be difficult to discern in the fossil record. Actually, Eldredge's trilobites in his eastern quarry did not exactly fit the bill, for he did not observe rapid *gradation* between 17- and 18-columned eyes. The fossils *either* had 18 *or* 17-columns; but it is, of course, difficult to see how they could have done otherwise.

WILLIAMSON'S EMPIRICAL EVIDENCE FOR THE THEORY OF PUNCTUATED EQUILIBRIUM

Given the interest generated by the work of Gould and Eldredge, the attention of readers of *Nature* was quickly attracted by Williamson's paper, published in 1981,[7] offering an account of detailed palaeontological and stratigraphical investigations in north-west Kenya. These researches revealed such vast quantities of fossil shells, occurring in sediments securely datable by physical determination of the ages of tuff (volcanic ash) marker bands, that the very process of morphological evolutionary change could be perceived with precision, as it were under a microscope. As J. S. Jones expressed the matter in an editorial in *Nature* (Jones, 1981), the Kenyan rocks constituted an "uncensored page of fossil history". And certainly the new results appeared to provide convincing empirical evidence in favour of a punctuationist model for evolutionary change. Thus, on the basis of his observations, Williamson suggested that he could discern several changes of type (e) in Figure 1.

I shall not seek to give a complete picture of Williamson's work, but in Figure 2, reproduced from his report in *Nature*, we see how he displayed his palaeontological findings, according to the general framework of the theory of punctuated equilibria. It will be seen that the claimed speciation events are represented as rapid 'punctuations' of periods of morphological stasis (occurring particularly at the stratigraphical horizons marked by the two thick, downward-pointing arrows). Williamson contended that he could discern in the fossil record the very occurrence of punctuation or speciation. This was seemingly an advance, since palaeontological records of the actual

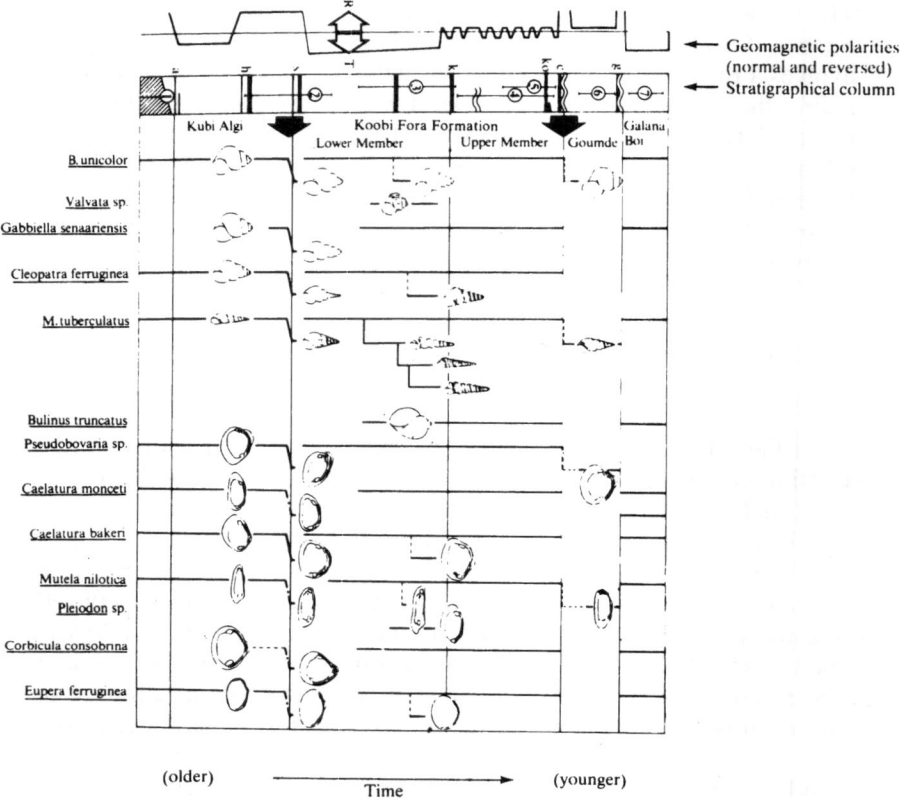

[g = Top Guomde Tuff; c = Chari/Karari Tuff; ko = Okote Tuff; k = KBS Tuff;
— = Tulu Bor Tuff; s = Suregei Tuff Complex; h = Hasuma Tuff; a = Allia Tuff.]

Fig. 2. Reprinted by permission from *Nature*, **293**, p. 442. Copyright © 1981,
Macmillan Journals Ltd.

process of very rapid evolutionary change had not been observed
exactly by Eldredge.

The gradual elucidation of the stratigraphical column that is represented at the top of Figure 2 will be outlined in what follows; also the
arguments leading to the dating of the sedimentary sequences. It will be
evident that accurate knowledge of the dating of the sedimentary layers,

together with their thickness and the completeness, is essential to determine rates of sedimentation at specific horizons, and hence speciation rates. The extent to which these requirements were met will be considered later in this paper. Here we should note that the chief horizon at which speciation was supposed to occur was that of the Suregei Tuff Complex. And Williamson suggested (1981, p. 441) that the changes that occurred at this horizon were continuous and unidirectional, and on the evidence of "general thickness estimates" would have taken between 5000 and 50000 years to accomplish. However, he did not at that time provide evidence on which this estimate was based, beyond referring readers to his doctoral dissertation (University of Bristol). I have subsequently been informed by Dr. Williamson (personal communication, 24 February 1986) that the estimate was in fact made by Professor Francis Brown of the University of Utah. To my knowledge no publication has revealed the exact line of reasoning by which Brown arrived at his conclusion. Thus the evidence on which the punctuated equilibrium hypothesis rested, so far as Williamson's 1981 Turkana paper was concerned, is somewhat obscure so far as the published record is concerned. We shall, however, say something further about the basis of Brown's argument below (p. 137). And we shall also examine briefly the evidence that has been adduced more recently by Williamson in relation to punctuations at the Suregei Tuff horizon.

THE HISTORICAL BACKGROUND TO WILLIAMSON'S CLAIMS:
THE KOOBI FORA RESEARCH PROJECT

Lake Turkana lies in north-west Kenya in arid country. Its eastern shores received very little attention from geologists before the 1960s,[8] but following a brief reconnaissance by Richard E. Leakey in 1967, which revealed the existence of numerous vertebrate fossils and some stone tools, a substantial expedition was launched the following year under Leakey's leadership. This eventually blossomed into the 'Koobi Fora Research Project' (Leakey and Leakey, 1978; Harris, 1983; Leakey, 1983), named after a locality, Koobi Fora, on the eastern margin of the lake (see Figure 3, based chiefly on Leakey and Leakey, 1978, p. 2). Work has proceeded actively in the area since 1969.

Early in the research programme, artefacts were discovered *in situ* on the Koobi Fora Ridge, near a prominent light-grey tuff band, about 1 metre thick and of characteristic 'salt-and-pepper' appearance. The dis-

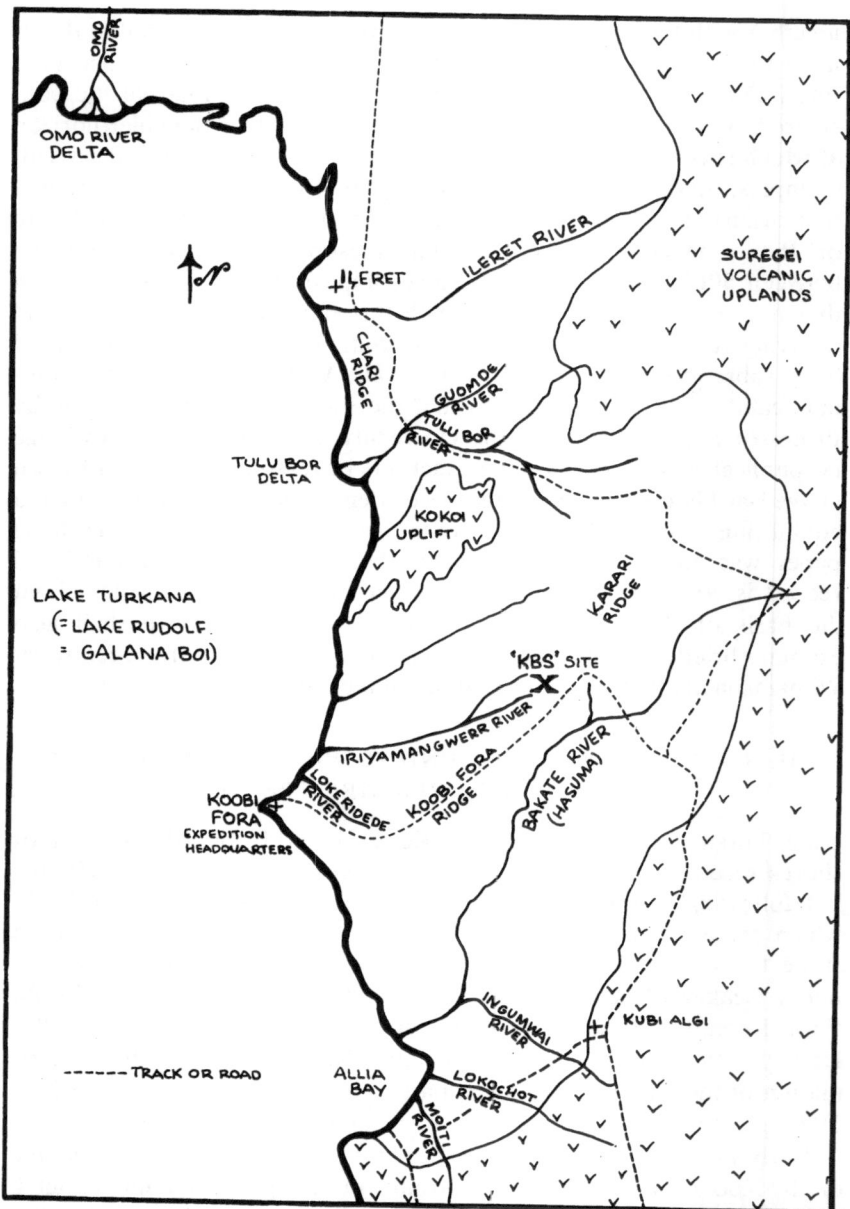

Fig. 3. Koobi Fora.

coverer was the expedition's first geologist, Miss A. K. Behrensmeyer, a graduate student of Harvard (later of Yale University and currently at the Smithsonian Institution), and the location was appropriately named the 'Kay Behrensmeyer Site' in her honour. The associated tuff thus became known as the 'KBS Tuff'. Its age was to become a major bone of contention in the story, and the whole episode has become known as the 'KBS Tuff Controversy'. Various hominid remains were also discovered.[9]

Samples of the KBS Tuff were despatched to England for dating, and were examined by Mr. Frank Fitch and Dr. John Miller of Birkbeck College, London, and Cambridge University respectively. In addition to their university work, these investigators also ran a private firm engaged in geological consulting, named F. M. Consultants Ltd., and initially they supplied their expertise on a commercial basis. Subsequently, they worked as associates to the Koobi Fora Expedition, with a research grant from the National Environment Research Council (U.K.) (Leakey, 1983, p. 134). Their preliminary investigations yielded a result of about 220 million years (220 Myr), which was clearly in error for rocks that were obviously young, geologically speaking, and the errors were attributed to impurities in the rock samples. Further analysis, however, yielded a value of 2.61 Myr (Fitch and Miller, 1970). This result caused much excitement, and considerable importance was attached to it (Isaac et al., 1971), since it suggested a significantly increased time-scale for the existence of tool-making hominids on this planet, and it hardly needs saying that there was reluctance to give up such a striking piece of evidence favouring the considerable extension of human or tool-making hominid history. We shall not, however, be concerned with the issue of human evolution in the present account.

EMPIRICAL METHODS: THE POTASSIUM/ARGON METHOD FOR DETERMINATION OF ROCK AGES

How were the Fitch and Miller results obtained, and what kinds of measurement were involved? To answer these questions, an outline of the potassium/argon method for the dating of rocks is required. There are two versions of the method in common use.

In both versions, use is made of the spontaneous conversions, with known half-lives, of ^{40}K into ^{40}Ar and ^{40}Ca, by electron capture[10] and electron emission (or beta-decay) respectively, for which the partial decay constants are symbolized as λ_e and λ_β. If the quantity of radiogenic argon in a specimen is symbolized by $^{40}Ar^*$, and the quantity of residual radioactive isotope of potassium by ^{40}K, and if a mineral specimen has existed for time t, during which the spontaneous conversion of ^{40}K to $^{40}Ar^*$ has occurred, then:[11]

$$(1) \quad t = [1/(\lambda_e + \lambda_\beta)] \ln \{1 + [(\lambda_e + \lambda_\beta)/\lambda_e]^{40}Ar^*/^{40}K\}$$

So the problem in the determination of the age of a rock specimen is to measure ^{40}K and $^{40}Ar^*$ — the decay constants, λ_e and λ_β, being assumed from information supplied by atomic physicists. Usually, potassium-rich minerals such as sanidine are extracted and measured. In the conventional K—Ar method, the potassium content is commonly determined with a flame photometer. A solution of the substance is prepared and introduced into the air of a gas burner in a fine spray. The intensity of the lilac light resulting from the presence of potassium atoms is determined with a sensitive light detector (phototube), calibrated against a standard potassium source.[12] Following the measurement of the total potassium, the ^{40}K is calculated as 1/8600th, based on the assumption that this is the fraction of the isotope occurring in nature at the present time. Potassium may also be determined in the foregoing manner for whole rocks if desired, rather than particular minerals in the specimens.

The determination of argon is more complex. It has to be driven out of the rock (or extracted mineral) by heat and then analyzed by some method. Early on, the argon was separated from the other inert gases and its pressure measured directly, from which the quantity of argon could be calculated. This method is now obsolete. In modern work, the argon driven off is analyzed with a mass spectrometer, which allows the proportions of ^{40}Ar (partly radiogenic and partly arising from air contamination), ^{38}Ar (tracer isotope, added in known quantity as 'spike'[13]), and ^{36}Ar (from air contamination) to be determined. From these measurements, the molar concentration of $^{40}Ar^*$ in the sample can be calculated from the equation:[14]

$$(2) \quad [^{40}Ar^*] = [^{38}Ar_{spike}]\{([^{40}Ar]/[^{38}Ar])_{mixture}$$
$$- ([^{40}Ar]/[^{36}Ar])_{air}/([^{38}Ar]/[^{36}Ar])_{mixture}$$
$$- ([^{40}Ar]/[^{38}Ar])_{spike}\}.$$

The technique for the extraction of argon is complicated and is shown in Figure 4 (based on Dalrymple and Lanphere, 1969, p. 63).

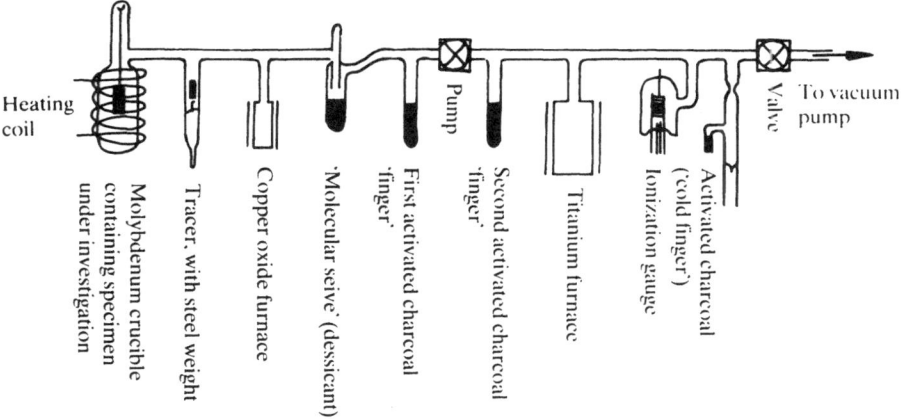

Fig. 4. The extraction of argon[15].

The extracted gas is introduced into a mass spectrometer. In the traditional form of this instrument, positive ions generated by an electron
beam acting on the different argon isotopes are accelerated into a
magnetic field, where they are separated according to their different
charge/mass ratios. Ions from the different isotopes are directed in turn
to a collector electrode, whence they leak to earth through a high resistance, and the voltage drop across the resistor is measured. This drop is
proportional to the number of ions hitting the electrode. Hence, the
relative quantities of the ^{40}Ar, ^{38}Ar and ^{36}Ar isotopes in the sample can
be determined. Usually the voltage is plotted against the deflecting
magnetic field, yielding three peaks whose heights correspond to the
relative quantities of the three isotopes. Or in modern machines the
data are fed directly into a computer. For the purposes of improved
accuracy, readings of argon concentrations should be taken at measured time intervals after the gas is introduced into the mass spectrometer, and by extrapolating backwards the relative concentrations of
the isotopes at the time of the first introduction of the gas may be
calculated. (Argon from a previous run may have been present in the
apparatus, adsorbed on solid surfaces, and on mixing or exchanging
with introduced gas it may cause errors. Such 'memory' errors are
overcome by the use of the extrapolation procedure.)

In the investigations we shall be describing, however, a substantially
different piece of apparatus was used: the 'omegatron' mass spectro-

meter (Grasty and Miller, 1965). This device, which works on a principle similar to that of the better-known cyclotron, was first devised by
workers at the National Bureau of Standards, Washington, D.C., to
measure fundamental physical constants such as the Faraday, the
charge/mass ratio of the proton, and the proton/electron mass ratio
(Sommer, Thomas and Hipple, 1951). Subsequently, it was adapted by
workers of the Westinghouse Laboratory, Pittsburgh, to measure very
low partial pressures (Alpert and Buritz, 1954), and later it was used by
Grasty and Miller at Cambridge for the analysis of argon.

Following the description of Sommer, Thomas and Hipple, the
apparatus may be represented as shown in Figure 5. (The ionizing
chamber is surrounded by a glass vessel containing the gas to be
analyzed, so that the argon is both within and without the omegatron
chamber.)

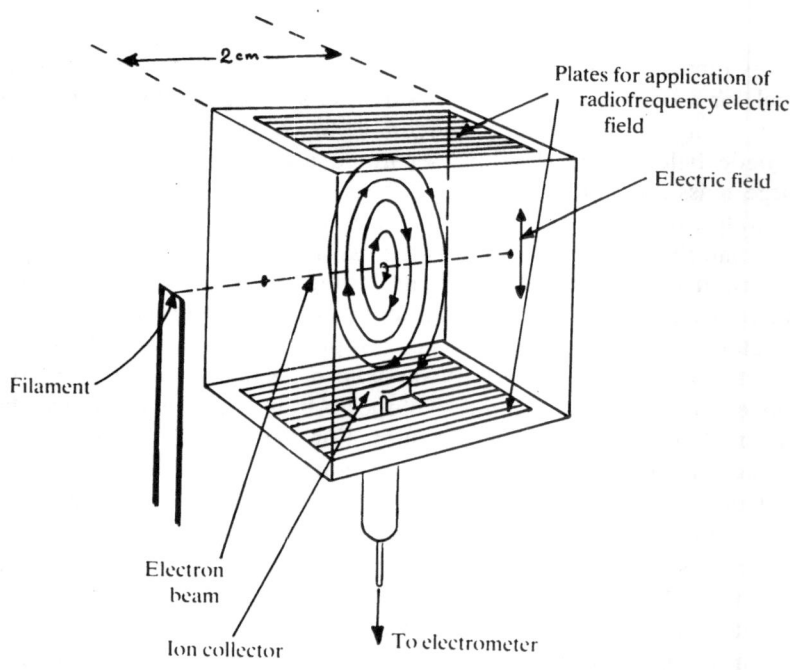

Fig. 5. The Omegatron mass spectrometer.

A narrow beam of electrons is directed into the apparatus from a heated filament, causing the production of positive ions from the gas contained within the small (2 cm square) omegatron chamber. These ions can move in circles perpendicular to the electron beam with a frequency given by $\omega = eH/M$, where e = electonic charge, H = magnetic field strength and M = mass of ion. In the presence of an impressed radiofrequency electric field, however, ions whose circulation frequency is the same as that of the applied radiofrequency are caused to spiral outwards where they may eventually be picked up by the collector, as shown in Figure 5. So ions of different mass in the chamber may be scanned by gradually altering the frequency of the applied electric field. As in the orthodox mass spectrometer, the current leaking away to earth from the ion collector is representative of the presence of ions of a particular mass in the chamber. Thus, on evacuation of the apparatus, followed by the admission of an argon sample, the relative proportions of the several argon isotopes may be determined.

It will be evident that in determinations such as those outlined above, limitations are presented by the necessary calibrations of the mass spectrometer, the purity of the ^{38}Ar spike (tracer), and errors in the physical constants used in the calculations. The combined effects of these uncertainties is likely to vary somewhat from laboratory to laboratory.

An important way of treating the data is to prepare so-called isochron diagrams. Various samples are extracted from the rock, and the several measured ^{40}Ar/^{36}Ar ratios are plotted against corresponding ^{40}K/^{36}Ar values. For a rock that has just been formed, the ^{40}Ar/^{36}Ar ratio will normally be uniform throughout, with a value of about 296 : 1, as in air. (See Figure 6.)

In time, however, the ^{40}K will convert to ^{40}Ar*, so that the proportion of ^{40}Ar will rise. The extent to which it does this will depend on the quantity of ^{40}K that happens to be in the sample. So after so many million years a plot will look like Figure 7. The slope of this graph (or isochron diagram) will be dependent on the age of the rock, which may thus be determined.[16]

All this may seem rather complicated, but it might have been better in the KBS Tuff controversy if only the conventional K—Ar technique had been used. Instead, recourse was had to another procedure developed in the 1960s at the independent suggestion of T. Sigurgeirsson

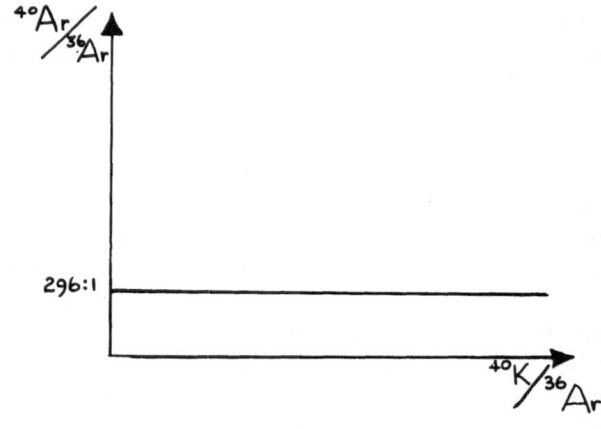

Fig. 6.

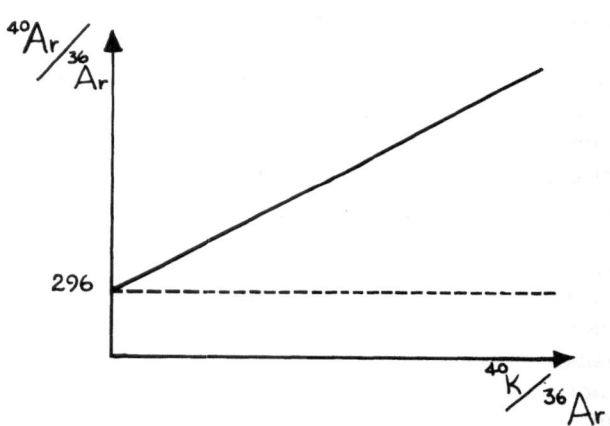

Fig. 7.

(1962) and C. M. Merrihue (1966). Thus we come to the second method mentioned above.

In this method, which obviates the necessity of measuring the potassium concentration with the flame photometer, some of the potassium (^{39}K) in the sample is converted to ^{39}Ar by exposure to neutron irradiation in an atomic reactor. A standard sample of known age is treated similarly, and the ^{39}Ar is found for the two specimens by driving it off by heat and measurement in the mass spectometer; also ^{40}Ar is determined in the same run. Clearly, ^{40}Ar/^{39}Ar will increase according to the age of the sample. Thus comparison of $^{40}Ar/^{39}Ar$ of the unknown with that of the standard of known age allows determination of the age of the sample. The determination of the absolute concentrations of K and Ar is not need therefore.

This seemingly simple method can be elaborated by driving off the argon at increasingly high temperatures, instead of proceeding to fusion immediately. In this manner, the *apparent* ages of the sample can be determined for different temperatures by measuring the $^{40}Ar/^{39}Ar$ ratio at each stage of the stepwise heating. The apparent age will not necessarily be constant, since argon at different sites in the crystal lattices may have different isotopic ratios, due to diffusion of gas over time, entrapment of argon at the time of original crystallization, and so on. If the rock has suffered a heating episode subsequent to its original formation, *part* of the argon may have been driven off at that time, and more may have been formed subsequently by radioactive decay of ^{40}K. In principle, then, examination of the age-spectrum may tell one something about the geological history of the rock — for example, the date of its original formation (by Ar from its securest sites in the crystal lattices) and its date of subsequent 'overprinting(s)'. Examples of step-heating diagrams are shown in Figure 8.

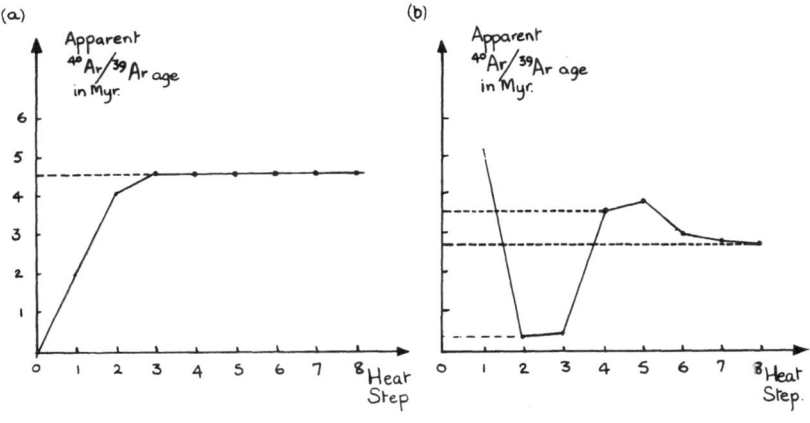

Fig. 8.[17]

In case (a), one age only for the sample is indicated — 4.5 Myr. (The early low values correspond to the initial displacement of loosely-held argon, usually atmospheric, but not necessarily so.) Case (b) is obviously more complicated, but one might venture an interpretation as follows.

The initial reading of 5 Myr is actually spurious, due to loosely-held argon from some external source. The probable age is 3.6 Myr, with 'overprintings' at 2.7 and 0.3 Myr. Obviously, however, interpretation of such diagrams in this manner is risky, given that each point in the graph has its own region of possible error. And in fact few authors would be willing to venture any definite interpretations on the basis of such data, for which no adequate theoretical model is available as to how the argon is held in the crystal lattices.

In yet another way of handling the data, if the $^{40}Ar/^{36}Ar$ and $^{39}Ar/^{36}Ar$ ratios are determined at each stage of the heating process, one may obtain an isochron graph, the slope of which should reveal the age of the rock. But one might, perhaps, find several lines, whose several slopes might be taken to indicate distinct episodes in the history of the material. Thus, Figure 9 shows isochron diagrams for the step-heating process: (a) for a theoretically ideal straight-line plot; (b) a plot revealing the incorporation of some ^{40}Ar at original crystallization; and (c) a specimen that has undergone overprinting since its formation.

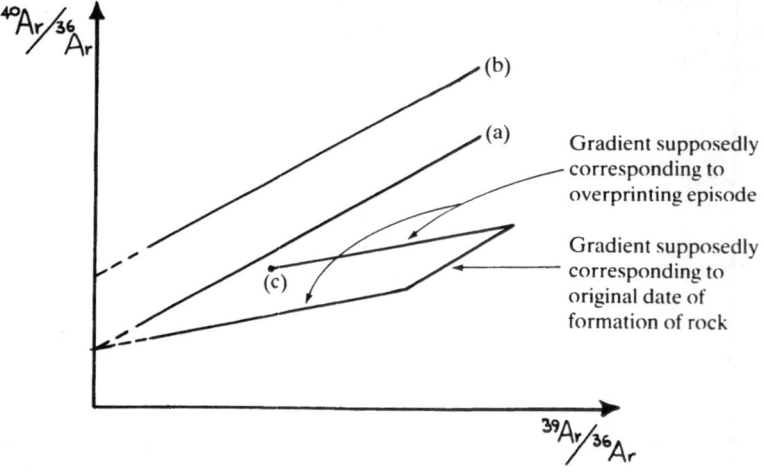

Fig. 9.

Clearly, the application of this method may have certain advantages for the elucidation of a rock's history. With the standard K—Ar method, using total fusion, one may obtain anomalous results owing to inherited argon in xenoliths,[18] from argon incorporated at the time of initial formation of a rock (which might not have had the normal isotopic ratio), or by argon introduced or lost during the subsequent geological history of the rock (overprinting). Also, when dealing with young rocks the quantity of $^{40}Ar^*$ will be minute, and even small errors in the components of the right-hand side of Equation 2 (p. 100) will become very significant in the taking of differences. However, some of these complications might, it seemed, be side-stepped with the help of the new method. One can see, therefore, why it attracted Fitch and Miller. They commended it publicly in a paper presented to the Volcanic Studies Group of the Geological Society of London in March 1971 (Fitch and Miller, 1971).[19]

However, the new method was not without problems for any discrepant points on a diagram can lead to a 'cobweb' of lines, and thereby stimulate an outpouring of *ad-hoc* hypotheses![20] One may add the thought that for the tuffs of the Koobi Fora region (as opposed to igneous lava flows, etc.) one was dealing with volcanic ash deposited as sediment. Such material could readily incorporate material from more than one source so that a composite age might be revealed, over and above what might be apparent owing to some period of subsequent heating (overprinting). So while the method appeared to have considerable attractions, in fact its use added to the controversies attending to the Turkana investigations, rather than making all things plain. Also, the omegatron spectrometer used by Fitch and Miller is now regarded with disfavour by nearly all K—Ar researchers, who maintain that it does not yield the precision that is possible with the standard apparatus. I am informed, however, that the Fitch/Miller team are still satisfied with the apparatus (Fitch, personal communication, 28 May 1986).

THE DEVELOPMENT OF THE STRATIGRAPHICAL COLUMN AND
ROCK DATING AT KOOBI FORA

Let us turn now to consider how some of these principles were deployed in the gradual elucidation of the geology of the Koobi Fora region through the 1970s. A geological reconnaissance survey was required which was undertaken by Kay Behrensmeyer (1970). She recognized several tuff bands in a sedimentary sequence of alternating lacustrine and fluvial deposits. An upper sequence, chiefly fluvial with mammalian remains and including what is now called the KBS Tuff, was called the Koobi Fora II unit. An underlying lacustrine sequence with fish and molluscs was designated Koobi Fora I. We shall not here attempt to describe all the steps leading to the establishment of the overall stratigraphical column. The task seems to have proceeded fairly smoothly at first, though problems were encountered due to mistaken tuff correlations. (For the most part, the tuffs do not offer extended outcrops.) And faulting also caused difficulties. So it is perhaps not surprising that some radical revisions where called for subsequently. (See p. 138.) The chief contributors in the early stages of the work were Carl Vondra and Bruce Bowen (Vondra *et al.*, 1971; Bowen and Vondra, 1973; Vondra and Bowen, 1976, 1978) of Iowa State University, and Ian Findlater of Birkbeck College and later of the Leakey Memorial Institute for African Prehistory at Nairobi (Findlater, 1976, 1978). By 1978, then, the generalized stratigraphical column was represented as shown in Figure 10 (Findlater, 1978, p. 15), which was more or less the picture used by Williamson in his 1981 paper.

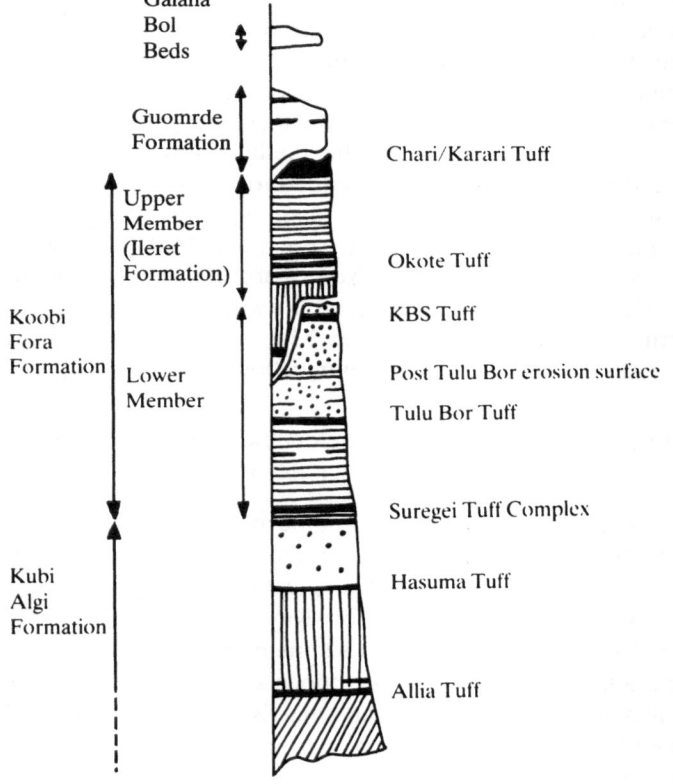

Fig. 10.

It should be noted that the gap in the stratigraphical column in the Koobi Fora Upper Member, above the KBS Tuff, was claimed on field evidence *before* it became a subject of controversy when the data on tuff datings became available (Isaac *et al.*, 1971, p. 1129).

Let us now examine in a little detail the story of the determination of the ages of the several tuffs. As mentioned above, the first effort in this direction was undertaken by Fitch and Miller (1970). As we have seen, in their initial investigations of the KBS Tuff they used the orthodox K—Ar method, obtaining values of about 220 Myr, which were obviously incorrect, due to extraneous argon. Using two fresher, and carefully prepared, samples, preliminary values of 3.02 and 2.37 Myr were obtained, again by the standard method, the first using whole rock, the second extracted feldspar crystals. They then proceeded to a ^{40}Ar/^{39}Ar determination, and using the step-heating process on the feldspars obtained the graph shown in Figure 11 (based on Fitch and Miller, 1970, pp. 227—228).

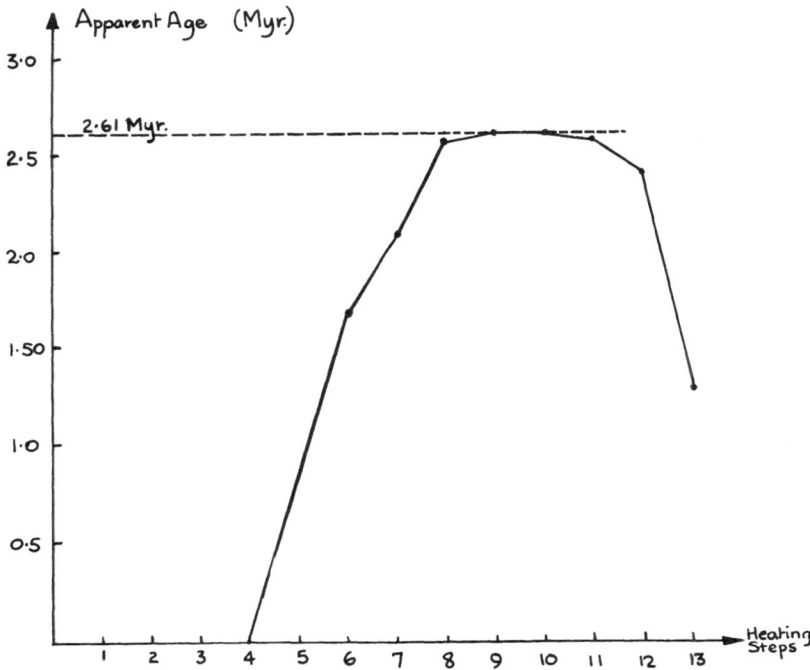

Fig. 11.

This *appeared* to be a satisfactory graph, and gave the value of 2.61 Myr, compared with 2.64 for $^{40}Ar/^{39}Ar$ total degassing and 2.37 for the standard K—Ar method. The value of 2.61 was accepted by many authors over the next few years, notably Leakey, Findlater (also of Birkbeck, it may be remembered) and Behrensmeyer, who was not a geophysicist. Fitch and Miller themselves clung tenaciously to their result. But their dating was never reproduced by other K—Ar researchers.

The establishment of a secure, dated stratigraphical column in a given area does not, however, depend on the results of physical measurements alone. Palaeontological evidence is also highly relevant, and it is necessary to achieve correlation with observations made in other regions. Also, as we shall see, geochemical and geomagnetic data may be utilized. So, following the preliminary classification of the rocks at Koobi Fora by Vondra, Behrensmeyer and others (Vondra *et al.*, 1971), palaeontologists sought to establish definitions of the different strata in terms of their fossil contents. For this purpose, the fossil remains of pigs and elephants proved to be of particular importance.

As early as 1969, Leakey invited Basil Cooke of Dalhousie University, Canada, to examine the fossil vertebrate specimens that had been recovered by the Koobi Fora Expedition, with a view to establishing palaeontological correlations with strata in adjacent regions. This Cooke did, in collaboration with Vincent Maglio, a young palaeontologist from Princeton University. Their results were first made public at a symposium held under the auspices of the Wenner-Gren Foundation for Anthropological Research at Burg Warterstein in Austria in July, 1971, but their results were not published until the following year (Cooke and Maglio, 1972), when Maglio also published another paper in *Nature* (Maglio, 1972). These early palaeontological investigations laboured under some difficulties, since the Leakey expedition had been somewhat careless in recording the locations of its fossils (Leakey, 1983). Nevertheless, despite the uncertain data, it was already apparent at the Wenner-Gren meeting that the palaeontological results were not meshing satisfactorily with the data furnished by the K—Ar daters. And as time went on, the difficulties seemed to become more acute, rather than being clarified by the presentation of more data with improved precision.

In his paper of 1972, Maglio employed four stratigraphical zones, characterized by four distinctive suites of fossils and named according to particular types. The faunal zones seemingly corresponded with four disinct lithological zones:

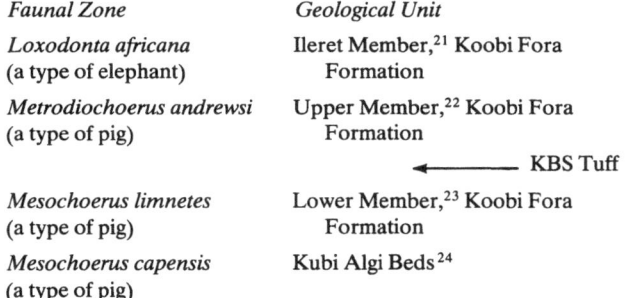

Faunal Zone	Geological Unit
Loxodonta africana (a type of elephant)	Ileret Member,[21] Koobi Fora Formation
Metrodiochoerus andrewsi (a type of pig)	Upper Member,[22] Koobi Fora Formation
	◄———— KBS Tuff
Mesochoerus limnetes (a type of pig)	Lower Member,[23] Koobi Fora Formation
Mesochoerus capensis (a type of pig)	Kubi Algi Beds[24]

Using these palaeontologically defined boundaries, Maglio sought to correlate the Koobi Fora succession with that found in the Omo River Valley to the north of Lake Turkana, which also had clear tuff marker beds, already dated (Butzer and Thurber, 1969; Brown and Lajoi, 1971) and separating highly fossiliferous strata. In examining the fossil pigs from the Omo River, Maglio contended that a clear gradation in the size of the third molars (M_3) of *M. limnetes* could be discerned through time — reckoned by reference to the dated tuff horizons. This trend is illustrated in Figure 12 (based on Maglio, 1972, p. 384).

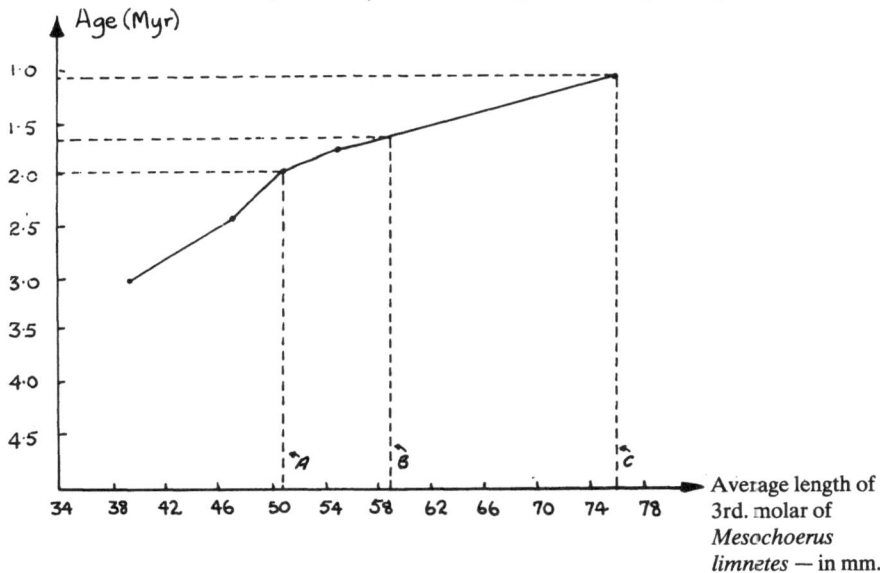

Fig. 12. Plot of age against M_3-length for fossils of *M. Limnetes* in Omo Valley.

Here, A, B, and C represent the average M_3 lengths of $M.$ *limnetes* for the three faunal zones of the Koobi Fora Formation.[25] One can, thereby, form a rough estimate of the age of the three zones, using the dated tuffs of the Omo River as standard of comparison. It will be seen that Zone A, which is *below* the KBS Tuff, has an estimated age of 2.0 Myr — considerably less than what might be expected according to the results of Fitch and Miller. In this *Nature* paper, Maglio supposed that the two sets of data were just about compatible (allowing for a range of dates of 2.7 to 1.8 Myr for the Omo deposit).[26] But in the paper of Cooke and Maglio of the same year, which was the published version of the Wenner-Gren presentation, it was stated that the suid (pig) data argued against an age of 2.6 Myr for the Koobi Fora II unit, and it was suggested that a 'best-fit' age for the fossils would be between 1.9 and 2.1 Myr (Cooke and Maglio, 1972, p. 316). In other words, the palaeontologists were willing to query the geophysicists' results: the research programme was beginning to display lack of coherence of results at a crucial point.[27] The work of Cooke and Maglio was subsequently supported by John Harris and Timothy White of the Kenya National Museum and Michigan University respectively (White and Harris, 1977), who had been invited by Leakey to give a second opinion on the palaeontological datings.

The following year (1972), a conference was held in Nairobi to assess the results that were emerging from the Lake Turkana researches, and reading between the lines of the published proceedings one can readily discern a considerable amount of intellectual tension between the participants. However, since these proceedings were not published until 1976 (Coppens *et al.*, 1976), I shall deal with them at a later stage, following here the order of the papers as they were published. Two additional physical techniques were now introduced in order to try to resolve the anomaly: magnetostratigraphy and fission-track dating.

EMPIRICAL METHODS: MAGNETOSTRATIGRAPHY

When rocks are laid down, they become magnetized according to the direction of the Earth's field at the time of deposition. Since this field reverses from time to time, the pattern of reversals may be recorded in the rocks as they are successively deposited, the particular pattern of reversals being unique for any given geological epoch. This pattern has been revealed by examination of the magnetism in unambiguously dated rock sequences (Harland *et al.*, 1982). Consequently, comparison of the palaeomagnetic history of a sequence of rocks of uncertain age with the standard scale may be of considerable assistance towards their dating. This technique was soon applied to the Koobi Fora region by Andrew Brock of the University of Nairobi and Glynn Isaac of the University of California, who was co-leader of the Koobi Fora Expedition,[28] assisted by Joab Ndombi (Brock and Isaac, 1974). They based their work on the best geomagnetic scale available at the time — that of Allan Cox of Stanford University (Cox, 1969).

To be of value in geochronology, the polarity acquired by a sample at its original deposition has to be determined, but this is likely to be obscured by magnetic vicissitudes suffered by the specimen since its first deposition. However, the obscuring secondary magnetism may be 'cleaned away', either by stepwise heating, or by stepwise demagnetization produced by an applied magnetic field. There are several methods for determining the remanent magnetism of rock samples. Brock and Isaac used a spinner magnetometer, with a so-called 'fluxgate' detector, the design being due to John H. Foster of the Lamont Geological Observatory, Columbia University (Foster, 1966).

A 'fluxgate' (which is used for many things besides the study of rock magnetism) consists of two parallel solenoids, with opposite windings on two ferrite cores. Another coil is wound round the two solenoids, and can be used to detect changes in the currents flowing therein. (See Figure 13.)

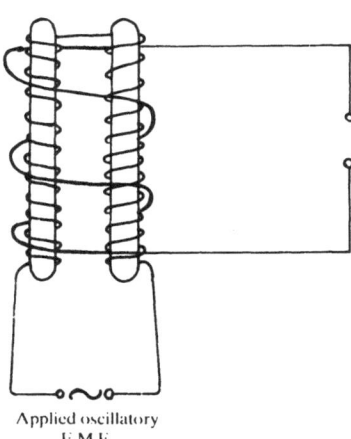

Applied oscillatory
E.M.F.

Fig. 13. Fluxgate.

In the absence of an external field, there will be no induced current in the secondary circuit (signal coil), but in the presence of a field there will be a bias and one core will change polarity before the other during magnetization and after it during demagnetization, which difference can be converted into an electrical signal related to the external field's magnitude and direction.

In Foster's apparatus, a rock sample is spun at a known rate adjacent to the fluxgate, and from the resulting signal at the output the magnetic moment and direction of the field can be inferred. Extraneous fields are eliminated by the presence of shielding Helmholtz coils. The Foster instrument was specially well suited to the examination of soft sediments, since it had a slow rate of spin which did not cause the specimens to break up.

Brock and Isaac collected cubic rock samples, measuring their alignments at the collection sites. The samples were then tested with the magnetometer, the strength and direction of the magnetism being determined for each[29] and plotted on a stereometric diagram. Each sample was then demagnetized in steps by the application of an increasing magnetic field and the remaining magnetism, measured at each step,[30] was plotted on the diagram. The supposition is that the final stage of the demagnetization process (before total demagnetization) reveals the field direction at the time when the rock sample was originally formed. In Figure 14, we see two examples in which shifts on the stereogram are displayed as demagnetization proceeds.

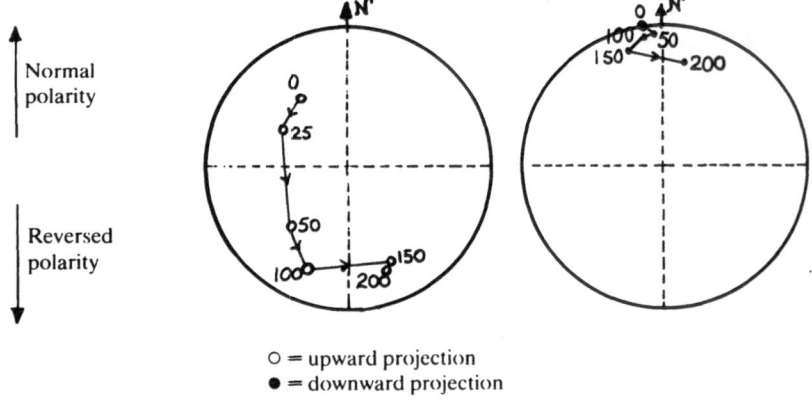

O = upward projection
● = downward projection

Fig. 14.

In Figure 14 (a), the sample began with normal polarity, but was eventually revealed as having had reversed polarity initially. In Figure 14 (b), the rock sample had the same polarity (normal) at the completion of the demagnetization as at the beginning. In this manner, then, the initial polarities of rock samples can be determined as aids to geochronological work.

In their 1974 paper, Brock and Isaac examined the magnetic polarities of rocks in four locations: Ileret, Karari, the KBS locale, and Koobi Fora. They *started* with the assumption that the age of the KBS Tuff was 2.61 Myr, thereby locating themselves, so to speak, on the international geomagnetic time scale (in what is called the 'Gauss Normal' epoch, or 'chron'). Then, to make the data as a whole fit, they found it necessary to propose a disconformity and a time gap of about 0.7 Myr above the KBS Tuff.[31] (See Figure 15.) As noted above, a disconformity had previously been postulated on the basis of field evidence.

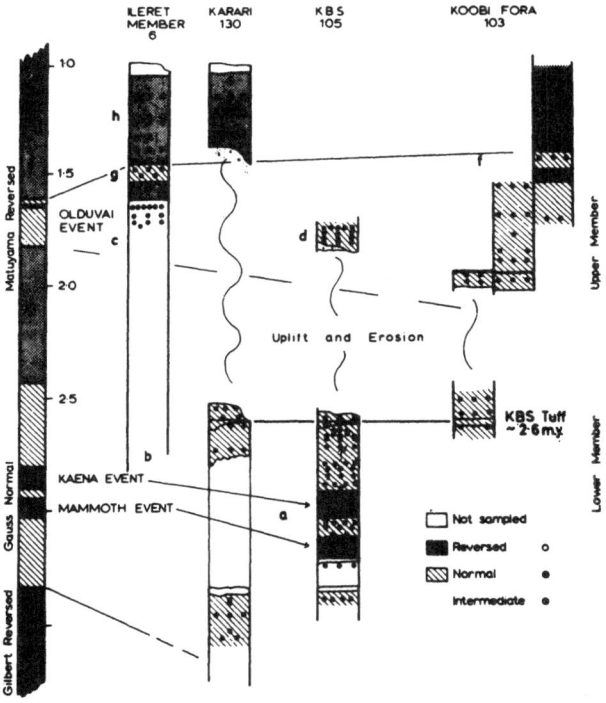

Fig. 15.

The authors claimed that their data *were* compatible with Maglio's palaeontological zones, given that their dating was not very precise, though they suggested that a better fit would be possible if the centre of the *Mesochoerus limnetes* zone were raised to about 2.7 Myr.

EMPIRICAL METHODS: FISSION-TRACK DATING

Useful support for the Fitch/Miller dating was furnished by the application of an independent technique for age determination: the so-called 'fission-track' method. This work was performed by Anthony Hurford of Birkbeck College, whose results were published in *Nature* in May, 1974 (Hurford, 1974). The method was originally introduced by P. B. Price and R. M. Walker (Price and Walker, 1963), and was subsequently described by Robert Fleischer and Howard Hart of the General Electric Research and Development Center, U.S.A., in the context of its application to dating of hominid evolution (Fleischer and Hart, 1972). In this technique, the tracks produced by the spontaneous fission of ^{238}U atoms within a crystal are examined. For each atomic fission, a small 'fission track' is produced in the surrounding crystal. So if a crystal is extracted, a clean face produced by grinding and polishing, and this face then etched with a suitable solvent such as potassium hydroxide or hydrofluoric acid, the tracks may clearly be seen through a microscope where they meet the crystal surface, and may be measured by direct counting. Knowing the rate of spontaneous fission of ^{238}U, the age of the crystal may in principle be calculated, provided that its uranium concentration is known. This may be found, after natural tracks have been counted, by placing the crystal in an atomic reactor and subjecting it to a known dose of slow neutrons. These bring about fission of ^{235}U atoms in the crystal, and the new crop of fission tracks may be counted. This allows calculation of the ^{235}U concentration, and hence estimation of ^{238}U, and the age of the sample. The neutron dose is determined by counting the tracks produced in a standard glass of known uranium concentration (called a dosimeter), which is irradiated alongside the crystals under investigation.

Then, if ρ_s = natural track density of sample; ρ_i = induced track density of sample; ρ_D = induced track density of dosimeter; A = age of sample; and ζ = constant,

$$A = \zeta(\rho_s/\rho_i)\rho_D$$

ζ is calculable from data concerning the decay constant of ^{238}U, the cross-section of ^{235}U for fission by slow neutrons, the number of atoms per unit volume of material, the fraction of these that are ^{235}U and the fraction that are ^{238}U in the sample and the dosimeter, the lengths of the etchable tracks for ^{235}U and ^{238}U, and the fraction of tracks crossing a surface that are revealed by etching. (See Fleischer and Hart, 1972, pp. 136—137.)

All this seems fairly straightforward in principle, and one might suppose that Hurford's result would settle the point at issue without undue controversy. It was not so, however. For if a mineral or glass suffers heating during its lifetime (after original formation) the fission tracks may be melted away, so to speak. That is, they may suffer erasure due to annealing. Consequently, the fission-track method may give anomalous results due to thermal 'overprinting', as does the K—Ar technique. Also, proponents of the K—Ar technique sometimes complain that fission track counting may be too subjective. It involves a nice judgment as to what is or is not a fission track. And for young rocks, with low track densities, such judgements may possibly 'inform' the results to a significant degree.

Be that as it may, in the Birkbeck/Cambridge group, the hypothesis was being developed that an overprinting event occurred in the Koobi Fora region about 1.75 Myr ago, which might account for the younger ages for the KBS Tuff that were being claimed by a research team at Berkeley, California. (See below, p. 118.) Hurford examined a glassy tuff from the Kubi Algi Formation (*below* the KBS Tuff) and obtained an age of 1.8 Myr. So it appeared *either* that the KBS Tuff date (2.61 Myr) was wrong, *or* that the vitric tuff had suffered a total annealing at 1.8 Myr. Given that the rock showed a complicated age spectrum, it seemed reasonable to assume that annealing had occurred and that the rock was older than 1.8 Myr. Thus an acceptable coherence of results (or interpretation) was maintained, but in a sense the new (anomalous) fission-track data were simply 'explained away'.

AN AMERICAN CHALLENGE TO THE FITCH/MILLER
GEOCHRONOLOGY

In 1975, two important papers were published by Professor Garniss Curtis of Berkeley, California,[32] and his co-workers, challenging the work of the British geochronologists (Curtis, 1975; Curtis *et al.*, 1975). Curtis, who was one of the founders of the K—Ar method, reviewed the procedures involved, drawing attention to the many technical difficulties that they presented. The discussion was illustrated by reference to the work of Fitch and Miller. Generously, Curtis emphasized the great labour involved in their determining dates for five tuffs, and the problems of contamination and alteration of samples that they had had to face. Nevertheless, with respect to the KBS Tuff "so much questioning . . . was going on privately in anthropological and palaeontological circles" (Curtis, 1975, p. 206) that a new run of datings was undertaken at Berkeley, with Fitch and Miller's approval. The results were written up in Curtis *et al.*, 1975.

The authors returned to the orthodox K—Ar dating method, with plots of isochrons. They examined both sanidine crystals and pieces of glass from the same pumice clasts in the tuff, obtaining concordant results. On this basis, they disputed the overprinting hypothesis, since one would not expect two such different substances as sanidine and glass to lose argon to the same proportional extent when heated. Criticizing the Birkbeck/Cambridge team, they wrote (p. 396):

[T]he interpretation of the $^{40}Ar/^{39}Ar$ isochrons that yield multiple apparent ages from a single phase, such as Fitch and Miller have obtained on KBS sanidines, is uncertain. Their interpretations rely on oversimplified and uncertain models for the diffusion of argon in solids, both in nature and in the laboratory procedure necessary to make the determination. For this reason we feel that their results, even when they are reproducible to high precision, may be an artefact of experimental procedure, and not geologically meaningful.

However, Curtis *et al.* had a problem of their own in that in collecting KBS specimens from three areas they found two distinct sets of dates: 1.60 Myr (near Ileret) and 1.82 Myr (near Karari Ridge). Hence they concluded that they were dealing with *two* tuffs, not one as previously supposed.

EMPIRICAL METHODS: ISOTOPE ANALYSES.
FURTHER PROBLEMS AND CONTROVERSIES

In 1976, the proceedings of the 1973 Nairobi Conference were eventually published (Coppens *et al.*, 1976), and in introducing the papers Isaac referred to the considerable controversy that had been generated at the meeting. Thure Cerling of the Berkeley group reported on his efforts to differentiate the tuffs by careful measurement of the isotopic composition of the constituent oxygen. Since the rates of diffusion of ^{18}O and ^{16}O are not quite the same, different conditions at the time of formation of different rocks may give rise to slightly different $^{18}O/^{16}O$ ratios. Thus exact determination of $^{18}O/^{16}O$ ratios may provide a suitable method for the 'fingerprinting' of otherwise similar rocks. To make a determination, the powdered substance is heated with cobalt fluoride, oxygen is driven off and its isotopic composition measured with a mass spectrometer. Unfortunately, however, the method lacks high precision, and while Cerling found it possible to discriminate between the KBS and the Tulu Bor Tuffs he could not differentiate all the younger tuffs, and the question of whether or not the KBS Tuff was composite was not settled.

Fitch and Miller (1976) explained the many difficulties attending the $^{40}Ar/^{39}Ar$ method, but still stood by the technique, and the work that they had done with it on the Koobi Fora deposits. They offered a date of 2.61 Myr, with overprinting at 2.42, about 1.75 and 1.07 and 1.02 Myr. It would appear that they were being led into a whole morass of hypotheses to explain their experimental findings.

Brock and Isaac (1976) redescribed their palaeomagnetic investigations, as published in 1974, and essentially the same interpretation was advanced as previously. They emphasized that the notion of a disconformity above the KBS Tuff, which they supported, had been proposed by Behrensmeyer on field evidence *before* the hypothesis seemed to be called for by the palaeomagnetic and K—Ar data.[33] They further claimed that reducing the age of the KBS Tuff would necessitate some unusually rapid rates of sedimentary deposition.

In October 1976, two papers were published in *Nature* offering a new date for the KBS Tuff. The first was by Hurford and Andrew Gleadow, of Birkbeck,[34] and C. W. Naeser of the U.S. Geological Survey, Denver, Colorado (Hurford *et al.*, 1976). These authors worked on extracted zircon crystals from pumice in the tuff, etching them according to the technique previously developed by Naeser (1969), with some recently devised etching techniques (Gleadow *et al.*, 1976) that were not perhaps fully developed and working satisfactorily at that time (Gleadow, personal communication, 23 May 1986). They obtained an average age of 2.44 Myr, in agreement with the new figure of 2.42 Myr now proposed by Fitch and Miller in the second of the two papers (Fitch *et al.*, 1976).

The second paper reveals the severe problems that were now beginning to beset the proponents of the $^{40}Ar/^{39}Ar$ method. The reduced figure of 2.42 Myr, for a sample originally investigated in 1969, was attributed to improvements in technique and instrumental calibrations, though the basis of these changes was not given explicity. And it was evident that severe problems were still being encountered. Consider for example the following results, obtained on some more recently collected material, using the age-spectrum technique:

from collecting area 130 . . . approx. 1.8 Myr,
overprinted at 1.07 Myr
"   "   "   105 . . . 1.90 and 1.02 Myr
"   "   "   10 . . . 2.46 and 1.94 Myr

One published correlation diagram is reproduced in Figure 16.

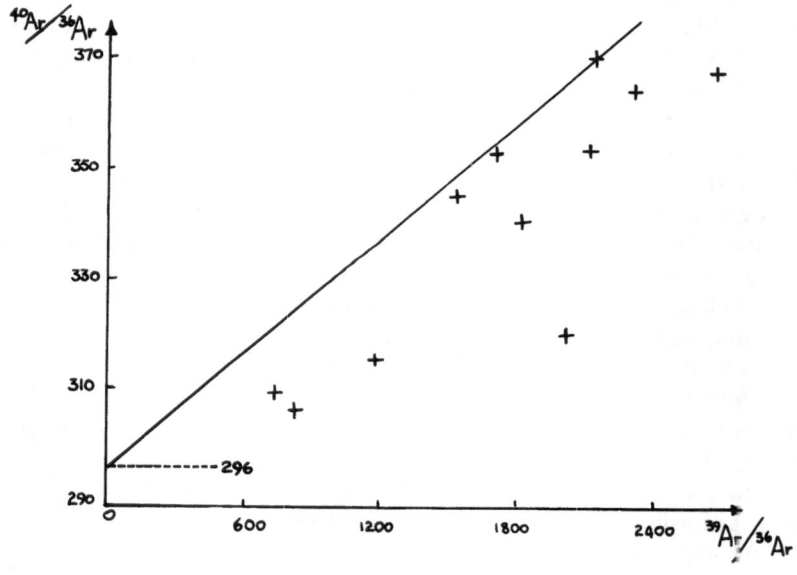

Fig. 16.

Little use could be made of such results, and others were only slightly less perplexing. Fitch and Miller nevertheless maintained the validity of their method, claiming that it had provided coherent results for the other tuffs. This claim was not, however, well supported by the results that they themselves published.[35] In the case of the KBS Tuff, there appeared to be initial argon contamination; possibly some older included detrital material; possibly slow mineralogical change of sanidine and anorthoclase minerals leading to argon loss; and very likely one or more periods of thermal overprinting. But their method was held to be suited to reveal such circumstances. The point made by Cerling *et al.* that overprinting was unlikely since concordant K—Ar results were obtained on glass and sanidine alike was countered by the claim, based on an age-spectrum analysis of a certain specimen of the rock, that it had an age of 2.47 Myr, with an *apparent* age of 1.8 Myr. Curtis *et al.*, it was alleged, had not corrected for initial argon. However, Fitch, Hooker and Miller's age of 2.47 Myr rested in this case, on only one point in the age-spectrum diagram, and the graph was by no means flat at 1.8 Myr, as was desirable. The point that *concordance* of ages was obtained using sanidine and glass was not dealt with satisfactorily.

The various puzzles and ambiguities notwithstanding, a figure of 2.42 Myr was recommended, and only one tuff was accepted at the KBS horizon. The agreement with the fission-track data from the Birkbeck laboratory was encouraging. But Dr. Gleadow informs me that the fission-track workers were under considerable pressure at the time, and were encouraged to publish as soon as possible when their results seemed to accord with the $^{40}Ar/^{39}Ar$ data. It is, of course, possible that they would have produced a different result if they had carried on with their work somewhat longer, and perhaps in a calmer atmosphere, before publishing. As we shall see, the fission-track workers did eventually produce considerably lower dates. The empirical basis of the change of Fitch and Miller's K—Ar KBS dating from 2.61 to 2.42 Myr remains obscure.

In 1977, a paper by J. W. Hillhouse of the U.S. Geological Survey, California, J. W. M. Ndombi and A. Cox of Stanford University, and Andrew Brock, who had moved to the University of Lesotho (Hillhouse *et al.*, 1977) was published in *Nature*, giving an account of further investigations of palaeomagnetism of Koobi Fora and revisions and extensions of earlier measurements. They showed how the palaeomagnetic data could be meshed with the datings for the KBS Tuff of *either* 2.61 Myr (Fitch *et al.*, 1974) *or* 1.6—1.8 Myr (Curtis *et al.*, 1975). (See Figure 17.)

Neither scheme of fit was entirely satisfactory, as can be seen. The second proposal placed a strain on the age of the Tulu Bor Tuff, about which there had been little previous disagreement, and they tended to favour the traditional scheme (Figure 15, left-hand). Hillhouse *et al.*

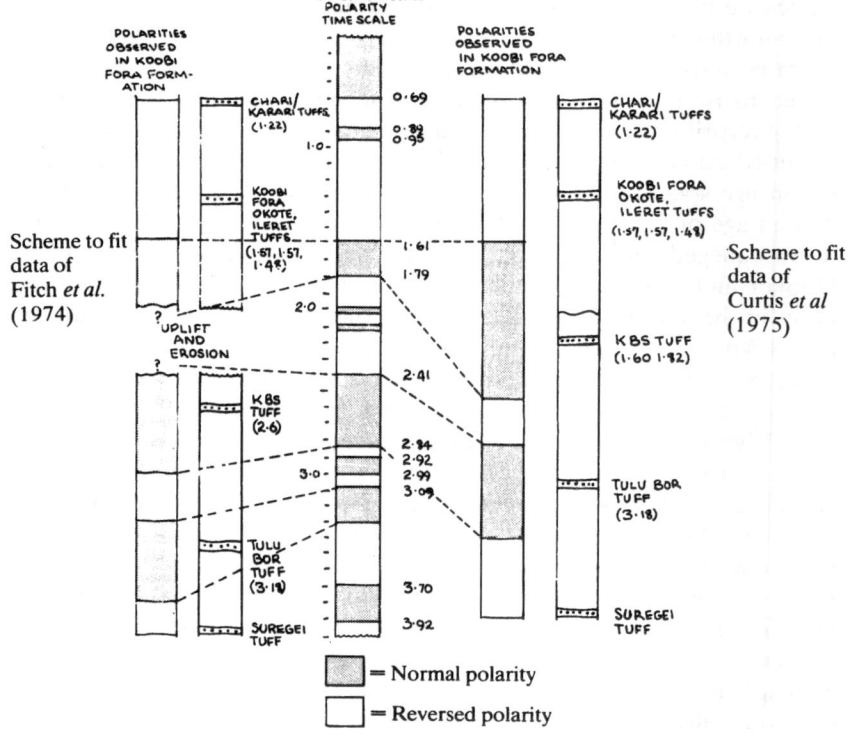

Fig. 17.

were not too concerned about the palaeontological evidence of Cooke and Maglio against Scheme 1. Maybe the "biostratigraphical range of a single species . . . [could] vary between regions depending on palaeo-ecological conditions" (Hillhouse *et al.*, 1977, p. 415). In other words, the rate of evolution of some species might not be constant in all parts of its geographical range.

SOME *AD HOC* THEORIZING AND EFFORTS
TOWARDS COHERENCE

The preceding line of argument was pursued in greater depth by Kay Behrensmeyer (1978) in an important volume edited by William W. Bishop, in which the various issues were thrashed out yet again in the light of the greatly increased empirical data by then available. (This book represented the proceedings of a three-day Symposium organized by the Geological Society of London in February, 1975.) Behrensmeyer reviewed the problem of correlation between the Koobi Fora and the Omo River exposures. Broadly speaking, the Omo River data presented no serious difficulties. Alternative fittings for the palaeomagnetic data were possible there but such changes would involve upsetting a lot of good correlations just for the sake of getting rid of the palaeontological problem raised by Cooke and Maglio. For her part, Behrensmeyer was willing to contemplate accepting Fitch and Miller's 2.61 Myr dating of the KBS Tuff, suggesting that the pigs of a certain tooth-length might have arisen 0.7 Myr earlier at Koobi Fora than at the Omo River, 100 km to the north, and could have been maintained thus for that period of time with a slightly differing anatomy. The ecological conditions could, she suggested, have been significantly different at the two sites, even if they are much the same today.[36] Thus, while most participants in the debate seemed to regard the controversy as a way of testing the merits of rival methods for the dating of rocks and minerals, Behrensmeyer was willing to take the geophysical data more at face value, and see what implications they had for evolutionary theory. In other words, the same measurements could be made to serve two roles: either the palaeontologists could whip the geochronologists into line, so to speak; or the physical data could offer evidence for new schemes of interpretation in evolutionary theory.

The Bishop volume contained a further statement on the KBS problem by the Birkbeck/Cambridge team (Fitch et al., 1978). The authors gave an account of their use of the $^{40}Ar/^{39}Ar$ technique and the results that they had obtained with it at Koobi Fora, applied to the several tuffs of the region. Concordant results were claimed in all cases except that of the KBS Tuff, which as we have seen (Figure 16) gave an almost meaningless set of data for the $^{40}Ar/^{36}Ar$—$^{39}Ar/^{36}Ar$ plot. Problems with the KBS Tuff were again attributed to overprinting, and for the first, and I believe the only, occasion a diagram was published showing the possible cause of the heat — a hypothetical basalt lava flow (possibly equivalent to that presently observable at the 'Bakate Gap') that erupted shortly after the deposition of the KBS Tuff, and was subsequently removed by erosion, leading to confusion in the data for the dating of the KBS Tuff. (See Figure 18, based on Fitch et al., 1978, p. 457.)

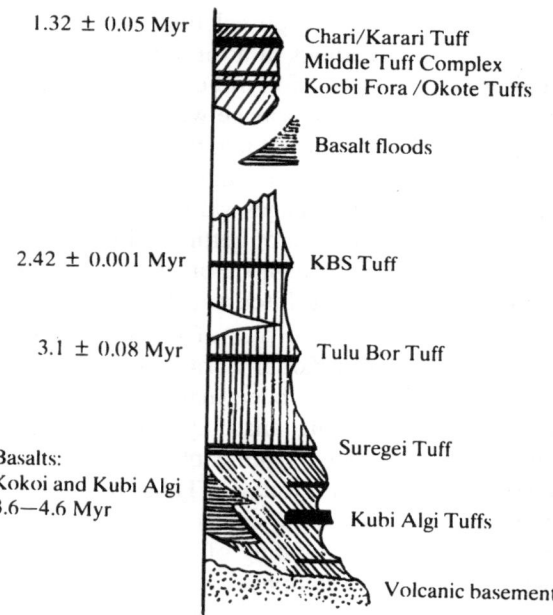

1.32 ± 0.05 Myr — Chari/Karari Tuff, Middle Tuff Complex, Kocbi Fora /Okote Tuffs

Basalt floods

2.42 ± 0.001 Myr — KBS Tuff

3.1 ± 0.08 Myr — Tulu Bor Tuff

Suregei Tuff

Basalts: Kokoi and Kubi Algi 3.6—4.6 Myr — Kubi Algi Tuffs

Volcanic basement

Fig. 18.

Such a proposal (due, we are told, to a suggestion by Ronald T. Watkins) had the marks of a classic *ad hoc* hypothesis.

Curtis and his co-workers also had their say in the Bishop volume (Curtis *et al.*, 1978), but essentially they rested their case on previous evidence and interpretations. A balanced survey of the points at issue was provided in the paper in the same volume by F. H. Brown, F. Clark Howell and G. G. Fleck (Brown *et al.*, 1978). By 1975 (the date of the Geological Society Symposium), problems were being found in the pre-KBS magnetostratigraphy of the Koobi Fora district, but satisfactory correlation was established between the several upper tuffs and the corresponding tuff beds of the Omo River sequence. And, using the Cerling data for the KBS Tuff, rather than that of Fitch and Miller, it appeared that the palaeontological difficulties raised by Cooke and Maglio could be avoided. Thus Brown and his co-authors preferred a revised tuff dating, rather than a rethinking of evolutionary theory and principles, such as was contemplated by Behrensmeyer.

Particularly interesting from the present perspective were the estimates now made for the *rates* of sedimentation for the several units (corresponding to the several magnetostratigraphical zones) of the East Turkana region.

	Thickness (in metres)	Duration (Myr)	Rates of sedimentation (cm/1000 yr)
Chari Tuff — Koobi Fora Tuff	36	0.35	10
Koobi Fora Tuff — KBS Tuff	30	0.24	13
KBS Tuff — bottom of sediments below KBS Tuff	34	0.06	57
Tulu Bor Tuff — Suregei Tuff	50	1.2	4

These figures offered reasonable agreement with those at Omo, except for the bottom unit. However, the several figures were only regarded as provisional and the continuity of sedimentation was not necessarily assured. This might account for the low value of 4 cm/1000 yr.

So far as I am aware, the final published statement on the KBS Tuff controversy by the Birkbeck/Cambridge group was provided in the Proceedings of a conference held at Snowmass-at-Aspen, Colorado, in August, 1978 (Fitch *et al.*, 1978). The various dates that had been published for the Tuff were summarized, and a frequency diagram for the different results was provided! These clustered round 2.4 and round 1.8 Myr — corresponding broadly to the views of the Birkbeck/ Cambridge and the California laboratories respectively. A number of possible explanations for the discrepancies were canvassed, and since agreement was claimed for the other tuffs in the sequence the explanation tentatively advanced was that tuffs from more than one horizon were being conflated. In the light of recent investigations at Lake Turkana (see p. 139), this seems more than likely, though as we shall see (p. 130) a date has now been arrived at that attracts general agreement.

EMPIRICAL METHODS: CHEMICAL FINGERPRINTING

Further efforts toward the correlation of the tuffs at Koobi Fora and at Omo River were made by workers at Berkeley, in collaboration with colleagues at the University of Utah, using geochemical techniques

(Cerling *et al.*, 1979). Glass was separated, crushed and purified, then pressed into pellets for examination by X-ray fluorescence analysis. In this technique, a substance to be analyzed is bombarded with X-rays, leading to the emission of radiation with wavelengths characteristic of the elements present in the sample. The concentrations of the several constituent elements may be determined by comparison of the intensities of the emitted X-radiation of particular wavelengths with the intensities produced by standard test substances of known chemical composition.

In the case under consideration, however, more data were obtained by means of an 'electron microprobe' using a technique of A. E. Bence and A. L. Albee (1968), which derived from the work of earlier investigators such as J. Hillier (1947), R. Castaing (1951), I. B. Borovskii (1953), V. E. Coslett and P. Duncumb (1956), L. S. Birks and E. J. Brooks (Birks, 1971, pp. 5—19). In this technique, the prepared specimen is bombarded with an electron beam and the emitted X-rays, characteristic of the atoms of each kind of element submitted to the bombardment, are analyzed with an X-ray spectrometer. The intensities of the emitted X-rays may be compared with those emitted by standard sources to allow estimation of the quantities of the different elements present in the specimens under test. While simple in principle, the method is rather complicated in practice, since there are interaction effects between the different elements. The empirical correction factors, allowing for these effects, can be determined using standard samples of known concentration. For example, if three elements are concerned, three different mixtures of these elements, of known composition, have to be prepared, and the relative X-ray intensities measured for the three elements in the three standard samples. Nine simultaneous equations then have to be solved to enable the nine empirical interaction coefficients (α_{AA}, α_{AB}, α_{AC}, α_{BA}, α_{BB}, α_{BC}, α_{CA}, α_{CB}, α_{CC}) to be determined (Birks, 1971, p. 104). This can be done 'by hand' where, say, only three elements are involved, but for more complicated cases the services of a computer are obviously required. Once the interaction coefficients have been determined, measurement of the relative intensities of emitted X-rays for unknown samples allows calculation of the concentrations of the constituent elements.

By such precise analytical techniques, the different glasses, pumices, etc., could each be chemically 'fingerprinted' so to speak; that is, their precise chemical composition with respect to trace elements such as

niobium or zirconium could be determined. This would facilitate correlation from one region to another, and would assist the recognition of
the individual tuffs within a tuff complex.

Using such techniques, Cerling *et al.* showed that the Chari and
Karari Tuffs were one and the same, as had long been supposed, and
were equivalent to the tuff labelled '*L*' in the Omo River Basin. But
things were not so simple with the KBS Tuff. Three different pumices
were distinguished in collecting Area 131. One of these correlated with
the tuff at Area 105 and with a tuff, 'H_2', at Omo River of age about
1.8 Myr. However, another tuff in Area 105 East, previously designated as KBS, had a significantly different composition, and correlated,
composition-wise, with 'H_4' at Omo River. It appeared, then, as previously suspected, that there were several components to what had
previously been designated the KBS Tuff. But they all had about the
same age, and one that agreed with that of the Berkeley researchers'
dating rather than that of the Birkbeck/Cambridge group.

Thus the Berkeley group now announced that they wished to date all
the KBS suite at around 1.8 Myr (Drake *et al.*, 1980). (In 1975, some
areas had been dated at 1.6 Myr.) Potassium concentrations were re-
determined, previous discrepancies being attributed to a faulty balance
at the Berkeley laboratory!

Arguing against the British team's results, the Berkeley group maintained that Fitch and Miller — experienced in the examination of
ancient plutonic rocks with overprinting — were over-sanguine in the
extension of their method and interpretation to young pumices (though
in fact "several of Fitch and Miller's better total degassing dates and at
least one ^{40}Ar—^{39}Ar age spectrum support[ed] a 1.8 Myr age" [p. 370]).
Hurford's 1976 date of 2.44 Myr by fission-track analysis was rejected
as a product of the statistical analysis used; for with low track densities
(for young rocks) the estimated error might be of the same magnitude
as the assessed age. The British team's results being thus dismissed,
Drake *et al.* felt confident to propose the following correlation for the
Omo River and Lake Turkana tuffs (Figure 19), correlations with
Olduvai beds and elsewhere also being given but omitted here.

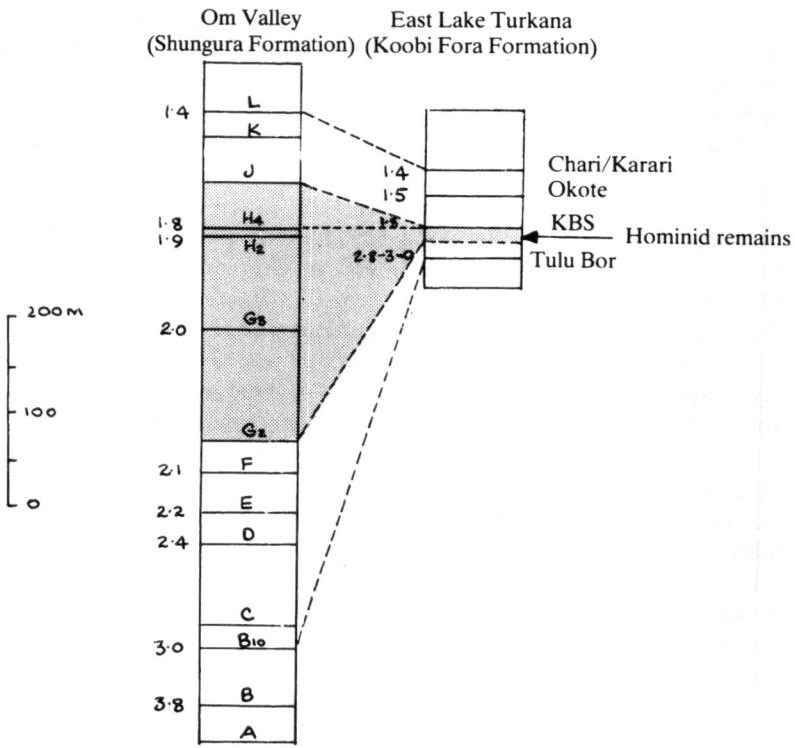

(The shaded area here represents a palaeontological correlation, using the suid data.)

Fig. 19.

IMPROVEMENTS IN PRECISION AND ACCURACY LEADING TO COHERENCE OF RESULTS AND A 'DECISION' ON THE KBS TUFF CONTROVERSY

Matters were now brought to a reasonable state of concordance, though the Birkbeck/Cambridge workers still resisted the reduced age of the KBS Tuff. A remaining difficulty was the high fission-track data of Hurford *et al.* (1976) for the age of the KBS Tuff. This problem was eliminated by the careful fission-track study of zircons from the KBS Tuff by Andrew Gleadow (1980), who obtained a date of 1.84 ± 0.04 Myr, using carefully controlled etching techniques and counting procedures. In the same issue of *Nature*, Ian McDougall — who had trained in the laboratory of Garniss Curtis in California (Glen, 1982) — and his co-workers from the Australian National University, Canberra, (McDougall *et al.*, 1980) reported an age of 1.89 ± 0.01 Myr for anorthoclase crystals extracted from pumices within the KBS Tuff, examined by the standard K—Ar procedure, with total degassing and an isochron diagram plotted using different extracts. A revised decay constant for ^{40}K was employed, which increased the results by 2.67% compared with previous calculations. The whole overprinting hypothesis was rejected, but so also was the claim of Cerling *et al.* (1979) that the tuffs of Areas 105 and 131 were different. Interestingly, one set of specimens was divided and analyzed separately in California and Australia. The Berkeley results were 6% lower, but the general concordance was acceptable. As we noted above (p. 103), variation in results from one laboratory to another may be anticipated due to differences in the calibration of the instruments and the different standards used. However, in the present case McDougall *et al.* simply suggested that the Berkeley group failed to extract the last puff of argon from their samples!

So, by 1980 McDougall *et al.* could claim (p. 230) that they had achieved the "best estimate currently available for the time of formation of [the KBS Tuff —] this important marker horizon within the East Turkana Basin." And in fact the KBS Tuff controversy was by this stage almost over. A fortnight after the publication of McDougall's claim, Professor Richard L. Hay of the Department of Geology and Geophysics, University of California, Berkeley, published a brief (invited) note in *Nature* (Hay, 1980), reviewing the chief published contributions to the controversy. He concluded that "the latest and perhaps the final statement on the age of the KBS Tuff are provided by McDougall *et al.*". Gleadow's work was also singled out for praise and the note concluded with the remark that "the KBS Tuff has been a testing ground for various geochronological methods, both new and old, and the science of geochronology has learned much from the 'KBS Tuff controversy'."

So far as I am able to determine, Hay's *ex cathedra* pronouncement did settle the matter. The scientific community had made up its collective mind on the question of the age of the KBS Tuff, and turned its attention to other problems at Lake Turkana. There was, however, one issue still to be attended to so far as the KBS Tuff was concerned. Why had the $^{40}Ar/^{39}Ar$ method apparently failed when applied to the dating of this rock?

McDougall dealt with this in a further paper, published in November, 1981 (McDougall, 1981). He examined crystals of anorthoclase from pumice in the Tuff with extreme care and attention to possible sources of error. Both total fusion and step-heating techniques, with analysis of the $^{40}Ar/^{39}Ar$ and $^{40}Ar/^{36}Ar$ ratios were used. The age spectra experiments yielded very satisfactory flat graphs, showing no evidence of overprinting, and isochron diagrams were obtained that were almost exactly linear. Ages were now quoted to four significant figures, and an average result of 1.88 ± 0.02 Myr was given. The argument was apparently so utterly convincing that the paper was accepted for publication just one month after receipt. With rather less restraint than is usually displayed in the scientific literature, McDougall wrote in triumph (McDougall, 1981, p. 124):

The $^{40}Ar/^{39}Ar$ total fusion ages measured on 10 different concentrates, as summarized by Fitch *et al.*, range from 0.53 to 2.48 Myr, typically with quoted errors between 0.1 and 0.5 Myr. The proportion of $^{40}Ar^*$ in these analyses is generally <20% of the total ^{40}Ar and commonly <10% .... On the basis of the large scatter in the ages and the small proportion of $^{40}Ar^*$ in the gas extracted from the anorthoclase concentrates, I suggest that the results are less analytically precise than given by these authors.

And with respect to the step-heating experiments and the hypothesis of overprinting:

[G]eological evidence for such thermal events is lacking. I suggest that unrecognized analytical difficulties and larger than quoted errors must be invoked to explain these earlier $^{40}Ar/^{39}Ar$ results.

THE SOCIAL CONSTRUCTION OF COHERING KNOWLEDGE

All this long controversy was involved, therefore, in determining the age of just one geological unit at Lake Turkana. The case was not typical, of course, but it displays in a most interesting manner what may on occasions be involved in the determination of geochronological data, and more generally the measurement of quantities in science. It is evident that the prime consideration was *coherence*: the truth of a scientific statement was to be gauged by the extent to which it agreed or disagreed (cohered or otherwise) with other statements.[37] That is to say, consistency was demanded for measurements of a quantity by methods that were independent of one another — in that they were utilizing distinct empirical methods and theories from different branches of physics and different branches of geology. Thereby, order was sustained in the total theoretical and empirical structure of science. So far as the scientists were concerned, this requirement provided the criterion of truth. The whole controversy blew up when lack of coherence (in this case revealed by lack of empirical consistency) was discovered between two sets of measurements (palaeontological and geophysical), between which there should have been no disagreement if all was well. And since the problem was of wider significance — for it had implications for the whole evolutionary history of man — a great deal of attention was devoted to the controversy and no effort was spared until the anomaly was eliminated.

In the case before us, it does not appear that the several participants had particular axes to grind so far as over-arching theory was concerned — the chief theoretical issue being the age of tool-making man which was not, in itself, a question on which the geophysicists and geochemists might be expected to have excessively strong feelings. There were, however, strong rivalries between the different laboratories, and between proponents of different techniques — though eventually all proved viable. (The omegatron spectrometer has not, however, come into general use for rock dating.) Personal *interests* were involved in that each disputant in the controversy wished to demonstrate the efficacy of the particular empirical practice with which he or she was associated, and obviously did not wish to lose face by having to admit inadequacies in practical capability. And Leakey, I suggest, had a considerable interest in an extended date for the Koobi Fora hominids, for if this were accepted his work would be of heightened significance by virtue of the extended time range of tool-makers on this planet. There was, therefore, as might be anticipated, a strong social component in the construction of the scientific knowledge concerning the age of the KBS Tuff, and the achievement of coherence has to be seen as a process with a strong social dimension. So again one can hardly say that coherence is, in itself, a sufficient criterion of truth or a guarantee of closure in a scientific controversy.

The whole episode may usefully be seen from a perspective similar to that employed in M. J. S. Rudwick's recent detailed examination of the 'Great Devonian Controversy' in the nineteenth-century geological community (Rudwick, 1985; Oldroyd, 1985). On this view — which stems from the writings of students of metascience such as Bruno

Latour and Steve Woolgar (Latour and Woolgar, 1979), which in turn derive from writings such as those of Pierre Bourdieu (1971; 1975) — the arena of scientific debate may be seen as an 'agonistic field'. In its most literal and limited sense, this simply means a 'field of contest', analogous, say, to the arena of an Olympic Games. But for Bourdieu, the scientific 'field' is a much more subtle affair, not just one of winners and losers. Indeed, one might liken it to an electromagnetic field in which activity in every constituent part influences all the other parts to a greater or lesser degree.[38] The participants in a scientific 'field' are engaged in enquiries about the world; but they are also engaged in a 'contest' in which they seek acceptance of *their* account of the world. They seek to become 'authorities'. The case is intriguing since there is, on this view, no source of authority for scientific truth external to the community of scientists. There is, as Bourdieu puts it (Bourdieu, 1975, p. 25), "no judge who is not also a party to the dispute". The scientist is, then, forever involved in a series of feed-back loops: nothing succeeds like success (and nothing fails like failure, one might add). Of course, extraneous factors may influence the field also. A scientist's 'field-strength', so to speak, is a function of his or her general social position, and this may be expected to contribute to the perception of the scientist's authority in the field. So this social (or political) strength will play its part alongside prowess as an experimentalist, mathematician, or whatever. Thus numerous factors can contribute to one's 'authority in a field'.

Rudwick's theoretical discussion also draws from the work of H. M. Collins and his notion of a 'core-set' (Collins, 1981). According to this notion, there can be, for any scientific specialty, a group of *authorities* whose opinions *count*, particularly in the matter of which empirical technique is to be deemed acceptable, and whose results or observations are to be relied on. However, functioning 'core-sets' are not so very common in science. Indeed, they will only be found to emerge when some significant controversy develops — some 'hot-spot' (Collins, 1981). Otherwise, the 'normal' processes of research and publication proceed in a relatively untroubled way, so most scientists do not have personal experience of 'hot-spots' or the overt activities of a 'core-set' during their whole working lives.

In keeping with these ideas, then, we find that field observations and laboratory measurements have to be made, analyzed and interpreted, and then brought to the bar of the scientific community for evaluation.

Normally, this is a fairly straightforward procedure, but it is not so when a 'hot-spot' develops. In the community that Rudwick studied, the final court of appeal and the judicial decision makers were to be found at the meetings of the Geological Society of London where the Devonian Controversy was keenly debated. And for 'victory' in the field contest the opinions of the more influential members of that society such as Charles Lyell and William Whewell had to be secured in support of one's position. (Participants in the debate — such as de la Beche, Murchison and Sedgwick in Rudwick's example — were also judges or members of the jury, as Bourdieu's thesis would require.) In the case we have been considering, the various conferences relating to the Koobi Fora project served as the arena for the contest, in which, as reported by F. H. Brown *et al.* (1978, p. 473), there was "substantial ante-chamber speculation". A further forum was provided by the pages of *Nature*. It is clear that publication in this prestigious journal was in itself guarantee neither of truth nor acceptance. Eventually, however, the geological community made up its collective mind on the matter, and its judgement was pronounced through the synopsis of the debate provided by Hay. His opinion commanded respect in that he was well informed in geochronological work, and had contributed to the dating of rocks at Olduvai Gorge. But he was not a direct protagonist in the KBS Tuff controversy so far as rock-dating procedures were concerned, and his judgement might therefore command general assent (even though he had links with the Berkeley team).

I find some further corroboration of Collins's views in the following points. Collins emphasizes that a very frequent concern of the protagonists of a scientific controversy is the relative efficacies of the contending empirical procedures, the relative adequacy of the apparatuses used, and the precision and accuracy that they allow. It is on these issues, therefore, that the core-set ultimately has to make a determination. This being so, it is to be expected that the issue on which the protagonists will be most sensitive is their technical procedures and the precision and accuracy of their measurements. One may, perhaps, make an implausible speculation or hypothesis, and this is tolerated by the scientific community, even if the hypothesis subsequently proves to be unsatisfactory and has to be withdrawn. One may even make empirical mistakes and be forgiven, as, for example, when Drake *et al.* (1980) changed some results that proved to be erroneous due to the use of a faulty balance at Berkeley. But it is one's reputation as a reliable

observer and/or experimentalist that is most central to an experimental scientist's authority. So if one's perferred empirical methods are
rejected by the 'core-set' then all credibility is lost. One may descend a
long snake and have many years of ladder climbing if one is ever to
recover. It is for this reason, I believe, that Fitch and Miller have
vehemently resisted the suggestion that they obtained an incorrect value
for the age of the KBS Tuff because of experimental incompetence.[39]
Moreover, I understand from Mr. Fitch (personal communication, 29
June 1986) that he and Dr. Miller still find the omegatron apparatus
satisfactory for their purposes. It is suggested that it never received
wide acceptance by K—Ar workers simply because it did not become a
commercially available package. It appears, however, in the view of
those experimenters who use the standard techniques for argon determination, that the omegatron apparatus does not offer sufficient precision for first-class work, owing to its limited resolution. It is, of course,
difficult for me to make a judgement on this point, and to know
whether the early difficulties encountered by the Fitch/Miller team
were due to the use of the omegatron or to other procedural problems.

   Given the foregoing discussion of the agonistic field of science,
authority, etc., does this mean that the outcome of scientific measurement can ultimately be reduced to the outcome of 'power-politics' in
science? Well, yes and no! The knowledge that we have is filtered
through the social system of science, which is the arbiter of what is true
or false. But the arbitration process clearly involves empirical information. In the case of the KBS Tuff controversy, the rocks presumably
have a *real* age;[40] the problem is to determine it, by the best means
available. The leading principle deployed for this purpose is, as mentioned previously, *coherence*, which, I believe, provides an accepted
guide to decisions reached in the agonistic field. This coherence is
achieved by the same result being reached by independent methods,
with gradual improvement in precision of measurement. In the case that
we have been examining, before coherence was achieved, with social
agreement concerning age determinations, the agonistic field was open
to speculation and the construction of multiple working hypotheses (*cf.*
Chamberlin, 1897), or less charitably, *ad hoc* hypotheses. For example,
while uncertainty was rife concerning the age of the KBS Tuff, Behrensmeyer could readily speculate about the possibility of different rates
of suid evolution in neighbouring regions under somewhat different
environmental conditions. With the precision of McDougall's measure-

ments, however, such speculations became untenable and were quietly dropped.

Thus, with increasing precision the 'space' available for differences of interpretation is reduced. With imprecision in the data, a range of 'multiple working hypotheses' may appear viable and attractive, but as precision increases some of them are squeezed out. Needless to say, that is why precision is such a sought-after commodity. It is the signpost to 'Truth', and helps to minimise the number of possible interpretations of the data, thereby promoting scientific objectivity and an approach to an understanding of the real world. Nevertheless, the social component of knowledge claims is always present. What *counts* as adequate precision is something that requires social negotiation in each case. There is no 'pure' observation, and the extension of concepts is always revisable. Further, the establishment (acceptance) of coherence occurs within the scientific community and is subject to the action of social forces within its agonistic field.

It should also be remarked that the 'field' must be seen extending beyond the loci of scientific meetings; beyond the acceptance, rejection and appraisal of contributions to scientific journals. It extends right into the thoughts of scientists and their proffered empirical observations. In making observations, the scientist is necessarily thinking about how they will be received by the community, and the social system of science will influence *what* he or she observes and what is overlooked or not investigated. It has for a good many years been accepted that observations may be 'theory-laden'. But one can also think of them as 'controversy-laden' (Rudwick, 1985, p. 431). For example, in the case we have been examining, the Birkbeck workers were actively seeking coherence between the results of fission-track and K—Ar determinations. Without the pressures produced by the KBS Tuff controversy, I do not think there would have been so much effort to achieve coherence, or indeed the need felt to investigate the matter so thoroughly, and to settle *who*, among the 'core-set' was to be regarded as the authority on the question. Thus social factors may enter into the very heart of the empirical enterprise.

Even so, and contrary to what may appear from the foregoing remarks, this does not mean that scientific knowledge is 'nothing but' politics and thereby lacks all objectivity. During the course of the KBS Tuff controversy, consensus amongst the majority of participants was eventually attained when *consistent* results, of acceptable precision, were achieved, using methods from several distinct branches of science. The *results* were consistent: the *knowledge* cohered. And this was taken by the community as a signature indicating objective knowledge or truth. This, I think, is the way science works, though to give a fuller account (which I shall not attempt here) one would obviously have to include a discussion of practical efficacy and its influences on the ongoing activities of the agonistic field.

In relation to his discussion of 'core-sets', Collins remarks on "the intransigence of defeated parties in the face of near consensus" (Collins, 1981, p. 13). It is interesting, then, that in the case before us, the Fitch/Miller group has not entirely given up the contest. In a letter to Roger Lewin dated 28 May 1986, Fitch has suggested that the original sample dated in 1969 was not *bona fide* KBS Tuff. That is, it is now suggested that the sample may have consisted of crystals of about 2.4 Myr that became *included* in the KBS Tuff when it was laid down about 1.9 Myr ago. Fitch points out that the crystals that he originally investigated were large and clean, and similar ones have never since been seen from the KBS horizon. He suggests, following a hypothesis of Cerling and Brown, that there may have been a "considerable thickness of sediment . . . deposited in the Koobi Fora region between the Tulu Bor and KBS times" (letter to Lewin), which has subsequently been removed by erosion. It appears, then, that it may *still* be possible to account for the earlier anomaly without invoking experimental deficiencies. Even so, were the new hypothesis accepted, we would still have consensus in the 'field'. Indeed it might be a stronger one, since all disputants would feel themselves vindicated.[41]

THE KBS TUFF PROBLEM A KUHNIAN ANOMALY?

The relationship of the KBS Tuff controversy to the work of Thomas Kuhn is worth some consideration. In the language of Kuhn (1962), the KBS Tuff presented an 'anomaly' in a process of 'puzzle solving' within a community working under the aegis of a common paradigm. And eventually the anomaly was resolved satisfactorily, without precipitating a scientific revolution. It will be apparent, however, that this account of the matter hardly does justice to the situation. We had something bigger than an anomaly. It was, as Collins puts it, a 'hot-spot'; or in the language of Rudwick (1985, p. 426) it was a 'focal problem'. The intense interest generated at the time, and the fact that the episode is already being noticed by historians, is an indication that we are dealing with something outside the personal experience of most scientists in their working lives. I suggest that the analysis of Collins is more apt than that of Kuhn for cases of this kind. However, a modified Kuhnian approach, such as that of Barnes (1982) can do reasonable justice to the case that we have been considering, and Collins's schema can be regarded as neo-Kuhnian.

THE KBS TUFF AND PUNCTUATED EQUILIBRIUM THEORY

The KBS Tuff controversy having been settled to our satisfaction, with an agreed age of the Tuff of about 1.9 Myr, what is the bearing of all this on the work of Williamson? The connection is not immediately obvious; but neither is the means whereby Williamson made his estimate of a punctuation episode as taking between 5000 and 50 000 years. As mentioned above, in his 1981 paper he footnoted this claim by referring to his Bristol Ph.D. thesis of 1979. Yet my inspection of this thesis reveals that it does not contain the evidence or argument whereby a reader might conclude that a punctuation event took between 5 and 50 thousand years. I therefore sought information from F. H. Brown of the University of Utah, who, according to Williamson, performed the original calculations. Professor Brown kindly furnished me with information as to how the argument might proceed (personal communication, 20 April 1986), but unfortunately he used recent data and referred to tuffs that had not been named as such at the time when the estimate was originally made. So I have some uncertainty as to exactly what was done at the time. However, he explains that if one takes the KBS Tuff as 1.89 Myr, the Toroto Tuff as 3.32 Myr,[42] and the intervening sediments as 223 metres thick, then we have a sedimentation rate of 15 cm per 1000 yr. It appears that the speciation event at the Suregei level occurred in a thickness of 4 metres, which is equivalent, therefore, to 27 000 yr. In confirmation, it is believed that in the Shungura Formation of the Omo Valley there are periodic cycles of fluvial deposition, which are thought to to take 30 000 yr for each cycle (using K—Ar datings of Omo Valley tuffs as the basis of this estimate). Williamson's punctuation event occurred, it seems, *within* one such analogous cycle at Lake Turkana, which is compatible with a time interval of 27 000 yr. The 5 to 50 thousand years allowed a reasonable spread around this figure. The lower limit was suggested by the fact that some of the sediments were finely laminated material, thought to have been deposited as rapidly as 1 metre per 1000 yr, by analogy with similar deposits at the Omo River and in the Koobi Fora Formation. Thus the published maximum rate of punctuation was perhaps founded on rather insecure evidence. In any case, the evidence was not revealed at the time in Williamson's paper.

It will be clear that much depends on the *completeness* of the sections measured in addition to the correct determinations of the ages of the marker tuffs. As to the completeness, this is a matter still under discussion, and until it is resolved satisfactorily we have no certain knowledge of the time represented by the 4 meters of sediment in which the speciation event is asserted to have taken place. As to the tuff ages, if we were to take the old Fitch/Miller dating for the KBS Tuff, we would have a sedimentation rate of 31 cm. per 1000 yr, so that 4 metres of sediments would represent about 13 000 yr. The speciation event would, then, have been even more dramatic! I believe that Brown and Williamson were using satisfactory ages for the marker tuffs; but it is clear that the whole basis of their estimate was dependent upon correct dating of these, including the KBS deposit. And, as I shall now show, one may still have doubt as to the security of the calculated punctuation interval: (a) because the stratigraphy of the Turkana region was revised shortly after the publication of Williamson's 1981 paper; and (b) because it is now becoming clear that Williamson is dealing with a *composite* stratigraphical column, built up from a synthesis of observations at a number of sites at Lake Turkana — not a single rock face. Whether these doubts should lead to skepticism is something for science rather than metascience to determine over the next few years. It is interesting to note, however, that if the Fitch/Miller dating had prevailed the punctuation would have been thought to be about twice as rapid as was estimated. This might (or might not!) have been too much for the community of evolutionary biologists to take seriously.

NEW EMPIRICAL EVIDENCE AND THE REWRITING OF
THE KOOBI FORA STRATIGRAPHY

With the firm dating of the KBS Tuff at 1.89 Myr one severe problem in the stratigraphy at Lake Turkana was eliminated, in that the need for a considerable disconformity above the KBS Tuff was removed. But a host of other problems soon became apparent. It will be observed from

Figure 2 that most of the asserted punctuations occurred at the horizon of the Suregei Tuff complex, and an unnamed tuff horizon, which I take, however, to be what was then called the Tulu Bor Tuff. On the other hand, there was stasis through Upper Koobi Fora time (then regarded as lasting for about 0.3 Myr, using the reduced age of the KBS Tuff). Such a period of stasis is certainly less remarkable than would be the case if the KBS Tuff were about 2.6 Myr old, which would involve about 0.7 Myr of stasis. With the elimination of the disconformity above the KBS Tuff (which Williamson shows indistinctly in Figure 2, but does not discuss), the stasis again becomes noteworthy. The change in shell form occurring betwen the Koobi Fora Upper Formation and the Guomde Formation is not particularly unexpected, given that there is an unconformity below the Guomde.

But let us concentrate attention for a moment on the lower part of the sequence, and the punctuations asserted near the Suregei Tuff and the Tulu Bor(?) Tuff. Eleven months after the appearance of Williamson's October 1981 paper, two papers by Cerling and Brown of the University of Utah appeared in *Nature* (Brown and Cerling, 1982; Cerling and Brown, 1982). These authors had been engaged in careful chemical 'fingerprinting' of the East Lake Turkana tuffs, many of which had been causing confusion, over and above that involved in the KBS Tuff controversy. The Tulu Bor Tuff was, we know, regarded as *above* the Suregei Tuff Complex, but Brown and Cerling found that in some collecting areas this was not so. Evidently, in mapping, two different tuffs had been conflated. In another place, rocks that had been construed as belonging to the Suregei Tuff Complex were now found overlying exposures of the Tulu Bor Tuff. In fact, the type section of the Koobi Fora Formation was found not to contain its defining marker beds! Thus radical revision of the whole rock sequence was clearly called for, allowing incidentally for a considerable disconformity that was now claimed to exist *below* the KBS Tuff. New names were introduced, and the sequence shown in Figure 20 was now proposed (Brown and Cerling, 1982, p. 214).

It will be noted that the whole sedimentary sequence was now extended by over 100 meters. A disconformity corresponding to over half a million years (by comparison with the dated Omo River sequence) was introduced into the Kubi Algi Formation below the Suregei Complex, which was now divided into two parts separated by upwards of $1\frac{1}{2}$ Myr![44] It must be remarked that Brown and Cerling claimed that their

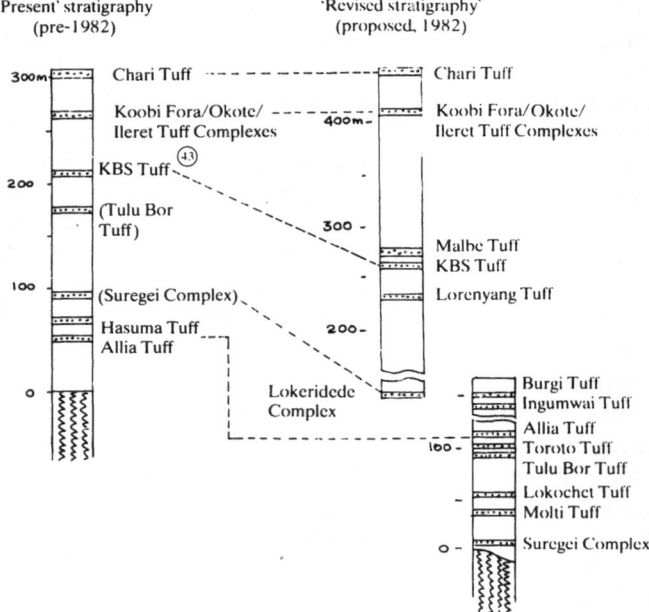

Fig. 20.

revisions accorded with ones that were proposed independently on the basis of palaeontological investigations, including those of Williamson (1982). And one can accept that Williamson might well observe rapid changes in shell morphology over very restricted vertical distances in a small region. Nevertheless, by 1982 the whole stratigraphical basis for his 1981 'observations' of punctuated equilibrium was thrown in doubt. The claim (Williamson, 1981, p. 438) that "the mollusc sequence [at Lake Turkana] can be studied within a well documented chronostratigraphic and palaeoenvironmental context" was evidently much too sanguine. Insofar as there was a "quasi-experimental context" it was, at the very least, one that was now radically altered. It should be noted, however, that recently published work by McDougall (1985) now gives precise dates for many of the tuffs described by Brown and Cerling, so that an agreed stratigraphical column has again emerged. And in a recent paper by Brown and Feibel (1986), we find a detailed revision of the whole stratigraphy of the Koobi Fora region, based on chemical characterization of the tuffs and palaeontological correlations. So we do have, it seems, a quite firm "quasi-experimental context" at last. The punctuation that Williamson formerly situated at the level of the Suregei Complex is now located at the level of the (newly-named) Burgi Tuff (Williamson, 1986).

RECENT CONTROVERSIES IN RELATION TO WILLIAMSON'S PUNCTUATIONIST INTERPRETATIONS

Criticism of Williamson's work has, however, come from a number of quarters. I do not have space here to mention all the issues that have been raised but two commentaries are worth particular consideration. One issue is, of course, the stratigraphical *completeness* of the sections examined, which remains an issue that is not fully resolved. Thus Lowell Dingus and Peter M. Sadler (1982) of the University of California have pointed out that the major punctuation claimed to occur at the level of the Suregei (= Burgi) Tuff Complex might represent not a response to environmental stress due to lacustrine regression and increase of salinity, as Williamson supposed, but rather a time gap in the record, as at the base of the Guomde Formation.

Further important comment on Williamson's claims has been made by G. Fryer of the Freshwater Biological Association, Windermere Laboratory (Cumbria, U. K.), and P. H. Greenwood and J. F. Peake of the British Museum (Natural History) (Fryer *et al.*, 1983). Their discussions take us into the realms of molluscan morphology and taxonomy, into which one may be ill-advised to enter here. The nub of the argument, however, lay in the question of whether Williamson was observing genetically-determined speciation events, or mere phenotypic changes occurring in response to altering ecological circumstances. Given that many of the new forms seemed to disappear relatively soon after their appearance (see Figure 2), one might well wonder whether phyletic change had in fact occurred. There was, perhaps, phenotypic change only, which was insufficient to protect the organisms against persistent environmental change, so that the new forms quite soon became extinct. Moreover, the critics argued, contrary to Williamson's assertions, modern forms, similar to those observed in Williamson's fossil collections, display considerable phenotypic variation from one lake to another, according to the local conditions. So the argument for sudden phyletic change loses weight on the basis of modern analogies.

Again, the general *recurrence* of ancestral forms recorded by Williamson above the Suregei Tuff Complex and above the Guomde Formation (see Figure 2) can well be explained by reversion to ancestral morphology on removal of environmental stress, and no phyletic change need be postulated just above the horizon of the Chari Tuff. In other words, Williamson's diagram (Figure 2) should be regarded as a theoretical *interpretation* of the data, rather than an 'objective' representation of the facts. It is, nevertheless, a perfectly 'feasible' interpretation, and the theory of punctuated equilibrium may well come to be accepted generally and regarded as true. Meanwhile, the debate continues (Williamson, 1985a; Fryer *et al.*, 1985; Williamson, 1985b; Hallam, 1985), with Williamson defending himself ably against his critics.

I should mention, however, that in the most recent statement of his views that I have available to me at the time of writing Williamson (1986) has revealed that his claim for punctuation at the Suregei/Burgi Tuff horizon is, as I have indicated earlier, based on a *composite* stratigraphical column, which has been pieced together from observations at several distinct exposures at Koobi Fora. No doubt, this was the procedure from the beginning. Be this as it may, it seems that we do *not* have a single exposure where several sudden morphological changes may by observed to occur in unison. Indeed, it appears from Williamson (1986) that at only *one* site is there evidence for punctuation, considering the evidence of a *single site* in isolation. All the other changes are infered by forming a composite stratigraphical column. However, Williamson does not allow any significant break in the sedimentary sequence (Williamson, personal communication, 4 June 1986) at the localities where he claims punctuations. And it is, of course, standard practice for the stratigrapher to seek to build up a historical picture of events in a region on the basis of observations at a number of separate collecting sites.

What the scientific community's final determination on all this will prove to be cannot be stated at the time of writing (1986). Again we have a 'hot-spot', or perhaps something of wider scope this time, for the interpretation of claimed punctuation sites is in dispute in several parts of the world. It is clear from the on-going debate in the agonistic field of the scientific community that more than one construction may be placed on the empirical evidence. And this state of affairs is likely to persist until greater precision is available for the data from Lake Turkana and elsewhere, particularly on the question of the rates of sedimentation. Meanwhile, the inferences drawn seem to be 'under-determined' by the data, and social factors are at work in their construal. However, work proceeds, and we may hope that in time 'truth will out'. In this process we may anticipate that empirical consistency and coherence between adjacent fields of science will be questions that players in the field will all regard as central issues.

ACKNOWLEDGEMENTS

I am much indebted to the following persons for their valued contributions to this paper: Randall Albury, Ron Amundson, Andrew Brock, Francis Brown, Frank Fitch, Diana Ford, John Forge, Andrew Gleadow, Richard Hay, Roger Lewin, Ian McDougall, David Miller, Phillip Schmidt, Peter Williamson and Susan Wright. It is probable that not all these people will agree with everything I have written. Nevertheless, I am extremely grateful to them for their help and I trust that none of them will be held responsible for any defects that may still be present in the paper.

University of New South Wales, N.S.W.

NOTES

[1] My historical account is necessarily incomplete, being largely dependent on printed sources. A great deal more might be said by the oral historian or one with access to letters and private papers. Such an inquiry, with respect to the KBS Tuff controversy is presently being conducted by Roger Lewin, as part of a book about the preconceptions in several areas of anthropology, two chapters of which pertain to the KBS affair. My goals in the present paper are, however, substantially different from those of Lewin.

[2] One author (Rhodes, 1983) has diligently seived through the Darwin texts to find a considerable number of examples which suggest that he would have been sympathetic to punctuated equilibrium theory.

[3] The classic case concerned changes in form of sea urchins collected at different horizons of the chalk cliffs of the southern English coast: A. W. Rowe, 'An analysis of the genus *Micraster*, as determined by rigid zonal collecting from the zone of *Rhynchonella cuvieri* to that of *Micraster cor-anguinum*', *The Quarterly Journal of the Geological Society of London*, 1899 **55**, 494–547. But on Gould and Eldredge's view, the "story is a sequence of three discrete points" (i.e. three discrete, fossil forms), not gradual transitions from one form to another.

[4] I have heard it suggested that the theory of punctuated equilibrium somehow exemplifies the process of 'quantitative' change giving rise to the 'qualitative', as might appeal to a dialectical materialist. But the punctuationists emphasize the significance of stasis

in species rather than gradual 'quantitative change'. Whether it is relevant or not I do not know, but Williamson has a quotation from Lenin at the front of his doctoral dissertation: "It is impossible to predict the time and progress of revolution. It is governed by its own more or less mysterious laws. But when it comes it moves irresistably." For an explicit attempt to relate Williamson's works to a Marxist viewpoint, see Howgate (1982). But this author does not suggest that Williamson himself had Marxist predilections.

[5] On this, see: Sober (1984, pp. 355—368).

[6] See: Stanley (1979, p. 27).

[7] This study drew on the author's doctoral dissertation (Williamson, 1979).

[8] The first European visitor to the region (1888) was Count Samuel Teleki, who named Lake Rudolf after the son of the emperor of the Austro-Hungarian Empire, Franz Joseph (Walker and Leakey, 1978). Teleki collected a few fossil shells from East Lake Turkana, but did not observe the mammal remains. In 1939, Vivian Fuchs reported that the eastern shores of the lake had been examined very briefly in the course of searching for two missing members of a geological expedition (Fuchs, 1939, pp. 246—247). The area was mapped in a very provisional manner. a description of the terrain, and the severe difficulties faced by fieldworkers in the district, is given in Reader (1981). This work also describes the archaeological investigations carried out at Turkana by Richard Leakey and his co-workers, and the inter-relationship of the archaeological and the geological research.

[9] It is noteworthy that a very similar debate attended the analysis in the 1960s of the remains of hominids in the Olduvai Gorge (Dalrymple and Lanphere, 1969, pp. 201—205). One may suspect that the participants in the KBS Tuff controversy saw themselves as involved in a kind of replay of the earlier contest.

[10] In electron capture (sometimes called K-capture), one of the atom's inner, K-shell electrons is captured by a nucleus proton to give a neutron, with emission of gamma radiation: $^{40}_{19}K + ^{0}_{1}e \rightarrow ^{40}_{18}Ar$. (The K symbol for the inner electron shell should not be confused with the symbol (K) for potassium.)

[11] For the simple derivation of this equation, see for example: Joplin et al. (1972, pp. 94—96). See also Dalrymple and Lanphere (1969).

[12] To allow for flame fluctuations, the basic technique is modified somewhat as follows. A known quantity of lithium is added to the test solution, and the resultant red colour is examined with a second phototube. The actual quantity measured, then, is the ratio of the K/Li intensities, using a differential amplifier applied to the two phototubes. Comparison is made with the effects produced by a standard K/Li mixture, and hence the potassium concentration of the unknown may be determined. As in other analytical work, measurement is achieved by comparison with arbitrarily stipulated standards.

[13] The 'spike' (^{38}Ar) will normally contain a small quantity of ^{40}Ar and ^{36}Ar, which should be known.

[14] In this equation, molar concentrations are represented by square brackets. For derivation, see Joplin et al. (1972, pp. 98 and 64). The $^{40}Ar/^{36}Ar$ in air is known to be 296 : 1. The equation shown here is a simplified version, involving some approximations. For a fuller expression, see: Dalrymple and Lanphere, 1969, p. 251.

[15] In this apparatus, the induction-heater coil induces current in the molybdenum crucible, and the heat generated melts the sample. Tracer argon is introduced by breaking the capsule with a steel weight moved by a magnet. The copper oxide furnace

oxidises any hydrogen or hydrocarbons. The activated charcoal, cooled with liquid nitrogen, absorbs gases, which are, however, evolved when the temperature is raised. The titanium furnace absorbs the gases nitrogen and oxygen. The ionization gauge estimates pressures in the apparatus by measurement of gaseous electrical conductivity. After absorption of all other gases, and step-wise transference of the evolved argon through the apparatus, this gas is collected on the cooled activated charcoal at the right-hand side of the apparatus. The take-off tube with its argon content, is eventually cut off the rest of the apparatus with a blow-pipe torch; or more commonly the gas is transferred directly to the mass spectrometer.

[16] The slope $= e^{\lambda t} - 1$, where λ is the decay constant of ^{40}K.

[17] In these diagrams, following Fitch and Miller, I merely indicate 'heat steps' on the horizontal axis. The usual practice is to plot the proportion of the total argon released.

[18] I.e., pieces of country rock incorporated in a solidifying magma.

[19] It is worth noting that not all listeners to the talk could have understood the basis of the method, for a question was asked as to why lines, such as those of Figure 8, could "fold back on themselves".

[20] This is not intended as a comment on Fitch and Miller's work. Rather, it is a personal comment on difficulties that seemed to attend the method, in the view of the present writer. According to A. J. Gleadow (personal communication, 23 May 1986), it was the "over-interpretation of data with no due consideration of its analytical accuracy which . . . [was] at the heart of the KBS controversy".

[21] These rocks were neither named nor divided by Behrensmeyer. Found in the Ileret, but not the Koobi Fora region, they were equivalent to the upper part of what came to be called the Upper Member of the Koobi Fora Formation.

[22] The rocks of the Okote Tuff down to the KBS Tuff.

[23] Rocks from the KBS Tuff down to the Suregei Tuff.

[24] The sequence subsequently recognized below the Koobi Fora Formation — from the Suregei Tuff Complex to the basal volcanics.

[25] Specimens of *M. limnetes* could in fact be found in all three zones.

[26] It is noteworthy that the whole argument involves an assumption of phyletic gradualism. Yet, when agreement was eventually reached about the KBS Tuff date, these data, which concurred with Cooke and Maglio's palaeontological arguments, were used in support of an argument for punctuated equilibrium theory by Williamson.

[27] In publications of 1971, 1972 and 1973, the expedition leader, Richard Leakey, was willing to accept a date of 2.61 Myr for the KBS Tuff. See Leakey, 1971, 1972, 1973a, and 1973b.

[28] Isaac has recently died, I am regretfully informed. Andrew Brock (personal communication, 18 September 1986) regards his contribution to the Koobi Fora Project as essential for the way in which he "held the scientific threads together".

[29] The field is determined with respect to the three rectangular co-ordinates of the cubical sample, and the resultant field is calculated. The apparatus is used to measure the field with respect to each co-ordinate twice: once; and then with the specimen turned over. For a detailed account of the empirical methods of palaeomagnetism, see Collinson, 1983. This text gives details of the considerable number of developments that have taken place in spinner magnetometry since the designs of the Foster fluxgate instrument used by Brock and Isaac.

[30] The constituent minerals in the rock, particularly weathered matter, do not all retain

their magnetism with the same tenacity. Hence a 'magnetic washing' procedure may be employed in order to display the original palaeomagnetism. Washing may be accomplished also by step-wise heating.

[31] It is interesting to note that this was about the amount of time by which the age of the KBS Tuff was eventually reduced.

[32] Curtis was an experienced geochronologist, who had done work with rocks from the important Olduvai Gorge excavations (Evernden and Curtis, 1965). For a photograph of Curtis at work with his argon determination equipment, see Reader, 1981, p. 159.

[33] However, the field evidence for a disconformity was queried at the conference.

[34] Dr. Gleadow is currently a Senior Research Fellow at Melbourne University.

[35] See Fitch and Miller, 1976. For example, ages of between 0.53 and 17.6 Myr were quoted for specimens from the Koobi Fora Tuff (pp. 129—131). Yet curiously one figure for an apparent date of crystallization of a sanidine sample was quoted as 1.57 ± 0.00 Myr.

[36] Behrensmeyer (1978, p. 436) openly questioned the uniformitarian hypothesis: "It may be valid, in this case, to question whether the present is a reliable key to the past".

[37] Philosophers generally contend that coherence is not a *sufficient* criterion of truth. The positivist, Moritz Schlick, for example, objected that any fabricated tale might cohere very well, but be quite false (Schlick, 1979, p. 376).

[38] I am indebted to W. R. Albury for this suggestive analogy.

[39] Frank Fitch has kindly furnished me with a copy of his response to Roger Lewin (dated 28 May 1986), who had been collecting evidence for his book dealing (*inter alia*) with the KBS Tuff controversy. Fitch's letter expresses particular pain at certain suggestions of experimental deficiencies, which were, it appears, circulating privately amongst the scientists involved in the controversy.

[40] Herein, however, lay a major part of the difficulty. The several components of the tuff might themselves have different ages. And even the *recognition* of the Tuff — away from its type area — was sometimes uncertain and open to controversy. Chemical composition and radiometric age were part of the definition of the Tuff, so it was not simply a question of recognizing the rock by its lithological appearance and then dating it.

[41] I am informed by Roger Lewin (personal communciation, 26 September 1986) that some of the original KBS crystals are still held at a dating laboratory in Berne, Switzerland, in the care of Fitch's former student, Tony Hurford. However, so far as I am aware at the time of writing, this material has not been reanalyzed.

[42] See Figure 20.

[43] The KBS Tuff was indicated on the figure, but (inadvertently?) not labelled as such.

[44] One part was renamed the Lokeridede Complex. The other part retained the name Suregei Complex, but was relocated at the bottom of the sequence.

BIBLIOGRAPHY

Alpert, D. and R. S. Buritz: 1954, 'Ultra-High Vacuum. II. Limiting Factors on the Attainment of Very Low Pressures', *Journal of Applied Physics* **25**, 202—209.

Barnes, B.: 1982, *T. S. Kuhn and Social Science*, Macmillan, London and Basingstoke.

Behrensmeyer, A. K.: 1970, 'Preliminary Geological Interpretation of a New Hominid Site in the Lake Rudolf Basin', *Nature* **226**, 225—226.

Behrensmeyer, A. K.: 1978, 'Correlations of Plio-Pleistocene Sequences in the Northern Lake Turkana Basin: A Summary of Evidence and Issues', in W. W. Bishop (ed.), *Geological Background to Fossil Man: Recent Research in the Gregory Rift Valley, East Africa*, Scottish Academic Press, Edinburgh and University of Toronto Press, Toronto, pp. 421—440.

Bence, A. E. and A. L. Albee: 1968, 'Empirical Correction Factors for the Electron Microanalysis of Silicates and Oxides', *Journal of Geology* **76**, 382—403.

Birks, L. S.: 1971, *Electron Probe Microanalysis*, Wiley-Interscience, New York.

Bishop, W. W. (ed.): 1978, *Geological Background to Fossil Man: Recent Research in the Gregory Rift Valley, East Africa*, Scottish Academic Press, Edinburgh and University of Toronto Press, Toronto.

Bishop, W. W., J. A. Miller and S. Cole (eds.), *Calibration of Hominoid Evolution: Recent Advances in Isotopic and Other Dating Methods Applicable to the Origin of Man*, Scottish Academic Press, Edinburgh and University of Toronto Press, Toronto.

Bourdieu, P.: 1971, 'Intellectual Field and Creative Project', in M. F. D. Young (ed.), *Knowledge and Control: New Directions for the Sociology of Education*, Collier-Macmillan, pp. 161—188 (originally published in French in 1966).

Bourdieu, P.: 1975, 'The Specificity of the Scientific Field and the Social Conditions of the Progress of Learning', *Social Science Information* **14**, 19—47.

Bowen, B. E. and C. F. Vondra: 1973, 'Stratigraphical Relationships of the Plio-Pleistocene Deposits, East Rudolf', *Nature* **242**, 391—393.

Brock, A. and G. Ll. Isaac: 1974, 'Paleomagnetic Stratigraphy and Chronology of Hominid-Bearing Sediments East of Lake Rudoff, Kenya', *Nature* **247**, 344—348.

Brock, A. and G. Ll. Isaac: 1976, 'Reversal Stratigraphy and its Application at East Rudolf', in Y. Coppens *et al.* (eds.), *Earliest Man and Environments in the Lake Rudolf Basin: Stratigraphy, Paleoecology, and Evolution*, University of Chicago Press, Chicago, pp. 148—162.

Brown, F. H. and T. E. Cerling: 1982, 'Stratigraphical Significance of the Tulu Bor Tuff of the Koobi Fora Formation', *Nature* **299**, 212—215.

Brown, F. H., F. Clark Howell and G. G. Fleck: 1978, 'Observations on Problems of Correlation of Late Cenozoic Hominid-Bearing Formations in the North Lake Turkana Basin', in W. W. Bishop (ed.), *Geological Background to Fossil Man: Recent Research in the Gregory Rift Valley, East Africa*, Scottish Academic Press, Edinburgh and University of Toronto Press for the Wenner-Gren Foundation, Toronto, pp. 473—498.

Brown, F. H. and C. S. Feibel: 1986, 'Revision ot the Lithostratigraphic Nomenclature in the Koobi Fora Region, Kenya', *Journal of the Geological Society, London* **143**, 297—310.

Brown, F. H. and K. R. Lajoi: 1971, 'Radiometric Age Determinations on Pliocene/Pleistocene Formations in the Lower Omo Basin, Ethiopia', *Nature* **229**, 483—495.

Butzer, K. W. and D. L. Thurber: 1969, 'Some Late Cenozoic Sedimentary Formations of the Lower Omo Basin', *Nature* **222**, 1138—1143.

Cerling, T. E. and F. H. Brown: 1982, 'Tuffaceous Marker Horizons in the Koobi Fora Region and the Lower Omo Valley', *Nature* **299**, 216—221.

Cerling, T. E., F. H. Brown, B. W. Cerling, G. H. Curtis and R. E. Drake: 1979, 'Preliminary Correlations Between the Koobi Fora and Shungura Formations, East Africa', *Nature* **279**, 118—121.

Chamberlin, T. C.: 1897, 'The Method of Multiple Working Hypotheses', *Journal of Geology* **5**, 837—848.

Collins, H. M.: 1981, 'The Place of the Core-set in Modern Science: Social Contingency with Methodological Propriety in Science', *History of Science* **19**, 6—19.

Collinson, D. W.: 1983, *Methods in Rock Magnetism and Palaeomagnetism: Techniques and Instrumentation*, Chapman and Hall, London and New York.

Cooke, H. B. S. and V. J. Maglio: 1972, 'Plio-Pleistocene Stratigraphy in East Africa in Relation to Proboscidean [Elephant] and Suid [Pig] Evolution', in W. W. Bishop *et al.* (eds.), *Calibration of Hominoid Evolution: Recent Advances in Isotopic and Other Dating Methods Applicable to the Origin of Man*, Scottish Academic Press, Edinburgh and University of Toronto Press for the Wenner-Gren Foundation, Toronto, pp. 303—329.

Coppens, Y., F. C. Howell, G. Ll. Isaac and R. E. F. Leakey (eds.), *Earliest Man and Environments in the Lake Rudolf Basin: Stratigraphy, Paleoecology, and Evolution*, University of Chicago Press, Chicago and London.

Cox, A.: 1969, 'Geomagnetic Reversals', *Science* **163**, 237—245.

Curtis, G. H.: 'Improvements in Potassium-Argon Dating: 1962—1975', *World Archaeology* **7**, 198—209.

Curtis, G. H., T. Drake, T. Cerling and J. Hampel: 1975, 'Age of KBS Tuff in Koobi Fora Formation, East Rudolf, Kenya', *Nature* **258**, 395—398.

Curtis, G. H., R. E. Drake, T. E. Cerling, B. W. Cerling and J. H. Hampel: 1978, 'Age of KBS Tuff in Koobi Fora Formation, East Lake, Turkana, Kenya, in W. W. Bishop (ed.), *Geological Background to Fossil Man: Recent Research in the Gregory Rift Valley, East Africa*, Scottish Academic Press, Edinburgh and University of Toronto Press, Toronto, 1978, pp. 463—469.

Dalrymple, G. B. and M. A. Lanphere: 1969, *Potassium-Argon Dating: Principles, Techniques and Applications to Geochronology*, W. H. Freeman and Co., San Francisco.

Darwin, C. R.: 1859, *On the Origin of Species by Means of Natural Selection, or the Preservation of Favoured Races in the Struggle for Life*, John Murray, London.

Drake, R. E., G. H. Curtis, T. E. Cerling, B. W. Cerling and J. Hampel: 1980, 'KBS Tuff Dating and Geochronology of Tuffaceous Sediments in the Koobi Fora and Shungura Formations, East Africa', *Nature* **283**, 368—372.

Eldredge, N.: 1972, 'Systematics and Evolution of *Phacops rana* (Green, 1832) and *Phacops iowensis* Delo. 1936 (Trilobita) from the Middle Devonian of North America', *Bulletin of the American Museum of Natural History* **147**, 45—114.

Eldredge, N.: 1985, *Time Frames: The Rethinking of Darwinian Evolution and the Theory of Punctuated Equilibrium*, Simon and Schuster, New York.

Eldredge, N.: 1986, 'Progress in Evolution?', *New Scientist*, 5 June, 54—57.

Eldredge, N. and S. J. Gould: 1972, 'Punctuated Equilibrium: An Alternative to Phyletic Gradualism', in T. J. M. Schopf (ed.), *Models in Paleobiology*, Freeman, Cooper & Co., San Francisco, pp. 82—115.

Evernden, J. F. and G. H. Curtis: 1965, 'The Potassium-Argon Dating of Late Cenzoic Rocks in East Africa and Italy', *Current Anthropology* **6**, 343—363.

Findlater, I. C.: 1976, 'Tuffs and the Recognition of Isochronous Mapping Units in the East Rudolf Succession', in Y. Coppens *et al.* (eds.), *Earliest Man and Environments in the Lake Rudolf Basin: Stratigraphy, Paleocology, and Evolution*, University of Chicago Press, Chicago and London, pp. 94–104.

Findlater, I. C.: 1978a, 'Isochronous Surfaces within the Plio-Pleistocene Sediments East of Lake Turkana', in W. W. Bishop (ed.) *Geological Background to Fossil Man: Recent Research in the Gregory Rift Valley, East Africa*, Scottish Academic Press, Edinburgh and University of Toronto Press, Toronto, pp. 415–420.

Findlater, I. C.: 1978b, 'Stratigraphy', in M. V. Leakey and R. E. Leakey (eds.), *Koobi: Fora Research Project Volume 1: The Fossil Hominids and an Introduction to their Context 1968–1974*, Oxford University Press, Oxford, pp. 14–31.

Fitch, F. J., I. C. Findlater and R. T. Watkins: 1974, 'Dating of the Rock Succession Containing Fossil Hominids at East Rudolf, Kenya', *Nature* **51**, 213–215.

Fitch, F. J., P. J. Hooker and J. A. Miller: 1976, '^{40}Ar/^{39}Ar Dating of the KBS Tuff in Koobi Fora Formation, East Rudolf, Kenya', *Nature* **263**, 740–744.

Fitch, F. J., P. J. Hooker and J. A. Miller: 1978, 'Geochronological Problems and Radioisotopic Dating in the Gregory Rift Valley', in W. W. Bishop (ed.), *Geological Background to Fossil Man: Recent Research in the Gregory Rift Valley, East Africa*, Scottish Academic Press, Edinburgh and University of Toronto Press, Toronto, 441–461.

Fitch, F. J. and J. A. Miller: 1970, 'Radioisotopic Age Determinations of Lake Rudolf Artefact Site', *Nature* **226**, 226–228.

Fitch, F. J. and J. A. Miller: 1971, 'Atmospheric Argon Correction in the K–Ar Dating of Young Volcanic Rocks', *Journal of the Geological Society, London* **127**, 277–280.

Fitch, F. J. and J. A. Miller: 1976, 'Conventional Potassium–Argon and Argon-40/Argon-39 Dating of Volcanic Rocks of East Rudolf', in Y. Coppens *et al.* (eds.), *Earliest Man and Environments in the Lake Rudolf Basin: Stratigraphy, Paleoecology, and Evolution*, University of Chicago Press, Chicago and London, pp. 123–147.

Fitch, F. J., A. J. Hurford, P. J. Hooker and J. A. Miller: 1978, 'The KBS Tuff Problem', in R. E. Zartman (ed.), *Short Papers of the Fourth International Conference, Geochronology, Cosmochronology, Isotope Geology 1978*, United States Department of the Interior Geological Survey, Geological Survey Open-File Report 78–701, 1978, pp. 114–117.

Fleischer, R. L. and H. R. Hart: 1972, 'Fission Track Dating: Techniques and Problems', in W. W. Bishop *et al.* (eds.), *Calibration of Hominoid Evolution. Recent Advances in Isotopic and Other Dating Methods Applicable to the Origin of Man*, Scottish Academic Press, Edinburgh and University of Toronto Press, Toronto, pp. 135–170.

Foster, J. H.: 1966, 'A Paleomagnetic Spinner Magnetometer using a Fluxgate Gradiometer', *Earth and Planetary Science Letters* **1**, 463–466.

Fryer, G., P. H. Greenwood and J. F. Peake: 1983, 'Punctuated Equilibria, Morphological Stasis and the Palaeontological Documentation of Speciation: A Biological Appraisal of a Case History in an African Lake', *Biological Journal of the Linnean Society* **20**, 195–205.

Fryer, G., P. H. Greenwood and J. F. Peake: 1985, 'The Demonstration of Speciation in

Fossil Molluscs and Living Fishes', *Biological Journal of the Linnean Society* **26**, 325—336.

Fuchs, V. E.: 1939, 'The Geological History of the Lake Rudolf Basin, Kenya Colony', *Philosophical Transactions of the Royal Society, London*, Series B, **229**, 219—274.

Gleadow, A. J. W.: 1980, 'A Fission Track Age of the KBS Tuff and Associated Hominid Remains in Northern Kenya', *Nature* **284**, 225—230.

Gleadow, A. J. W., A. J. Hurford and R. D. Quaife: 1976 'Fission Track Dating of Zircon: Improved Etching Techniques', *Earth and Planetary Science Letters* **33**, 273—276.

Glen, W.: 1982, *The Road to Jamarillo: Critical Years in the Revolution in Earth Science*, Stanford University Press, Stanford.

Gould, S. J.: 1969, 'An Evolutionary Microcosm: Pleistocene and Recent History of the Land Snail P. (*Poecilozonites*) in Bermuda', *Bulletin of the Museum of Comparative Zoology Harvard University* **138**, 407—532.

Gould, S. J.: 1977, *Ontogeny and Phylogeny*, Harvard University Press, Cambridge (Mass.).

Gould, S. J.: 1983, *The Panda's Thumb: More Reflections in Natural History*, Pelican Books, Harmondsworth.

Gould, S. J.: 1984, *Hen's Teeth and Horse's Toes: Further Reflections in Natural History*, Pelican Books, Harmondsworth.

Gould, S. J. and N. Eldredge: 1977, 'Punctuated Equilibrium: The Tempo and Mode of Evolution Reconsidered', *Paleobiology* **3**, 115—151.

Grasty, R. L. and J. A. Miller: 1965, 'The Omegatron: A Useful Tool for Argon Isotope Investigation', *Nature* **207**, 1146—1148.

Hallam, A.: 1985, 'Comment', *Biological Journal of the Linnean Society* **26**, 341—343.

Harland, W. B., A. V. Cox, P. G. Llewellyn, C. A. G. Pickton, A. G. Smith, and R. Walters, and K.E. Fancett, *A Geologic Time Scale*, Cambridge University Press, Cambridge.

Harris, J. M. (ed.): 1983, *Koobi Fora: Researches into Geology, Palaeontology, and Human Origins. Volume 2 The Fossil Ungulates: Proboscidea, Perissodactyla, and Suidae*, Oxford University Press, Oxford.

Hay, R. L.: 1980, 'The KBS Tuff Controversy May be Ended', *Nature* **284**, 401.

Hillhouse, J. W., J. W. M. Ndombi, A. Cox and A. Brock: 1977, 'Additional Results on Palaeomagnetic Stratigraphy of the Koobi Fora Formation, East of Lake Turkana (Lake Rudolf), Kenya', *Nature* **265**, 411—415.

Howgate, M.: 1982, 'Marxism and Evolution: The New Synthesis', *Labour Review*, Nov. 1982, pp. 8—14.

Hurford, A. J.: 1974, 'Fission Track Dating of a Vitric Tuff from East Rudolf, North Kenya', *Nature* **249**, 236—237.

Hurford, A. J., A. J. W. Gleadow and C. W. Naeser: 1976, 'Fission-Track Dating of Pumice from the KBS Tuff, East Rudolf, Kenya', *Nature* **263**, 738—740.

Huxley, A.: 1982 'Anniversary Address, 1981', *Proceedings of the Royal Society, London*, Series A, **379**, ix—xx.

Isaac, G. Ll., R. E. F. Leakey and A. K. Behrensmeyer: 1971, 'Activities, East of Lake Rudolf, Kenya', *Science* **173**, 1129—1134.

Jones, J. S.: 1981, 'An Uncensored Page of Fossil History', *Nature* **293**, 427—428.

Joplin, G. A., J. R. Richards and C. A. Joplin: 1972, *Finding the Age of Rocks*, Angus and Robertson, Sydney.

Kuhn, T. S.: 1962, *The Structure of Scientific Revolutions*, University of Chicago Press, Chicago and London.

Latour, B. and S. Woolgar: 1979, *Laboratory Life: The Social Construction of Scientific Facts*, Sage Publications, Beverly Hills and London.

Leakey, M. G. and R. E. Leakey (eds.): 1978, *Koobi Fora Research Project Volume 1: The Fossil Hominids and an Introduction to their Context 1968—1974*, Oxford University Press, Oxford.

Leakey, R. E. F.: 1971, 'Further Evidence of Lower Pleistocene Hominids from East Rudolf, North Kenya', *Nature* **231**, 241—245.

Leakey, R. E. F.: 1972, 'Further Evidence of Lower Pleistocene Hominids from East Rudolf, North Kenya, 1971', *Nature* **237**, 264—269.

Leakey, R. E. F.: 1973a, 'Further Evidence of Lower Pleistocene Hominids from East Rudolf, North Kenya, 1972', *Nature* **242**, 170—173.

Leakey, R. E. F.: 1973b, 'Evidence for an Advanced Plio-Pleistocene Hominid from East Rudolf, Kenya', *Nature* **242**, 447—450.

Leakey, R. E.: 1983, *One Life: An Autobiography*, Michael Joseph, London.

Maglio, V. J.: 1972, 'Vertebrate Faunas and Chronology of Hominid-Bearing Sediments East of Lake Rudolf, Kenya', *Nature* **239**, 379—385.

Maynard Smith, J.: 1981, 'Did Darwin Get it Right?', *London Review of Books*, 18 June — 1 July, pp. 10—11.

Mayr, E.: 1963, *Animal Species and Evolution*, Harvard University Press, Cambridge (Mass.).

McDougall, I.: 1981, '^{40}Ar/^{39}Ar Age Spectra from the KBS Tuff, Koobi Fora Formation', *Nature* **294**, 120—124.

McDougall, I.: 1985, 'K—Ar and ^{40}Ar/^{39}Ar Dating of the Hominid-Bearing Pliocene-Pleistocene Sequence at Koobi Fora, Lake Turkana, Northern Kenya', *Geological Society of America Bulletin* **96**, 159—175.

McDougall, I., R. Maier, P. Sutherland-Hawkes and A. J. W. Gleadow: 1980, 'K—Ar Age Estimate for the KBS Tuff, East Turkana, Kenya', *Nature* **248**, 230—234.

Merrihue, C. M. and G. Turner: 1966, 'Potassium—Argon Dating with Fast Neutrons', *Journal of Geophysical Research* **71**, 2852—2857.

Naeser, C. W.: 1969, 'Etching Fission Tracks in Zircons', *Science* **165**, 388.

Oldroyd, D. R.: 1985, 'An Episode in Geology', *Science* **230**, 432—433.

Price, P. B. and R. M. Walker: 1963, 'Fossil Tracks of Charged Particles in Mica and the Age of Minerals', *Journal of Geophysical Research* **68**, 4847—4862.

Reader, J.: 1981, *Missing Links: The Hunt for Earliest Man*, Book Club Associates, London.

Rhodes, F. H. T.: 1983, 'Gradualism, Punctuated Equilibrium and the *Origin of Species*', *Nature* **305**, 269—272.

Ridley, M.: *The Problems of Evolution*, Oxford University Press, Oxford and New York.

Rowe, A. W.: 1899, 'An Analysis of the Genus *Micraster*, as determined by rigid zonal collecting from the zone of *Rhynchonella cuvieri* to that of *Micraster cor-anguinam*', *The Quarterly Journal of the Geological Society of London* **55**, 494—547.

Rudwick, M. J. S.,: 1985, *The Great Devonian Controversy: The Shaping of Scientific Knowledge among Gentlemanly specialists*, University of Chicago Press, Chicago and London.

Schlick, M.: 1979, *Philosophical Papers*, Vol. II (1925—36), D. Reidel Publ. Co., Dordrecht.

Sigurgeisson, T.: 1962, 'Age Dating of Young Basalts with the Potassium-Argon Method', University of Iceland Physics Laboratory Report (in Icelandic, not seen).

Simpson, G. G.: *Tempo and Mode in Evolution*, Columbia University Press, New York, ' Mornington Heights.

Sober, E.: 1984, *The Nature of Selection: Evolutionary Theory in Philosophical Focus*, MIT Press, Cambridge (Mass.) and London.

Sommer, H., H. A. Thomas and J. A. Hipple: 1951, 'The Measurement of e/M by Cyclotron Resonance', *Physical Review* **82**, 697—702.

Stanley, S. M.: 1975, 'A Theory of Evolution above the Species Level', *Proceedings of the National Academy of Science, U.S.A.* **72**, 646—650.

Stanley, S. M.: 1979, *Macroevolution: Pattern and Process*, W. H. Freeman & Co.: San Francisco.

Stanley, S. M.: 1981, *The New Evolutionary Timetable: Fossils, Genes, and the Origin of Species*, Basic Books, New York.

Vondra, C. F. and B. E. Bowen: 1976, 'Plio-Pleistocene Deposits and Environments, East Rudolf, Kenya', in Y. Coppens *et al.* (eds.), *Earliest Man and Environments in the Lake Rudolf Basin: Stratigraphy, Paleoecology, and Evolution*, University of Chicago Press, Chicago and London, pp. 79—93.

Vondra, C. F. and B. E. Bowen: 1978, 'Stratigraphy, Sedimentary Facies and Palaeo-environments, East Lake Turkana, Kenya', in W. W., Bishop (ed.), *Geological Background to Fossil Man: Recent Research in the Gregory Rift Valley, East Africa*, Scottish Academic Press, Edinburgh and University of Toronto Press, Toronto, pp. 395—414.

Vondra, C. F., G. D. Johnson, B. E. Bowen and A. K. Behrensmeyer: 1971, 'Preliminary Stratigraphical Studies of the East Rudolf Basin, Kenya', *Nature* **231**, 245—248.

Walker, A. and R. E. F. Leakey: 1978, 'The Hominids of East Lake Turkana', *Scientific American* **239** (2), 54—66.

White, T. D. and J. M. Harris: 1977, 'Suid Evolution and Correlation of African Hominid Localities', *Science* **198**, 13—21.

Williamson, P. G.: 1979, 'Evolution Implications of Late Cenozoic Mollusc Assemblages from the Turkana Basin, North Kenya', Ph.D. Dissertation, Bristol University.

Williamson, P. G.: 1981, 'Palaeontological Documentation of Speciation in Cenozoic Molluscs from Turkana Basin', *Nature* **293**, 437—443.

Williamson, P. G.: 1982, 'Molluscan Biostratigraphy of the Koobi Fora Hominid-Bearing Deposits', *Nature* **295**, 140—142.

Williamson, P. G.: 1985a, 'Punctuated Equilibrium, Morphological Stasis and the Palaeontological Documentation of Speciation: A Reply to Fryer, Greenwood and Peake's Critique of the Turkana Basin Mollusc Sequence', *Biological Journal of the Linnean Society* **26**, 307—324.

Williamson, P. G.: 1985b, 'In Reply to Fryer, Greenwood and Peake', *Biological Journal of the Linnean Society* **26**, 337—340.

Williamson P. G.: 1986, 'Fine Structure of the "Suregei Level" Speciation Event in Late Cenozoic Molluscs from the Turkana Basin', forthcoming.

JOHN NORTON

EINSTEIN, THE HOLE ARGUMENT AND THE
REALITY OF SPACE

1. INTRODUCTION

In November 1915, Einstein put the finishing touches to his general
theory of relativity. Then he proclaimed that the theory, through its
general covariance, "robbed time and space of the last trace of objective
reality" (Einstein, 1915, p. 831). This triumphant proclamation was
repeated early in 1916 in a review of the theory. The requirement of
general covariance "takes away from space and time the last remnant of
physical objectivity." (Einstein, 1916, p. 117).

This case, as Fine has reminded us, is one of a number of embarrass-
ments for scientific realists who like to think that progress in science
has depended at least in some measure on the realist orientation of
scientific investigators (Fine, 1984, pp. 91—92). But Einstein's work on
special and general relativity owed a great debt to Machian positivism
and in particular Mach's non-realist[1] attitude towards Newton's absolute space and time. It is striking, for example, that Einstein's earliest
and still best known exposition of the complete general theory of
relativity does not begin by revealing some empirical deficiency of
earlier theories. Rather he launches the theory by pointing out an
"epistemological defect" in special relativity and classical mechanics,
which, he tells us, was first noticed by Mach. (Einstein, 1916, p. 112.)

In the simplest of glosses, Einstein's work on relativity theory is
portrayed as the relentless pursuit of the implications of Mach's non-
realist view of space and time. Assertions about motion with respect to
space are to be rendered meaningless unless they can be reinterpreted
solely in terms of the relative motion of bodies. Special relativity was
the first step. In it motion with respect to some absolute state of
rest was eliminated and with it went the aether of electromagnetism.
General relativity completed the process by removing the unacceptable
intrinsic distinction between inertial and accelerated motion which had
still lingered in special relativity. The result was a complete victory for
Leibniz's relational view of space and time. In so far as one talked
about space and time within general relativity (and this is done

153

John Forge (Ed.), Measurement, Realism and Objectivity, 153—188.

frequently!), the talk must be understood entirely instrumentally. According to general relativity, the terms space, time and spacetime do not refer to any entities in the world.

This account portrays the development of relativity theory as driven by a naive positivism and non-realism and as such it does not capture the subtlety and depth of Einstein's work towards the discovery of relativity theory.[2] To illustrate my claim I am going to tell the story of the origin and significance of an argument of Einstein which seems at first glance to be quite naively positivistic in outlook and somewhat trivial in import. I follow John Stachel in calling the argument the "point-coincidence argument" and quote the well known 1916 presentations of it. The argument is given in the wake of Einstein's statement of the requirement of general covariance, the requirement that the laws of nature must "hold good for all systems of coordinates." It reads:

That this requirement of general co-variance, which takes away from space and time the last remnant of physical objectivity, is a natural one, will be seen from the following reflexion. All our space-time verifications invariably amount to a determination of space-time coincidences. If, for example, events consisted merely in the motion of material points, then ultimately nothing would be observable but the meetings of two or more of these points. Moreover, the results of our measurings are nothing but verifications of such meetings of the material points of our measuring instruments with other material points, coincidences between the hands of a clock and points on the clock dial, and the observed point-events happening at the same place at the same time.

The introduction of a system of reference serves no other purpose than to facilitate the description of the totality of such coincidences. We allot to the universe four space-time variables x_1, x_2, x_3, x_4 in such a way that for every point-event there is a corresponding system of values of the variables $x_1 \ldots x_4$. To two coincident point-events there corresponds one system of values of the variables $x_1 \ldots x_4$, i.e. coincidence is characterized by the identity of the coordinates. If, in place of the variables $x_1 \ldots x_4$, we introduce functions of them x'_1, x'_2, x'_3, x'_4, as a new system of coordinates, so that the systems of values are made to correspond to one another without ambiguity, the equality of all four coordinates in the new system will also serve as an expression for the space-time coincidence of the two point-events. As all our physical experience can be ultimately reduced to such coincidences, there is no immediate reason for preferring certain systems of coordinates to others, that is to say, we arrive at the requirement of general covariance.   (Einstein, 1916, pp. 117—118)

Presented in this form, the argument has little force. Its conclusion is not the most interesting claim of the passage, which is that space and time have lost the last remnant of physical objectivity. Fine (1984, p. 91) quite rightly calls it a "suspicious-looking verificationist argument". What I think makes the argument look suspicious is not its verificationism. Rather it is the caution with which Einstein pursues a quite trivial conclusion: that there should be no preferred spacetime coordinate systems, which acquires the lofty title of postulate or requirement of general covariance. If spacetime coordinate system has its usual meaning — a smooth but otherwise arbitrary numerical labelling of spacetime events — then it is hard to see how we could require otherwise. Certainly this requirement is a commonplace of differential geometry, in which any well formulated spacetime theory is automatically expressible in coordinate free (= generally covariant) terms. This makes the requirement essentially useless as a criterion for selecting between competing theories. For example, generally covariant formulations of Newtonian spacetime theories and of special relativity are well known.

Nevertheless Einstein's point-coincidence argument, with its verificationist turn of phrase, fascinated contemporary philosophers such as Reichenbach and Schlick, both of whom studied closely the new-born general theory of relativity. (I follow here the discussion of Friedman, 1983, Ch. 1 and Howard, 1984, Sect. 3.) Briefly, they saw in it a perfect example of the relation between theory and fact proposed by the soon to emerge logical positivist movement and in a manner essentially related to the non-realistic view of theoretical terms. For Reichenbach, for example, our freedom in choosing coordinate systems was another instance of the conventionality inherent in theory, which would surface elsewhere as the conventionality of geometry and distant simultaneity. Schlick applauded the point-coincidence argument as illustrating how we can eliminate elements which are superfluous to our theory in the sense that they are arbitrary and thus cannot correspond to anything real. For Schlick space and time were the arbitrary elements while the coincidences of the point-coincidence argument were non-arbitary. Thus Friedman identifies the arbitrariness of choice of coordinate system as "the genesis of Reichenbach's notion of '*coordinative definition*'" (p. 19) and in Einstein's 1916 statement above of the point-coincidence argument he sees "the birth of the modern observational/theoretical distinction." (p. 24)

We owe a great debt to John Stachel (1980), who discovered the key to a proper understanding of the point-coincidence argument. With the help of Einstein's correspondence from that period, he was able to identify the argument as Einstein's resolution of a grave difficulty which had helped delay the completion of the general theory of relativity by as much as three years. In the so-called "hole argument" of late 1913,

Einstein had convinced himself that generally covariant gravitational field equations were incompatible with physical determinism. Late in 1915, in order to be able to readmit generally covariant gravitational field equations into general relativity, Einstein had to find an answer to the hole argument. That answer was the point-coincidence argument, which Einstein then interpreted as establishing that space and time must forfeit the "last remnant of physical objectivity."

In this paper, I shall review the hole and point-coincidence arguments and the circumstances surrounding their origins. We shall see that the really important conclusion Einstein drew from this episode was a result which I label "Leibniz equivalence". It asserts that in a generally covariant theory such as general relativity, a single gravitational field cannot be represented by a single mathematical field, but must be represented by an equivalence class of diffeomorphic fields.

We shall see that without excursions into Einstein's earlier publications and his correspondence it is impossible for readers of Einstein (1916) to understand that this result was the issue or, for that matter, precisely how the verificationism of the point-coincidence argument was to be applied. Moreover we shall see the establishment of Leibniz equivalence did not depend on the verificationist argument, but was already forced on us by a stronger consideration not mentioned in Einstein (1916). It is clear from retrospective appraisal of the hole argument, that unless we accept Leibniz equivalence, we will commit ourselves to an altogether unacceptable variety of indeterminism when we come to formulate generally covariant field theories such as general relativity.

Finally we shall see that non-realism about space and time played only a secondary role in the episode, entering only at the denouement of a much longer story. I can see no way for the point-coincidence argument to support the non-realist view. Rather it establishes forcefully an antisubstantivalist view of spacetime, which asserts not that spacetime has no reality, but no reality independent of the fields it contains. We shall see that, within four years of 1916, Einstein retracted his non-realist statements in favour of explicit antisubstantivalism.

2. THE *ENTWURF* THEORY

In August 1912 Einstein returned to Zurich. Over the preceding five years he had worked intermittently on the problem of relativizing gravitation theory and extending the principle of relativity to accelerated motion. Within less than a year, with the mathematical assistance of his friend Marcel Grossmann, he was able to sketch out virtually all the essential components of his general theory of relativity. We now call that theory the *Entwurf* ("Outline") theory after the first word of the title of Einstein and Grossmann (1913a and 1913b), in which the theory first appeared.

I now describe some elements of this theory in modern terms, terms somewhat different to those used by Einstein and Grossmann. Without this more precise terminology it would be very difficult to explicate adequately the hole and point-coincidence arguments. The theory proposed that spacetime was a four dimensional differentiable manifold on which certain fields were defined, the most important of these being the metric tensor field.

A *four dimensional differentiable manifold* is a set whose members are identified with the points or events of spacetime in the standard developments of spacetime theories. If the manifold were just a set of events, then we would have no idea of which events neighbor on which others. This information is provided by the topological structure of the manifold, which specifies which subsets of events are the open sets (neighbourhoods). Thus we have a notion of locality through which we can identify the neighbourhoods containing each event. The manifold looks locally like a four dimensional Cartesian space — that is, for any event we can always find a neighbourhood containing the event which can be mapped one—one onto some open subset of R^4.

A *coordinate system* or coordinate chart is just such a map. It labels each point p of the relevant neighbourhood with some unique four-tuple of reals, $x^i(p)$ ($i = 0, 1, 2, 3$). If one such coordinate system K is possible, then from K, it is easy to define a second coordinate system K' which assigns a different four-tuple $x^{i'}(p)$ to p. All coordinate systems, which are related by continuous, infinitely differentiable transformation equations where they overlap, belong to the manifold's altas of coordinate systems.

In terms of manifold structure alone, it is possible to define curves (smooth maps from an interval of the reals into the manifold) and their tangent vectors. But a bare manifold is unlike a Euclidean space in the sense that we cannot define length along the curves and thus have no notion of a straight line. Moreover we cannot single out preferred coordinate systems. In a Euclidean space we could distinguish preferred

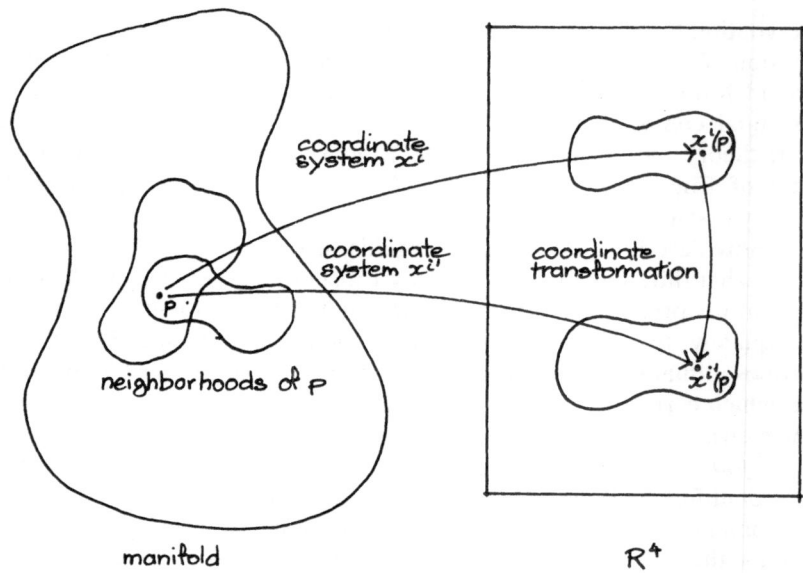

Fig. 1. A manifold with coordinate systems.

Cartesian coordinates, whose coordinate differences correspond to length, only because Euclidean spaces have extra structure enabling the defining of the length of curves.

A covariant, second rank, symmetric, Lorentz signature *metric field*, g_{ab} provides this notion of length (usually called "interval") in the *Entwurf* theory. Its Lorentz signature means that it does not assign lengths isotropically, unlike a Euclidean metric. It assigns positive lengths to curves in one direction, now identifiable as the "time-like" direction, and negative lengths to the curves of the other three "space-like" directions. Thus the Lorentz signature metric gives spacetime its light cone structure. Time-like curves are the possible trajectories or world lines of real non-zero rest mass particles. The null length curves forming the light cones are possible trajectories of light. The metrical length of a curve is given by the integral

$$\int \sqrt{g_{ac} V^a V^c} \, dl$$

along the curve, where V_a and V_c are the curve's tangent vector and l the associated path parameter.

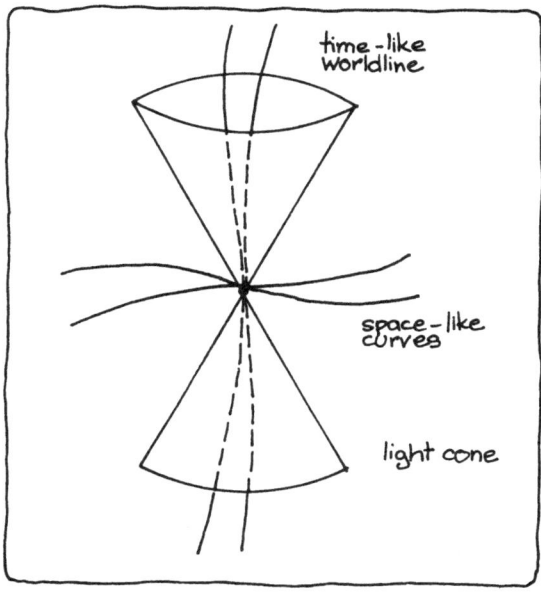

spacetime manifold

Fig. 2. Light cone structure of spacetime.

Gravitation and metrical curvature. The world lines of particles in free fall are time-like geodesics, curves of extremal interval, the analogue of Euclidean straights. If the metric is flat, we have a Minkowski spacetime, the case of special relativity. Particles in a Minkowski spacetime with initially parallel world lines never approach or diverge, analogous to the behaviour of parallel straights in a Euclidean space. This will no longer be the case if the metric tensor has non-vanishing curvature. Free particles with initially parallel world lines might now approach one another. This would be taken to be due to a gravitational action and the non-vanishing curvature of the metric associated with the presence of a gravitational field.[3]

Transformation law for tensor components. A second rank, covariant tensor, such as the metric tensor, can be represented uniquely in a given coordinate system by a 4×4 matrix of its components, g_{im}, where i, $m = 0, 1, 2, 3$.[4] The components of all such tensors obey the following transformation law under change of coordinate system. If any such tensor G_{ab} has components G_{im} in coordinate system x^i and com-

Minkowski spacetime.
Vanishing metrical curvature.
No gravitation

Spacetime with non-vanishing
metrical curvature.
Gravitation

Fig. 3. Gravitation and curvature.

ponents $G_{i'm'}$ in the new coordinate system $x'^{i'}$ (i', $m' = 0, 1, 2, 3$) *at
the same point in the manifold*, then

$$G_{i'm'} = \frac{\partial x^i}{\partial x^{i'}} \frac{\partial x^m}{\partial x^{m'}} G_{im} \tag{1}$$

where summation over repeated indices i' and m' on the right hand
side is implied in accord with the Einstein convention introduced in
Einstein (1916, p. 122).

This transformation law will figure prominently in the story to
follow. Notice in particular that if a tensor G_{ab} has all zero components
in one coordinate system x^i at p:

$$G_{im} = 0$$

then it follows immediately from the above law that it will have zero
valued components in any other coordinate system $x^{i'}$ at p:

$$G_{i'm'} = 0$$

Such a tensor is a zero tensor.

The *stress energy tensor* T_{ab} is another field in spacetime which we
need consider for what follows. It is a second rank, covariant tensor like
g_{ab} and represents the energy and momentum of all non-gravitational
forms of matter, such as electromagnetic fields or dust clouds.

As far as the above details are concerned, the *Entwurf* theory did not differ from the completed general theory of relativity. In fact the two theories agree in all but one essential aspect. The exception is crucial. The metric tensor takes the place of the scalar gravitational potential φ of Newtonian gravitation theory. As a result Einstein required in both the *Entwurf* and his final general theory that the metric tensor enter into a field equation analogous to Poisson's equation

$$\Delta \varphi = 4 \pi G \rho$$

in Newtonian theory. That equation was required to have the form

$$G_{ab} = k T_{ab} \tag{2}$$

where k is a constant, analogous to the Newtonian gravitation constant G; T_{ab} is the stress energy tensor, analogous to the source mass density ρ of Poisson's equation; and G_{ab} is the gravitation tensor, which is constructed out of any combination of the metric tensor and its first and second derivatives, and which is linear in the second derivatives. This tensor is the analogue of the Laplacian of the Newtonian gravitational potential, $\Delta \varphi$.

The derivatives in question here are derivative of the components of the metric tensor with respect to the coordinates. Notice that the above constraint appears to allow very many possible gravitation tensors. This freedom is illusory, however, for if g_{im} is a tensor, then it does not follow for example that its derivatives with respect to coordinate x^n, that is $g_{im,\,n}$, will also be a tensor. $g_{im,\,n}$ cannot represent a tensor since it can readily be confirmed that it does not satisfy a transformation law analogous to (1). It turns out to be very hard to combine the coordinate derivatives of the metric tensor to yield a new tensor. The only relevant possibilities are the metric tensor itself, the Riemann curvature tensor $R^a{}_{bcd}$, its contractions R_{ab} (the Ricci tensor) and R, and tensors formed from them. The last include the Einstein tensor, $R_{ab} - \frac{1}{2} g_{ab} R$. It is now well known that the addition of the requirement of energy momentum conservation is sufficient to force the choice of G_{ab} as the Einstein tensor, the gravitation tensor of Einstein's final theory of November 1915. (He then ignored the possibility of an additive cosmological term proportional to g_{ab}.)

Einstein and Grossmann clearly knew in 1912 and 1913 that the obvious place to look for a gravitation tensor was in the contractions of the Riemann curvature tensor. They even considered the Ricci tensor as a gravitation tensor, which would have given the final field equations at least in the source free case. But, they decided, this choice did not yield the correct Newtonian limit in the case of weak, static fields and, worse than that, they could find no acceptable gravitation tensor. An acceptable account of how they arrived at this conclusion has only recently become available. See Norton (1984) and in shortened form Norton (1985a) and (1985b). There I also describe the three year odyssey of compounded error upon which Einstein embarked and which culminated in a breathless and dramatic discovery of the final field equations in November 1915. The hole argument arose as one of the episodes of this odyssey in the following way.

To deal with their failure to find a gravitation tensor, Einstein and Grossmann took a desperate measure. They had found what they believed to be an acceptable quantity to stand for G_{ab} in equation (2). But that quantity was not a tensor, since the matrix of its components did not transform according to (1) for all coordinate transformations. As a result they distinguished two types of tensor:

(a) those whose components transformed as tensors under arbitrary coordinate transformations;
(b) those whose components transformed as tensors under some limited set of coordinate transformations.

To be generally covariant, the theory would have had to have a gravitation tensor of the first type, but Einstein and Grossmann offered a tensor of the second type. It followed that the field equations of the theory held only in a restricted set of coordinate systems. They soon began work on the problem of determining precisely how large this set was. After April 1914, with his move to Berlin, Einstein had to work on this problem alone. I have conjectured (Norton, 1984, p. 295) that it was in the process of the variational calculations involved that Einstein hit upon a marvellous way of converting failure into success. That was the hole argument.

3. THE HOLE ARGUMENT

The hole argument purported to demonstrate that any generally co-variant gravitational field equations in the context of the *Entwurf* theory would violate physical determinism in a severe and striking manner.[5] It aimed to show that if one had a matter distribution with a matter free spacetime neighbourhood (which Einstein called the "hole") and with the gravitational field specified everywhere outside the hole, then generally covariant field equations would be unable to determine uniquely the gravitational field within the hole, no matter how small the hole. Naturally this provided much comfort to Einstein, who could now regard his failure to find generally covariant field equations as unimportant. He need not doubt that such field equations were possible, but there was no point in pursuing them since they would be physically uninteresting.[6]

The hole argument was published four times by Einstein. In order of publication dates, they were Einstein and Grossman (1913b), pp. 260—261;[7] Einstein (1914a), p. 178; Einstein and Grossmann (1914), pp. 217—218; and Einstein (1914b), pp. 1066—1067. The first three of these were essentially the same. I quote the second:[8]

If the reference system is chosen quite arbitrarily, then in general the g_{mn} cannot be completely determined by the $\mathbf{T}_{sn}$. For, think of the $\mathbf{T}_{sn}$ and g_{mn} as given everywhere and let all $\mathbf{T}_{sn}$ vanish in a region Φ of four dimensional space. I can now introduce a new reference system, which coincides completely with the original outside Φ, but is different to it inside Φ (without violation of continuity). One now relates everything to this new reference system, in which matter is represented by $\mathbf{T}'_{sn}$ and the gravitational field by g'_{mn}. Then it is certainly true that

$$\mathbf{T}_{sn} = \mathbf{T}'_{sn}$$

everywhere, but unlike them the equations

$$g'_{mn} \neq g_{mn} \tag{3}$$

will definitely not all be satisfied inside Φ. The assertion follows from this.

If one wants a complete determination of the g_{mn} (gravitational field) by the $\mathbf{T}_{sn}$ (matter) to be possible, then this can only be achieved by a limitation on the choice of reference systems.   [Einstein's italics]

Reduced to its essentials, Einstein's argument appears to run as follows:

1. Consider a metric within the matter-free hole with components g_{mn} in some coordinate system. The metric satisfies the generally covariant source free field equations $G_{mn} = 0$, for some boundary condition specification of the metric and source matter distribution everywhere outside the hole.

2. Introduce a new coordinate system within the hole which agrees smoothly with the original coordinate system outside the hold. In the new coordinate system, the components of the gravitation tensor still vanish, i.e. $G'_{mn} = 0$, since G_{mn} is a zero tensor (in accord with

the discussion in Section 2). Thus the new components of the metric tensor g'_{mn} still satisfy the source free field equations.

3. Therefore we have a case of a unique boundary condition outside the hole but two distinct fields within, both satisfying the field equations.

Einstein's argument seems to rest on a simple beginner's blunder. It is certainly the case that the *components* of the metric tensor will differ in the new and old coordinate system within the hole, so that the equality of (3) will fail within the hole. But the failure of this equality does not mean that one has arrived at a different metric tensor as is claimed in step 3. Rather we only repeat the well known result that different matrices of components can represent the same metric in different coordinate systems.

Whilst it is difficult to imagine that Einstein could commit such a beginner's blunder repeatedly on a question which had his devoted attention for nearly three years, many commentators have been unable to resist convicting him of it. The most recent is Pais, 1982, pp. 221—222. What makes this 'blunder account' untenable is the footnote Einstein appended to the sentence containing equation (3). (An equivalent footnote did not appear in the first or third versions of the argument cited.) It read

The equations are to be understood in such a way that each of the independent variables x'_n on the left-hand side are to be given the same numerical values as the variables x_n on the right-hand side.

In the blunder account there is simply no good reason for Einstein to insist on this perverse way of reading the equation. The only reading of the hole argument compatible with it is one in which the transformation introduced in step 2 is understood in the active sense, in which case Einstein's argument becomes far from trivial.

I now review the active and passive view of transformations. The coordinate transformations discussed in the last section are generated from a smooth map from R^4 to R^4 which assigns the 4-tuple x'^n to the 4-tuple x^n. This map can be used in two ways:

Passive view: Coordinate transformation. The map is used to generate a new spacetime coordinate system x'^n from x^n. That is, it is used to relabel the points of the manifold with different coordinates.

Active view: Point transformation. The map is used to generate a another map *in the manifold* which will smoothly assign points in the manifold to other points in the manifold. Represent this induced map by h. Then the point p with coordinates x^n will be mapped onto the point hp with coordinates x'^n *in the same coordinate system*. If h is invertible and both it and its inverse are continuous and infinitely differentiable — which is the case usually dealt with — then h is called a diffeomorphism.

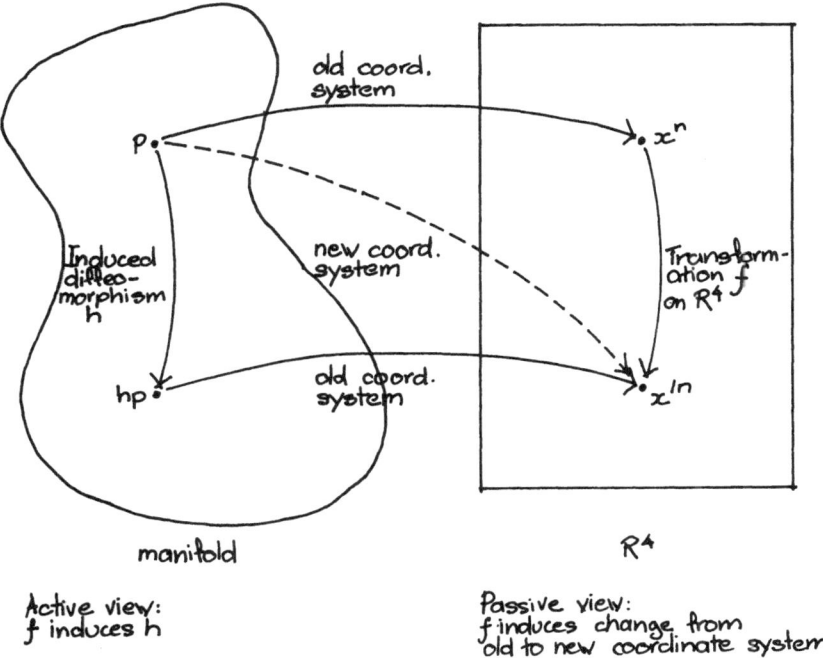

Fig. 4. The active and passive view of transformation.

Each h induces another map, h^*, the *carry along*, which maps structures defined on the manifold at a point p to structures defined on the manifold at hp. Thus h^* defines a carried along coordinate system. The carried along coordinates h^*x^n at hp are defined naturally by the requirement that they be numerically equal to the coordinates of x^n of p. Similarly the carried along metric h^*g_{ab} is defined by the requirement that:

The components of the carried along metric h^*g_{ab} at hp in the carried along coordinate system at hp are numerically equal to the components of the original metric g_{ab} at p in the original coordinate system.

Thus if primed indices represent the carried along coordinate system and unprimed indices the original coordinate system, this amounts to requiring

$$(h^*g)_{m'n'}(hp) = g_{mn}(p) \tag{4}$$

The inverse of h^* is the "pull back".

Diffeomorphism represents the gauge freedom of tensor field equations. Recall the earlier result that if a tensor has all zero valued components in one coordinate system, then it has all zero valued components in all coordinate systems and is the zero tensor. Thus it follows from the above rule that the carry along of a zero tensor will still be a zero tensor. Therefore if a metric tensor g_{ab} satisfies a tensorial gravitational field equation $G_{ab} = 0$, it then follows that the carry along of g_{ab} will also satisfy the field equation. For the carry along of G_{ab} will still vanish.

We now ask how to test whether h^*g_{ab}, the carry along of a metric tensor g_{ab}, will be the same tensor as g_{ab}, the original metric tensor. We begin with equation (4), which does not allow immediate comparison because the matrices of components on either side of the equation belong to different coordinate systems. The easiest way to compare the carry along h^*g_{ab} and the original g_{ab} is to transform the components of the carry along in (4) from the carried along coordinate system back to the original coordinate system. To do this, we must carry out the following operation:

> *Algorithm for comparing components of g_{ab} and h^*g_{ab} in the same coordinate system.* We take the matrix $g_{mn}(p)$, which comprises the components of g_{ab} in the original coordinate system, transform it to the new coordinate system x'^n and compare the resulting matrix of components with the components of the original metric at hp. To ensure that we compare metrics at the same point in the manifold (which here is hp), we recall that the matrix of components will be a function of the coordinates and must insist that the comparison be carried out for matrices with equal coordinate values (here the coordinate values of hp).

But this operation is precisely the 'perverse' reading of equation (3) upon which Einstein insisted in the footnote to the hole argument quoted above! Thus the only reasonable conclusion is that Einstein viewed the transformation of the hole argument actively and that this recipe did genuinely yield a new metric, the carry along of the original, and that the failure of the equality in equation (4) shows that the new metric does differ from the original within the hole. Finally it follows from the above discussion that the carried along metric will satisfy the field equations if the original metric already does, so completing Einstein's argument.

Einstein seemed to realize that his first three versions of the hole argument were not transparent. In his fourth and final version, he went to great pains to remedy this defect and in particular to show that he did intend the transformation to be viewed actively.

We consider a finite region of the continuum Σ, in which no material process takes place. Physical occurrences in Σ are then fully determined, if the quantities g_{mn} are given as functions of the x_n in relation to the coordinate system K used for description. The totality of these functions will be symbolically denoted by $G(x)$.

Let a new coordinate system K' be introduced, which coincides with K outside Σ, but deviates from it inside Σ in such a way that the g'_{mn} related to the K' are continuous everywhere like the g_{mn} (together with their derivatives). We denote the totality of the g'_{mn} symbolically with $G'(x')$. $G'(x')$ and $G(x)$ describe the same gravitational field. In the functions g'_{mn} we replace the coordinates x'_n with the coordinates x_n i.e. we form $G'(x)$. Then, likewise, $G'(x)$ describes a gravitational field with respect to K, which however does not correspond with the real (or originally given) gravitational field.

We now assume that the differential equations of the gravitational field are generally covariant. Then they are satisfied by $G'(x')$ (relative to K'), if they are satisfied by $G(x)$ relative to K. Then they are also satisfied by $G'(x)$ relative to K. Then relative to K there exist the solutions $G(x)$ and $G'(x)$, which are different from one another, in spite of the fact that both solutions coincide in the boundary region, i.e. *occurrences in the gravitational field cannot be uniquely determined by generally covariant differential equations for the gravitational field.*

Einstein (1914b, pp. 1066—1067) [Einstein's italics.]

Einstein's use of the "$G(x)$" notation is not standard, but his purpose is clear enough. He carefully acknowledges that a mere coordinate transformation cannot produce a new field — "$G'(x')$ and $G(x)$ describe the same gravitational field." Rather he uses the transformation to generate a new field with the required property. That new field is $G'(x)$, which we can identify as the carry along of the original metric, or, more precisely, the components of the carry along in the original coordinate system. The comparison of $G'(x)$ with $G(x)$ implements exactly the above algorithm for comparing h^*g_{ab} and g_{ab}.

Thus in summary Einstein's hole argument, when read actively, has the force Einstein claimed. It amounts to the following:

1. Consider a metric g_{ab} within the matter free hole. The metric satisfies the generally covariant source free field equations $G_{ab} = 0$, for some boundary condition specification of the metric and source matter distribution everywhere outside the hole.
2. Let h be a diffeomorphism which maps points within the hole to different points within the hole and which smoothly becomes the identity map everywhere outside the hole. Because of the tensor nature (general covariance) of the field equations, the carry along h^*g_{ab} will still satisfy the field equations. But in the general case, the carry along will differ within the hole from the original metric.
3. Therefore we have a case of a unique boundary condition outside the hole but two distinct fields within, both satisfying the field equations.

Einstein concluded that such a violation of physical determinism was unacceptable and that the only way out was to deny use of generally covariant field equations.

4. THE POINT-COINCIDENCE ARGUMENT

Einstein's fourth version of the hole argument was communicated to the Prussian academy in October 1914. A year later Einstein had completely lost confidence in the *Entwurf* field equations and returned in desperation to the search for generally covariant field equations. He communicated a new set of field equations to the Prussian academy on November 4. He submitted a modified version on November 11. The following week on November 18 he submitted his celebrated explanation of the then anomalous motion of Mercury. But he still did not have the modern field equations with the Einstein tensor as gravitation tensor. These field equations — the third set to be offered by him in the course of a month — were communicated to the Prussian academy on November 25.[9]

The following month Einstein wrote to Ehrenfest and reflected upon the momentous events of the past month.[10]

It is comfortable for Einstein. Each year he retracts what he wrote the previous year; now my duty is the extremely sad business of justifying my most recent retraction.

Nowhere in Einstein's frantic communications of November 1915 to the Prussian academy had he explained how the hole argument could be reconciled with his new generally covariant field equations. This task was the "sad business" to which Einstein now turned. He addressed the fourth statement of the argument. (It had appeared as section 12 of Einstein (1914b) and its first three paragraphs were quote above.) Einstein continued:

In §12 of my work of last year, everything is correct (in the first three paragraphs) up to the italics at the end of the third paragraph. One can deduce no contradiction at all with the uniqueness of occurrences from the fact that both systems $G(x)$ and $G'(x)$, related to the same reference system, satisfy the conditions of the grav. field. The apparent force of this consideration is lost immediately if one considers that

(1) the reference system signifies nothing real
(2) that the (simultaneous) realization of two different g-systems (better said, two different grav. fields) in the same region of the continuum is impossible according to the nature of the theory.

In the place of §12 steps the following consideration. The reality of the world-occurrence (in opposition to that dependent on the choice of reference system) subsists *in spacetime coincidence.** For example the intersection

[Footnote] *and in nothing else!

points of two different world lines are real, as is the assertion that they do *not* intersect one another. Those assertions, which refer to physical reality, are not lost then through any (unambiguous) coordinate transformation. If two systems of $g_{\mu\nu}$ (or [more] gen.[erally], variables used for describing the world) are so constituted, that one can obtain the second from the first merely through a space-time-transformation, then they refer to exactly the same thing [voellig gleichbedeutend]. For they have all timespace coincidences in common, i.e. all that is observable. This consideration shows immediately how natural is the requirement of general covariance.   [Italics in the original.]

The argument developed above is the point-coincidence argument, which we can now identify as Einstein's answer to his hole argument. Seen in this context it is clear why it is nearly impossible for modern readers to understand the point-coincidence argument if they only read the well known version given in Einstein (1916), which was quoted above in the introduction.

First, modern readers will not be able to see why there is a need for any such argument at all for general covariance, for it is a minimum requirement of any well formulated modern spacetime theory. In particular there is no hint in Einstein (1916) of the long period of doubt about general covariance which preceded the paper, let alone any mention of the hole argument.[11] Second, it is by no means clear that the argument seeks to establish anything more than the following: a spacetime coordinate system is a smooth but otherwise arbitrary labelling of events with four numbers; therefore no theory can suppose that any coordinate system is distinguished or preferred independently of the other structures defined on the manifold. Why, we must ask, would Einstein seek to derive this entirely straightforward result from contentious assertions about reality being constituted of spacetime coincidences? Finally, it is not clear even after reading Einstein's letter to Ehrenfest, that the point-coincidence does reconcile the hole argument with general covariance.

In short I will show that we can retain these objections only as long as we read the transformation it invoked passively. If we read it actively — and I shall urge that there is good reason to do so — then Einstein's argument becomes cogent and makes a strong case for the results claimed. The argument will be broken up into two steps. The first argues for what I call "Leibniz equivalence"; from it, the second seeks to establish the naturalness of general covariance, now understood in an active and non-trivial sense.

I begin by making the argument more precise. Represent a model of a generally covariant gravitation theory as the ordered triple $\langle M, g_{ab}, T_{ab} \rangle$, where M is a four dimensional manifold, g_{ab} a Lorentz signature metric and T_{ab} a stress energy tensor. In accord with the usual convention, g_{ab} and T_{ab} represent two tensor fields in coordinate free fashion. Of course each tensor can be represented by a matrix of components g_{ik} and T_{ik}. The corresponding model of the theory based on components in coordinate system K is represented by $\langle M, K, g_{ik}, T_{ik} \rangle$.

The basic assertion of the point-coincidence argument is made most clearly towards the end of the passage quoted above from Einstein's letter to Ehrenfest: two systems of spacetime quantities represent the same physical system if they are related by a coordinate transformation, for then each yields identical observables, that is, spacetime coincidences.

Point-Coincidence Argument (Passive Reading)

Thesis: Two models $T_1 = \langle M, K_1, g_{ik}, T_{ik} \rangle$ and $T_2 = \langle M, K_2, g'_{ik}, T'_{ik} \rangle$

represent the same physical system in case T_1 becomes T_2 under coordinate transformation from K_1 to K_2.

Justification: The transformation from K_1 to K_2 preserves spacetime coincidences, which are the only observables.

Under this passive reading, the thesis becomes trivially true. It merely reminds us that coordinate transformations do not alter quantities, but only the matrices of components which represent them. The justification offered does not establish this trivial thesis. It is just irrelevant to it.

If we read the transformation actively — that is, as a diffeomorphism induced by the coordinate transformation — we have:

Point-Coincidence Argument (*Active Reading*)

Thesis (*Leibniz equivalence*): Two models $T_1 = \langle M, g_{ab}, T_{ab} \rangle$ and $T_2 = \langle M, g'_{ab}, T'_{ab} \rangle$ represent the same physical system in case there exists a diffeomorphism h such that the carry along of T_1 is T_2. Then we have

$$\langle M, g_{ab}, T_{ab} \rangle = \langle hM, h^* g_{ab}, h^* T_{ab} \rangle$$

Justification: The diffeomorphism h preserves spacetime coincidences, which are the only observables of the system.

T_1 and T_2 are said to be diffeomorphic. Thus a convenient expression for the thesis of the active point-coincidence argument is

Leibniz equivalence: Diffeomorphic models represent the same physical system.

On this active reading, the argument is far from trivial. A model and its carry along are quite distinct mathematical structures. That they represent the same physical system is a claim which requires justification. A verificationist justification is provided. Observables are claimed to be preserved under the carry along, so that a model and its carry along agree on all observables.[12] To insist that a model and its carry along represent different physical systems, is to insist that there are physical systems which differ in some property, even though there can be no possible observational verification of the difference.[13]

Einstein supports the claim that observables are preserved under carry along by asserting that all observables can be reduced to coincidences, that is, to the relative points of intersection of physical systems. What is not preserved is the locations of these coincidences in the manifold. But these locations are in principle unobservable. Consider, for example, a model in which the system of fields representing a ray of light strikes a system of fields representing a photographic plate at its midpoint. Then that coincidence will be preserved in an arbitrary carry along of the model, even though the coincidence will be located at a different point in the manifold. Moreover that it is located at a different point has no observational consequences at all.

The active reading of the argument (but *not* the passive reading) does release Einstein from the problem of the hole argument. Recall that the indeterminism established by the hole argument was a consequence of our ability to take a solution of a tensor field equation and produce arbitrarily many diffeomorphic replicas, which still satisfied the field equation for the same boundary conditions but were nevertheless distinct from the original. Leibniz equivalence eradicates such indeterminism by asserting that all these diffeomorphic replicas represent the same physical system. The fields within the hole are *mathematically* underdetermined, but not *physically*, underdetermined since the allowed fields all represent the same physical situation. Thus the generation of diffeomorphic copies of the original solution within the hole amounts to the exercising of a gauge freedom akin to that of electrodynamics. Given any solution of Maxwell's equations in terms of a scalar and a vector potential, we can generate arbitrarily many more mathematically distinct solutions by a change of gauge, but each solution still represents the same electric and magnetic field.

What grounds are there for reading the point-coincidence argument activity? There are several. To begin, coherence points directly to the active reading. Only the active reading is not trivial and the justification offered actually relevant to the thesis. Only the active reading does what Einstein claimed, namely, resolve the hole argument. Some of Einstein's contemporaries — Schlick for example[14] — clearly read the point-coincidence argument actively. The two versions of the point-coincidence argument quoted so far are most naturally read passively by modern readers. We cannot allow this to rule out the active reading as Einstein's intended reading. We have already seen in the case of the hole argument that Einstein simply failed to make clear in two of four versions that he intended the transformation to be viewed actively. In another, as we have seen, he flagged this fact just by an opaque footnote. In the fourth and final version he made his intention clear only by use of a clumsy and non-standard notation.[15]

The strongest evidence for the active reading arises from the scepticism of another of Einstein's contemporaries. Ehrenfest was not convinced by Einstein's letter of 26 December 1915 of the admissibility of general covariance. He presented Einstein with a counterexample in a letter which I believe is no longer extant. Einstein responded in a letter to Ehrenfest of 5 January 1916 (EA 9 372). Reconstructing the counterexample from Einstein's response, it dealt with the system of a star, an aperture and a photographic plate, illuminated through the aperture by the star. Einstein explained that "Your difficulty has its root in the fact that you instinctively treat the reference system as something 'real'." He then continued:[16]

Your example somewhat simplified: you consider two solutions with the same boundary conditions at infinity, in which the coordinates of the star, the material points of the aperture and of the plate are the same. You ask whether "the direction of the wave normal" at the aperture always comes out the same. As soon as you speak of "the direction of the wave normal *at the aperture*," you treat this space with respect to the functions $g_{\mu\nu}$ as an infinitely small space. *From this and the determinateness of the coordinates of the aperture it follows that the direction of the wave normal AT THE APERTURE for all solutions* are the same.[17]

Fig. 5.

This is my thesis. For more detailed explanation [I offer] the following. In the following way you recover all solutions allowed by general covariance in the above special case. Trace the above little figure onto completely deformable tracing paper. Then deform the tracing paper arbitrarily in the plane of the paper. Then make a carbon copy back onto the writing paper. Then you recover e.g. the figure

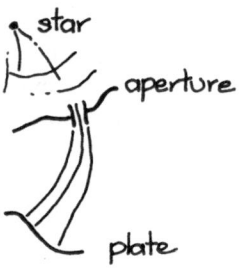

Fig. 6.

When you relate the figure once again to orthogonal writing paper coordinates, the solution is mathematically different from the original, and naturally also with respect to the $g_{\mu\nu}$. But physically it is exactly the same, since the writing paper coordinate system is only something imaginary. The same point of the plate always receives the light. If you carry out the distortion of the tracing paper only in the finite and in such a way that the picture of the star, the aperture and the plate do not lose continuity, then you recover the special case to which your question relates.

The essential thing is: as long as the drawing paper, i.e. "space", has no reality, then there is no difference whatever between the two figures. It [all] depends on coincidences
. . .   [Italics in the original.]

Einstein makes clear here that he intends the transformation used in the point-coincidence argument to be read actively as a diffeomorphism. This diffeomorphism is represented appropriately by a distortion of the tracing paper. Its carrying along of structures is represented very graphically by the carrying along of lines of a drawing by the distortion. The comparison of the two structures in question is clearly intended to be carried out in the same coordinate system, the orthogonal system of the writing paper, as required in an active (but not passive) reading of the argument. Einstein then continues in the following paragraph to give an *active* account of what it is for a theory to lack general covariance:

If the equations of physics were not generally covariant, then you certainly could not carry out the above argument; but relative to the writing paper system the same laws would not hold in the *second* figure as in the first. Then to this extent both would still not be equally justified. This difference falls away with general covariance.
                                                                [Italics in the original.]

In more modern terms we would say that a theory is generally covariant in the active sense if it satisfies the following condition:

General covariance of a theory (active reading): If T is a model of the theory, then a carry along of T under an arbitrary diffeomorphism is also a model of the theory.

Since the relation of being diffeomorphic is an equivalence relation, it follows that the set of models of a generally covariant theory can be divided into equivalence classes of diffeomorphic models. It is important to see that the requirement of general covariance is not the same as Leibniz equivalence. The former provided for the existence of equivalence classes of diffeomorphic models within the set of models of the theory. The latter requires that each member of a given equivalence class represents the same physical system.

For comparison, we can formulate the general covariance of a theory in the passive sense as follows. The formulation is specific to theories with models of the form $\langle M, g_{ab}, T_{ab} \rangle$. but its generalization is obvious.

General covariance of a theory (passive reading). If $T_1 = \langle M, K_1, g_{ik}, T_{ik} \rangle$ is a model of a theory, then so is any model $T_2 = \langle M, K_2, g_{i'k'}, T_{i'k'} \rangle$ where $g_{i'k'}$ and $T_{i'k'}$ are the matrices of components produced by transforming g_{ik} and T_{ik} from K_1 to K_2.

The two requirements of general covariance are not equivalent. It is easy to find examples of theories which satisfy one requirement but not the other. Consider, for example, a version of special relativity whose sole model is a particular Minkowski spacetime $\langle M, n_{ab} \rangle$, where M is a four dimensional manifold and n_{ab} a Minkowski metric. In component terms it has a set of models, each of the form $\langle M, K, n_{ik} \rangle$, which contains just all coordinate systems K defined on M and all the component representations of n_{ab}. This theory is generally covariant in the passive sense. But the theory is not generally covariant in the active sense since by stipulation none of the diffeomorphic replicas of $\langle M, n_{ab} \rangle$ are models of the theory.

Fortunately the two requirements agree in a number of important cases. For example, consider a general relativity-like gravitation theory whose models are the set of all triples $\langle M, g_{ab}, T_{ab} \rangle$ which satisfy the field equation

$$H_{ab} = G_{ab} - kT_{ab} = 0 \tag{5}$$

were G_{ab} is some generally covariant gravitation tensor. The theory is obviously generally covariant in the passive sense. It is also generally covariant in the active sense. For the H_{ab} tensor of any model is the zero tensor. Therefore the H_{ab} tensor of the carry along under arbitrary diffeomorphism of the model will also be a zero tensor, given our earlier result that the carry along of a zero tensor is always a zero tensor. Note that the active general covariance of this class of theories, restricted to the case in which g_{ab} is source free, was the central result discussed in Section 3 in the context of the hole argument. Writing the field equation in the form of (5), enables us to drop the restriction to the source free case.

If we read general covariance actively, we can now take the final step and see how Einstein's argument proceeds from Leibniz equivalence to general covariance. From Leibniz equivalence we have: given a model of a theory which represents some physical system, we can generate arbitrarily many diffeomorphic replicas which represent the observables of the same system equally well. The requirement of general covariance "is a natural one," to quote Einstein (1916, p. 117), since it just allows that all of these diffeomorphic replicas are also models of the theory. Without Leibniz equivalence, the requirement of general covariance is not at all natural. It is simply disastrous, as the Einstein of 1915 and 1916 well knew. For the diffeomorphic copies admitted by general covariance need no longer represent the same physical system and one arrives immediately at the radical indeterminism of the hole argument.

5. WHAT DO THE HOLE AND POINT-COINCIDENCE ARGUMENTS ESTABLISH?

Leaving aside the historical issues, let us ask what are we warranted to conclude from these two arguments. I think there is only one clear and unambiguous conclusion which we can draw and which has direct impact on the application of general relativity: Leibniz equivalence. This equivalence is now incorporated as a matter of course into some of the better modern texts on general relativity, although there is no acknowledgement of Einstein's original adoption of it. (See Hawking and Ellis (1973, p. 56), Sachs and Wu (1977, p. 27) and Wald (1984, p. 438).)

But Einstein clearly felt in 1915 and 1916 that some kind of conclusion about the reality of spacetime could also be recovered. I consider three possibilities. The first is literally Einstein's proposal of 1915 and 1916, the non-realist proposal, which I will argue is *not* established by Einstein's arguments. The second is a proposal by John Stachel and the third is due to John Earman and myself.

Non-Realism About Spacetime

Einstein urges (1916, p. 117) that we conclude that general covariance "takes away from space and time the last remnant of physical objectivity." Non-realism about spacetime is the direct reading of Einstein's assertion and presumably the one he indended. It claims that the term "spacetime" (or correspondingly "space" and "time") have no referen⁺ in the physical world. The claim should be tightened just a little, since what is really at issue is not whether the English word "spacetime" has a referent, but a theoretical structure known as "spacetime" in general relativity. Here I take the manifold to be that structure and the non-realist claim about spacetime to be that the manifold refers to nothing in the physical world.

But what supports the claim is not clear. General relativity read literally posits that the actual world and other possible worlds are represented by spacetime manifolds with fields, such as g_{ab} and T_{ab}, defined on them. In the model representing the actual world, these fields refer to real physical fields. Correspondingly the points of the manifold refer to real physical events.

The hole and point-coincidence arguments complicate the issue by making it impossible to determine which point of the manifold refers to which physical event without considering the fields defined on the manifold. The trouble is that according to Leibniz equivalence the same manifold can figure in two different (diffeomorphic) models which represent the same physical system. Imagine for example that the point p of the manifold M refers to some event — say the collision of two cars — in the model $\langle M, g_{ab}, T_{ab} \rangle$. Then the same point p will in general not refer to the same event in the diffeomorphic model $\langle M, h^*g_{ab}, h^*T_{ab} \rangle$. Rather the different point hp will refer to that event.

This difficulty however does not establish the non-realist claim. The relativization of reference is not the same as the elimination of reference entirely.

Spacetime Events Lose Individuation

The loss of individuation of spacetime events is the basic conclusion which Stachel urges we draw from the hole and point-coincidence arguments.[17a] As I understand it, Stachel's conclusion amounts to the difficulty mentioned above: the physical events to which points of the manifold refer cannot be determined from manifold structure alone. Their reference is determined by the fields defined on the manifold.

Recall that a spacetime coordinate system gives us a numerical labelling which enables us to distinguish points of the manifold from one another. So presumably Einstein had in mind such a loss of individuation when he sought to clarify his view to his correspondents in 1915 and 1916. He stressed to Ehrenfest (26 December 1915) that "the reference system signifies nothing real" and similarly to Besso (3 January 1916) that the coordinate "system K has no physical reality."[18] (Speziali, 1972, pp. 63—64) He returned to precisely this point in his letter of 6 January 1916 to Ehrenfest, describing a failure to grasp it as the root of Ehrenfest's objection to general covariance.

Thus Stachel (1985, Section 4) writes of the hole argument:

The main difficulty here was to see that the points of the space-time manifold (the "events" in the physical interpretation) are not individuated *a priori* but inherit their individuation, so to speak, from the metric field.

Stachel's response (1985, Section 6) to this loss of individuation is to cease representing physical events by points of the manifold in the case of spacetime theories without absolute objects, such as general relativity. In this case he represents physical events by structures in the fibre bundle formed from the manifold and geometric objects definable on it. (Specifically they are a set of maps from the manifold into cross-sections of the bundle.) In this way, the ambiguity of reference can be avoided since the geometric objects which determine that reference are automatically incorporated in the structure. I refer the reader to Stachel's paper for details of this construction and of his general proposal concerning spacetime theories with and without absolute objects.

Refutation of Spacetime Substantivalism

John Earman and I (Earman and Norton, forthcoming) have argued that the hole and point-coincidence arguments amount to a decisive refutation of the doctrine of spacetime substantivalism for a large class of spacetime theories. When the arguments are suitably generalized, that class includes general relativity as well as spacetime formulations of Newtonian theory and special relativity. I limit the discussion here to the case of general relativity.

According to this doctrine, spacetime is held to have an existence independent of anything it contains. The doctrine is best known through Newton's views towards absolute space and absolute time, whose properties are asserted to be entirely independent of the matter they contain. In particular one can have Newtonian absolute spaces and times devoid of matter. An exactly analogous formulation of the doctrine is not possible in the spacetime case within general relativity. For general relativity posits that every spacetime has both a manifold and a metric. By hypothesis, we cannot have a spacetime, understood to be the manifold, without the metric field it must contain. This renders spacetime substantivalism false by hypothesis in general relativity. Clearly this analysis resolves the question too cheaply, for spacetime substantivalism is not usually regarded as analytically false in general relativity.[19]

At this point the natural move is to seek an acceptable reformulation of the doctrine of spacetime substantivalism. Is it captured, we might ask, by the assertion that spacetime is not reducible to other structures; or that we must quantify unavoidably over spacetime events; or in Stachel's notion of no independent individuation of the points of the manifold? Fortunately we do not need to embark on this laborious quest. For our purposes it suffices that spacetime substantivalists must all agree on a simple acid test. The test is best known through Leibniz's challenge to Newtonian space substantivalists: would God have created a different universe if he had placed all the masses in it reversed East to West, but otherwise preserving all relations between them? Newtonian space substantivalists must concede that the new universe would be different to the old one, since the bodies in the new one are at quite different spatial locations, even though there would be no observable difference between the two universes.

The spacetime analogue of reflecting systems of masses East-West in space is a carry along by diffeomorphism over the manifold. Correspondingly, spacetime substantivalists, irrespective of the precise formulation of their views, must agree that diffeomorphic models of a spacetime theory represent different physical systems, for the fields are now located at different points in the manifold. We can express this by:

Leibniz test for spacetime substantivalism: Spacetime substantivalists must deny Leibniz equivalence.

But if spacetime substantivalists must deny Leibniz equivalence, then they face dire consequences. The hole argument forces them to agree that general relativity, with generally covariant gravitational field equations, is subject to what Earman and I call "radical local indeterminism". That is, the metric field within any neighbourhood of the spacetime manifold, no matter how small, is not uniquely determined by even the most complete specification of the fields outside that neighbourhood.[20] And in the case of the point-coincidence argument, they must insist that it is possible for there to be distinct systems which no possible observation could distinguish.

Thus we can summarize the import of the hole and point coincidence arguments for spacetime substantivalists in the form of two dilemmas:

Indeterminism dilemma (*Hole argument*): Spacetime substantivalists must either
(a) accept radical local indeterminism in general relativity, or
(b) deny their substantivalism.

Verificationist dilemma (*Point-coincidence argument*): Spacetime substantivalists must either
(a) accept that there are distinct systems which are observationally indistinguishable, or
(b) deny their substantivalism.

It is hard to imagine that even the most hardened of spacetime substantivalists could cling onto their doctrine in the face of these dilemmas. Perhaps they may do so in the case of the verificationist dilemma, given that verificationism is no longer fashionable. But surely the spectre of radical local indeterminism in the other dilemma is far too high a price to pay for a doctrine that adds nothing predictively to general relativity.

6. FROM NON-REALISM TO REALISM

If the Einstein of 1915 and 1916 held to non-realism about spacetime, he did not retain this belief for very long. By 1920 he had clearly shifted from non-realism (spacetime has no existence) to antisubstantivalism (spacetime has no existence *independent* of the fields it contains). This was a fortunate development since, as we have seen, the former view was not supported by his arguments, whereas the latter is most strongly supported. He wrote:

There can be no space [spacetime] nor any part of space without gravitational potentials; for these confer upon space its metrical qualities, without which it cannot be imagined at all. The existence of the gravitational field is inseparably bound up with the existence of space. (Einstein, 1920, p. 21)

Since non-realist claims about spacetime disappeared from Einstein's writings from this time onwards, I conjecture that this antisubstantivalism was the conclusion drawn ultimately by him from the hole and point-coincidence arguments and general covariance — and perhaps even what was intended all along in his 1915 and 1916 non-realist remarks.

Antisubstantivalism appears frequently in Einstein's writings of the 1950's, even though sometimes it appears in the form of the slogan, no space without metric field. (See Einstein (1953) and, in his correspondence, letters to D. W. Sciama, 28 December 1950 (EA 20 469), to G. Sandri, 24 June 1950 (EA 20 449) and to M. Fischler, 9 September 1954 (EA 11 023).) The best known version of the claim is in the 1952 appendix, "Relativity and the Problem of Space," to Einstein (1917), his popular exposition of relativity theory (p. 155):

In accordance with classical mechanics and according to the special theory of relativity, space (space-time) has an existence independent of matter or field . . . . On the basis of the general theory of relativity, on the other hand, space as opposed to "what fills space," which is dependent on the coordinates, has no separate existence . . . If we imagine the gravitational field, i.e., the functions g_{ik}, to be removed, there does not remain a space of the the type (1) [Minkowski spacetime], but absolutely *nothing*, and also no "topological space." For the functions g_{ik} describe not only the field, but at the same time also the topological and metrical structural properties of the manifold . . . . There is no such thing as an empty space, i.e. a space without field. Space-time does not claim existence on its own, but only as a structural quality of the field.
[Einstein's italics.]

Einstein's stress here on viewing spacetime as a property of the metric field rather than an independent entity makes it possible for us to characterize the view as something slightly stronger than antisubstantivalism. The view is a relational view of spacetime. That is, spacetime arises as an abstraction from the spatiotemporal properties of other things.

Readers of Einstein (1916) may well be able to see how he could modify his non-realism about spacetime to antisubstantivalism. But surely they will be surprised by the hopes expressed in 1930 by an Einstein deeply embroiled in the search for a unified field theory (Einstein, 1930, p. 184):

We may summarize in symbolical language. Space, brought to light by the corporeal object, made a physical reality by Newton, has in the last few decades swallowed aether and time and seems about to swallow also the field and the corpuscles, so that it remains as the sole medium of reality.

What complicates the whole discussion and lends an aura of contradiction to it is the fact that the term "spacetime" (or else "space" or "time") refer to different theoretical structures in different contexts.

In the analysis of spacetime substantivalism given by Earman and myself, spacetime is identified with the manifold. But it is harder to determine precisely what theoretical structure stands for "spacetime" in various of Einstein's writings. Presumably the Einstein of 1916 took spacetime to be the manifold. But the Einstein of 1930, who expresses the hope that space would become "the sole medium of reality," surely took space to be manifold plus metric or manifold plus the geometric structure of his unified field theory.

In the context of his slogan, no space without metric field, Einstein seems to take "space" to be certain spatiotemporal properties of a manifold with metric. Consider for example Einstein's (1953) remark that in a generally covariant field theory, "that which constitutes the spatial character of reality is then simply the four-dimensionality of the field"; and his remark in a letter to D. W. Sciama of 28 December 1953 (EA 469): "'Space' exists only as the continuum property of physical reality (field), not as a kind of container with independent existence, into which physical things are placed:" The difficulty with this reading of "space" is to make precise exactly which properties are in question. Stachel (Stachel, 1985, Section 6) makes the only attempt of which I am aware to deal with this problem.

The story of Einstein's change of viewpoint from non-realist (or perhaps just antisubstantivalist) to realist about spacetime is a fascinating one. Since it involves issues well beyond the scope of this paper, I can only mention a few of its highlights here.

The Einstein of 1915 and 1916, who rejoiced in the loss of objectivity of space and time, had by his own later admission (Einstein, 1946, p. 27) not appreciated fully the picture of reality demanded by a true field theory. Then he had sought to explain the origin of inertial forces solely in the interactions of bodies, alloting to fields a purely intermediate role. (Einstein, 1916, Section 2) The crucial insight, which he attributed to Mach, was that an epistemologically satisfactory mechanics could not admit inertial spaces as causes.

H. A. Lorentz, whom Einstein revered as a father figure, must have played some role in changing Einstein's mind. They corresponded extensively in the 1910's over relativity. Einstein conceded to him (15 November 1919, EA 16 494) that he had been hasty in concluding the non-existence of the aether from special relativity. He should only have concluded the non-reality of an aether *velocity*. The aether belonged in general relativity in so far as that theory posited spacetime as a bearer of physical qualities. Those qualities are the metric field. Thus Einstein began to portray general relativity as an aether theory and the term aether figured prominently in some of the titles of Einstein's papers. (See Einstein (1918, p. 702; 1920; and 1924).)

In particular, in a 1920 lecture at Leiden read before Lorentz, Einstein conceded that the Machian analysis of the origin of inertia no longer leads us to seek an account of inertia solely in the interactions of distant bodies, since we should no longer be prepared to posit action at a distance. Rather we are led to an aether, which "not only *conditions* the behavior of inert masses, but *is also conditioned* by them. Mach's idea finds its fullest expression in the aether of the general theory of relativity." (Einstein, 1920, p. 18; Einstein's italics)

Gradually Einstein replaced the term "aether" by "space" and with it the shift from non-realist to realist view of spacetime completed. Einstein now allowed that his Machian critique did not require a non-realist view of spacetime but the elimination of its preferred causal status. He summarized his changed viewpoint (Einstein, 1927, p. 260):

The general theory of relativity formed the last step in the development of the programme of the field-theory ... Space and time were thereby divested not of their reality but of their causal absoluteness — i.e. affecting but not affected — which Newton had been compelled to ascribe to them in order to formulate the laws then known.

7. CONCLUSION

My story began in 1916 with a rather unconvincing verificationist argument from Einstein for general covariance and an associated non-realist claim about space and time. Even though the argument and claim are much cited and quoted, we found that they are rarely understood, largely because Einstein failed to include in his presentation virtually all the ingredients necessary for this understanding.

, The notion of general covariance at issue was not merely the passive form of invariance of laws under arbitrary coordinate transformation. It was general covariance in the active sense under which any model of a theory belongs to an equivalence class of all possible models diffeomorphic to it. The crucial result was what I called Leibniz equivalence, that each member of one such equivalence class represents the same physical system. Einstein's verificationist argument makes good sense as an argument for Leibniz equivalence. Einstein did not mention the vital link which connected Leibniz equivalence to general covariance. Leibniz equivalence released him from the conclusion of his earlier hole argument, which was that general covariance would lead to a radical and unacceptable form of indeterminism in his gravitation theory. He also did not mention that the threat of this indeterminism in the hole argument could now be turned into an argument for Leibniz equivalence, which to modern eyes is stronger than the verificationist argument he offered. Perhaps Einstein felt that in 1916 an appeal to verificationism would be more readily accepted. Certainly any such expectation was vindicated by the enthusiastic response of such contemporary philosophers as Reichenbach and Schlick.

The non-realist claim about space and time entered only at the last moment of this episode and was dropped by Einstein within five years. We could find no argument within the episode to support this non-realism. Rather we extracted two dilemmas for those who hold to a related view, spacetime substantivalism, the view that spacetime has an existence independent of the fields it contains. These dilemmas force the rejection of that view. We found antisubstantivalist claims concerning spacetime common in Einstein's later work.

Einstein strove to express his ideas as simply and clearly as possible. Unfortunately sometimes his efforts backfired on him and he simplified his ideas to the point that they become unintelligible to even a diligent reader, as we have seen in the case here. I have not addressed the question of whether Reichenbach's or Schlick's reading of Einstein's work suffered from this difficulty. My story ends with a broader challenge to historians and philosophers of science: which other of Einstein's claims and arguments have been misunderstood for this reason?

University of Pittsburgh, U.S.A.

NOTES

* I wish to thank the Hebrew University of Jerusalem, Israel for its kind permission to quote the material in this paper from Einstein's unpublished writings, and Don Howard for discussion and comments on an earlier draft of this paper.

[1] A non-realist holds that "space" and "time" have no referents in the physical world. I would have preferred to use the term "anti-realist", but it has already been used by van Fraassen (van Fraassen, 1980, pp. 9—11) for a view which is agnostic about the existence of these referents.

[2] Einstein objected to this gloss also. For example he corrected Ehrenfest in correspondence of 1919 by noting that the novelty of special relativity in 1905 was not epistemological (non-existence of a resting aether) but empirical (equivalence of all inertial systems with respect to light). He allowed that epistemological demands came into play in 1907 when he commenced work on general relativity. But there too the empirically determined equality of inertial and gravitational mass played a significant role. A. Einstein to P. Ehrenfest, 4 December 1919, EA 9 451. (EA 9 451 refers to the document with control number 9 451 in the duplicate Einstein Archive, Mudd Manuscript Library, Princeton.)

[3] I have argued at length (Norton, 1985c) that this modern view was not Einstein's. He did not associate the presence of a gravitational field just with non-vanishing metrical curvature, but with the presence of a metric of any curvature. Thus a Minkowski spacetime for Einstein was already a special case of a gravitational field bearing spacetime. This was one of the crucial insights gleaned by Einstein from his principle of equivalence.

[4] I follow the usual modern conventions concerning indices. A sub- or superscripted $a, b, c, d, \ldots$ is used to represent the rank and type of a geometric object according to the abstract index convention. Thus g_{ab} represents a second rank covariant tensor. Sub- or superscripted $i, k, m, n, \ldots$ take values 0, 1, 2, 3 and are used to represent matrices of components of geometric objects. Thus a second rank, covariant tensor g_{ab} has components $g_{00}, g_{01}, \ldots, g_{23}, g_{33}$ in some coordinate system. These components are represented by g_{im}, where i, m are understood to take all values 0, 1, 2, 3.

[5] This was not Einstein's first argument against the physical acceptibility of general covariance. He had already argued against it as early as August 1913 on the basis of the limited covariance of the stress energy tensor of the gravitational field. See Norton, 1984, pp. 284—86.

[6] He did insist however that if his *Entwurf* field equations had any physical content, then they must have a generally covariant generalization. Einstein, 1914a, pp. 177—178.

[7] But the argument did not appear in the original separatum of this article, Einstein and Grossmann (1913a).

[8] For notational continuity, I have replaced Einstein's Greek indices by Latin indices both here and in the later version of the argument. Similarly T_{mn} is the stress energy tensor *density*, $\sqrt{-g}\ T_{mn}$, which Einstein denoted with a Gothic $\mathfrak{T}$.

[9] This episode is outlined in Norton (1984), (1985a) and (1985b) in which the first explanation of Einstein's need for three separate version of the field equations in this month is offered.

[10] A. Einstein to P. Ehrenfest, 26 December 1915, EA 9—363. Einstein presents

essentially the same arguments to Besso in slightly briefer form in A. Einstein to M. Besso, 3 January 1916, in Speziali (1972, pp. 63—64).

[11] Obviously I do not mean that Einstein was trying to suppress this episode. It would have been well known to any contemporary who had been following his theory in the literature. There was just no need for him to remind his readers of the embarrassing confusions of the previous three years.

[12] Thus the active reading makes sense of the remark of Einstein to Ehrenfest quoted above, where the passive reading does not: "Those assertions, which refer to physical reality, are not lost then through any (unambiguous) coordinate transformation."

[13] The denial that such observationally indistinguishable systems are different was called "Leibniz equivalence" above, since this was precisely the point Leibniz made to Clark in their celebrated correspondence when he asked how the world would differ if God had placed the bodies of our world in space some other way, only changing for example East into West. (Alexander, 1956, p. 26)

[14] Schlick's version, as quoted in Friedman (1983, p. 23), is very clear and simple — and unambiguously active.

[15] We can draw a useful moral here. If Einstein talks of coordinate transformation but his discussion is incoherent, it is worth considering the possibility that he may really mean the corresponding point transformation.

[16] The two figures shown have been redrawn after the sketches included in the original letter.

[17] I render the double underlining of "at the aperture" by uppercase italics.

[17a] For more details see Stachel 1985. Torretti (1983, Section 5.6) reviews the hole argument, the point-coincidence argument, Stachel's proposal and then offers what I believe amounts to Leibniz equivalence as a prefered alternative.

[18] The other point stressed in both letters was that it is impossible to realize simultaneously two different gravitational fields in the same neighbourhood of the manifold. If Einstein intends that the two fields are diffeomorphic (which is not clear), then I read this remark as a somewhat awkward statement of Leibniz equivalence.

[19] For general discussion of space, time and spacetime substantivalism, see Sklar (1977).

[20] Note that the argument has been generalized by dropping the requirement that the hole be matter free. The construction now requires that both g_{ab} and T_{ab} be carried along by the diffeomorphism and the field equations of form (5) used.

BIBLIOGRAPHY

Alexander, H. R. (ed.): 1956, *The Leibniz—Clark Correspondence*, Manchester Univ. Press.

Earman, J. and Norton, J.: (forthcoming), 'What Price Spacetime Substantivalism? The Hole Story', *British Journal for the Philosophy of Science*.

Einstein, A.: 1913, 'Zum gegenwaertigen Stande des Gravitationsproblems,' *Physikalische Zeitschrift* **14**, pp. 1249—66.

Einstein, A.: 1914a, 'Prinzipielles zur verallgemeinerten Relativitaetstheorie,' *Physikalische Zeitschrift* **15**, pp. 176—80.

Einstein, A.: 1914b, 'Die formale Grundlage der allgemeinen Relativitaetstheorie,' *Preuss. Akad. der Wiss., Sitz.*, pp. 1030—1085.
Einstein, A.: 1915, 'Erklaerung der Perihelbewegung des Merkur aus der allgemeinen Relativitaetstheorie,' *Preuss. Akad. der Wiss., Sitz.*, pp. 831—39.
Einstein, A.: 1916, 'The Foundation of the General Theory of Relativity,' in *The Principle of Relativity*, Dover, 1952.
Einstein, A.: 1917, *Relativity: The Special and the General Theory*, London, Methuen, 1977.
Einstein, A.: 1918, 'Dialog ueber Einwaende gegen die Relativitaetstheorie,' *Naturwissenschaften* 6, pp. 697—702.
Einstein, A.: 1920, 'Ether and the Theory of Relativity' in *Sidelights on Relativity*, New York, Dover, 1983.
Einstein, A.: 1924, 'Uber den Aether', *Schweizerische Naturforschendende Gesellschaft, Ver.* 105, pp. 85—93.
Einstein, A.: 1927, 'The Mechanics of Newton and their Influence on the Development of Theoretical Physics' in *Ideas and Opinions*, New York, Bonanza, pp. 253—261.
Einstein, A.: 1930, 'Space, Ether and the Field in Physics,' *Forum Philosophicum* 1, pp. 180—184.
Einstein, A.: 1946, *Autobiographical Notes*, La Salle and Chicago, Open Court, 1979.
Einstein, A.: 1953, Foreword to Jammer, M., *Concepts of Space*, Cambridge, Harvard Univ. Press, 1954.
Einstein, A. and Grossmann, M.: 1913a, *Entwurf einer verallgemeinerten Relativitaetstheorie und einer Theorie der Gravitation*, Leipzig, B. G. Teubner.
Einstein, A. and Grossmann, M.: 1913b, 'Entwurf einer verallgemeinerten Relativitaetstheorie und einer Theorie der Gravitation,' *Zeitschrift fuer Mathematik und Physik* 62, pp. 225—61.
Einstein, A. and Grossmann, M.: 1914, 'Kovarianzeigenschaften der Feldgleichungen der auf die verallgemeinerte Relativitaetstheorie gegruendeten Gravitationstheorie,' *Zeitschrift fuer Mathematik und Physik* 63, pp. 215—225.
Fine, A.: 1984, 'The Natural Ontological Attitude', in J. Leplin (ed.), *Scientific Realism*, Univ. of California Press, Berkeley, pp. 83—107.
Friedman, M.: 1983, *Foundations of Space-Time Theories*, Princeton Univ. Press.
Hawking, S. and Eillis, G. F. R.: 1973, *The Large Scale Structure of Space-Time*, Cambridge Univ. Press.
Howard, D.: 1984, 'Realism and Conventionalism in Einstein's Philosophy of Science: The Einstein—Schlick Correspondence,' *Philosophia Naturalis* 21, pp. 616—29.
Norton, J.: 1984, 'How Einstein Found his Field Equations,' *Historical Studies in the Physical Sciences* 14, pp. 253—316.
Norton, J.: 1985a, 'Einstein's Struggle with General Covariance,' *Rivista di Storia della Scienza* 2, pp. 181—205.
Norton, J.: 1985b, 'Einstein's Discovery of the Field Equations of General Relativity: Some Milestones,' *Proceedings, 4th Marcel Grossmann Meeting on Recent Developments in General Relativity*, Rome, Italy, 17—21 June, 1985.
Norton, J.: 1985c, 'What was Einstein's Principle of Equivalence?' *Studies in History and Philosophy of Science* 16, pp. 203—246.
Pais, A.: 1982, *Subtle is the Lord . . .: The Science and Life of Albert Einstein*, Oxford, Clarendon.

Sachs, R. K. and Wu, H.: 1977, *General Relativity for Mathematicians*, New York, Springer.

Sklar, L.: 1977, *Space, Time, and Spacetime*, Berkeley, Univ. of California Press.

Speziali, P., ed.: 1972, *Albert Einstein Michele Besso: Correspondance 1903—1955*, Paris, Hermann.

Stachel, J.: 1980, 'Einstein's Search for General Covariance.' paper read at the Ninth International Conference on General Relativity and Gravitation, Jena.

Stachel, J.: 1985, 'What a Physicist Can Learn from the Discovery of General Relativity.' *Proceedings, 4th Marcel Grossmann Meeting on Recent Developments in General Relativity*, Rome, Italy, 17—21 June, 1985.

Torretti, R.: *Relativity and Geometry*, Oxford, Pergamon.

van Fraassen, B. C.: 1980, *Scientific Image*, Oxford, Clarendon.

Wald, R.: 1984, *General Relativity*, Univ. of Chicago Press.

IAN LOWE

MEASUREMENT AND OBJECTIVITY:
SOME PROBLEMS OF ENERGY TECHNOLOGY

It has been almost an axiom of practising scientists and technologists that they deal in objective knowledge, with the arbiter between competing theories being the impartial measurement of a specified variable. In what follows, a range of examples from the broad field of energy technology are considered. The examples chosen are the hazards of low-level ionising radiation, the operating safety of nuclear reactors, the containment of radioactive waste, the life-time of energy resources, and the viability of "alternative" energy sources. Brief references to other controversial measurements — causal links with incidence of cancer, the testing of pharmaceuticals, the safety of the irradiation of food and the efficacy of fluoridating public water supplies — establish that the conclusions drawn from these examples have a much wider generality.

The examples lend little support to the idea of objective measurement. One interpretation, after Weinberg, is that there is a class of complex problems which transcend science, and cannot therefore be answered within the canons of science. According to this view, lapses from objectivity occur only in observations of complex phenomena, or where there is no possibility of a controlled experiment. Even this limited critique poses serious questions about entire scientific disciplines, such as astrophysics. Development of our understanding of the social process of science, however, suggests that the problem is more fundamental. Analysis of the problems of measurement leads inexorably to a questioning of the whole idea of science as an objective exercise.

SCIENCE AND OBJECTIVITY

Science has been referred to[1] as "the true exemplar of authentic knowledge". This view implies that scientific laws and measurements have a degree of permanence setting them apart from other forms of knowledge. Implicit, in its turn, in this view of science is the notion of objectivity; if human subjectivity were to influence measurements, they could not claim this special level of authenticity and permanence. Many scientists still see their profession as engaged in the objective, rational,

189

John Forge (Ed.), Measurement, Realism and Objectivity, 189—206.

socially-neutral and value-free accumulation of truth. I have heard a scientist say in admiration of a distinguished colleague that "anything 'measured by him stays measured".

There are different ways in which the concept of objective measurement can be scrutinised. This essay is not concerned with the incidence of scientific fraud, which has been extensively documented elsewhere[2]. It addresses the more fundamental question of whether it is possible even in principle for scientific measurement to be "objective", that is unbiased or free from personal prejudice. The general question is approached by way of a series of concrete examples in the field of energy technology.

HAZARDS OF LOW-LEVEL IONISING RADIATION

There is, unfortunately, ample documentation of the hazards to the human body of large doses of ionising radiation. At the other end of the dose spectrum, all humans are exposed daily to small quantities of radiation from the earth, the air and even from other humans. The question of the health risks from different low levels of ionising radiation is an obviously-important public health question. It bears on such issues as the acceptability of nuclear power, the safety of medical x-rays and the desirability of permitting the irradiation of food as a pest control measure.

Clearly this is not an area in which it is morally acceptable to conduct controlled experiments. While it might be possible in principle to irradiate different large groups of otherwise matched humans with different doses and observe over time the differences in their suffering, there appears to be little chance of such an experiment being regarded as ethically acceptable.

Even if a society were to emerge in which a minority group were regarded as sufficiently expendable to be the subjects of such experiments, that group would by definition not be a typical cross-section of the society, and so the results could not be validly extrapolated to the majority group.

Even were such an experiment ethically acceptable, it is not clear that it could be carried out satisfactorily. There are different theories of the relationship between radiation dose and damage to the body. Weinberg has shown[3] that an experimental protocol to test on laboratory rodents, the most straightforward theory — that the probability of one specific sort of damage, genetic mutation, is directly proportional to dose — would require the testing of 8×10^9 rodents! Thus the question of the relationship between radiation dose and damage to the body is one which must be measured indirectly or in consequence of unintended exposure, but even then it is not clear that it will be possible to obtain enough data to enable predictions to be made with acceptable degrees of certainty.

Bertell[4] tabulates health effects from various high-level doses of radiation before quoting the criteria used by the U.S. National Council and Radiation Protection and Measurement in developing radiation protection standards:

1. A value judgement which reflects, as it were, a measure of psychological acceptability to an individual of bearing slightly more than a normal share of radiation-induced defective genes.
2. A value judgement representing society's acceptance of incremental damage to the population gene pool, when weighted by the total of occupationally exposed persons, or rather those of reproductive capacity as involved in Genetically Significant Dose calculation.
3. A value judgement derived from past experience of the somatic effects of occupational exposure, supplemented by such biomedical and biological experimentation and theory as has relevance.

As Bertell observes, these criteria implicitly recognise that there is no absolutely safe level of exposure to ionising radiation, and the occupational standards established represent an explicit value judgement in which the unquantifiable health hazards are set against the unquantified benefits to society of having medical isotopes, nuclear power or nuclear weapons. Not only is this value judgement something over which different individuals can legitimately have different views, it is a value judgement which has varied over time. Comparisons are complicated by the advent of the unit of the rem, or roentgen equivalent man, but for these purposes the rem can be regarded as equivalent to the earlier unit of the roentgen. Bertell[5] points out that permissible occupational exposure to ionising radiation in the United States "was set at 52 roentgen (X-ray) per year in 1925, 36 roentgen per year in 1934, 15 rem per year in 1949 and 5 to 12 rem per year from 1959 . . .". The conclusion is clear; there is no possibility even in principle of an objective measurement of the health risk from low levels of ionising radiation. In its absence, standards are set which explicitly represent the collective value judgement of the particular society at a particular time.

In the absence of an objective measurement, different scientists reach different conclusions on such questions as whether the low levels of radiation from the British nuclear establishment at Windscale (now called "Sellafield") could be responsible for the local high rates of leukaemia and Down's Syndrome.

At the extreme end of the spectrum of contributions to this vexed question are activities which have been described as "cooking the books"; Collingridge quotes[6] Hinton's story of his involvement in the British decision to build an experimental breeder reactor at a remote site on the Scottish north coast:

Because of the unknowns we had, from the outset, planned to build the reactor on a remote site ... But why, if we were giving the reactor containment, were we putting it on a remote site? This could only be logical if we assumed that the sphere was not absolutely free from leaks. So we assumed, generously, that there would be 1% leakage from the sphere, and dividing the country around the sites into sectors, we counted the number of houses in each sector and calculated the number of inhabitants. To our dismay this showed that the site did not comply with the safety distances specified by the health physicists. *That was easily put right; with the assumption of a 99% containment the site was unsatisfactory so we assumed, more realistically, a 99.9% containment and by doing this we established the fact that the site was perfect.*   [my emphasis.]

This is admittedly an extreme example, but it allows little room for arguing that the scientific measurement of risk was objective. Since the answer was unacceptably high, it was effectively divided by ten to give a lower figure which "established that the site was perfect".

OPERATING SAFETY OF NUCLEAR REACTORS

Attempts to quantify the operating risk of nuclear power stations have usually come from sources within the industry, and have (perhaps not surprisingly) suggested that the hazards are minimal. Lowrance[7] quotes from the U.S. Atomic Energy Commission report[8] WASH—1400:

From the viewpoint of a person living in the general vicinity of a reactor, the likelihood of being killed in any one year in a reactor accident is one chance in 300 000 000 and the likelihood of being injured in any one year in a reactor accident is one chance in 150 000 000 ...

This impressively precise estimate can be scaled up to give the overall hazard for an entire nation; in this case, it was suggested that a country like the U.S.A. with a projected future complement of 100 reactors might have one person killed and two injured every twenty-five years. The calculation was also claimed to be capable of further refinement to give the risk from any particular type of accident; for example, it was stated that "the most likely core melt accident would occur on the average of one every 17 000 years per plant". As if that heroic precision were not stretching credibility, the report went on to give quantitative estimates of the differing probabilities of various degrees of severity in such accidents:

If we consider a group of 100 similar plants then the chance of an accident causing 10 or more fatalities is 1 in 2500 per year or, on the average, one such accident every 25 centuries. For accidents involving 1000 or more fatalities the number is 1 in 1 000 000 or once in a million years.

It is scarcely necessary to remark that the basis of such astoundingly precise assertions has been challenged; more importantly, the basis was challenged long before such accidents as Chernobyl demonstrated the absurdity of the claims. The whole technique of fault-tree analysis has been thoroughly scrutinised[9] and shown to have two fundamental flaws:

(i)   the technique inevitably assumes that failure will occur as a result of a given sequence of events, whereas actual failures are — arguably, by definition — more likely to result from a sequence of events which was not foreseen;

(ii)   the likelihood of failure is computed by combining the various probabilities for designated events, such as specified failure modes, but many of these separate probabilities are not actually known with the sort of precision implied by the quoted results.

Thus there are serious doubts about the underlying bases of such calculations. The bizarre sequences of events leading to serious nuclear reactor accidents have been documented elsewhere[10]; they include human "errors" so startling that they could, in the words of one leading advocate of nuclear power[11], almost be described as sabotage, as well as unforeseeable equipment failures. There is no possibility even in principle of quantifying all the factors involved, but the desire for a numerical estimate leads to arbitrary assumptions being made. A recent book[12], referring to a 1979 study by the U.K. Nuclear Installations Inspectorate, quotes the then Director of Nuclear Safety as saying that the worst accident which should be considered for planning purposes would involve the release of one per cent of the volatile material in the core. No justification has been devised for this assumption; why not two per cent, or 0.5 per cent?

It is possible, in principle, for the experience of catastrophic reactor accidents eventually to give accurate estimates of the various probabilities of releasing different proportions of the radioactive inventory. Until such experience is obtained, the "estimates" will continue to be more accurately described as "guesstimates". Of course, those giving the assurances of safety often have careers dependent on the future of the nuclear industry, and do have some incentive to be optimistic.

CONTAINMENT OF RADIOACTIVE WASTE

One of the crucial questions in the debate about the safety of nuclear power has been the long-term containment of high-level radioactive waste. Quite large quantities of waste are produced by operating reactors, so that the world now has a large stockpile awaiting treatment and disposal. In 1978, the U.S. House of Representatives Committee on Government Operations noted[13] that the U.S. would have about 20 000 tonnes of spent nuclear fuel in store by 1990, with "no demonstrated technology for permanently and safely disposing of this waste". The waste is a complex mixture of many different radioactive substances, each of which has its own characteristic level of activity and rate of radioactive decay. Because of the long-lived transuranic elements, such as plutonium, it is generally agreed that the waste needs to be isolated from the biosphere for a period greater than 100 000 years. The enormity of this task is clear. The period for which the waste must be contained is not only orders of magnitude longer than any human civilisation has ever endured; it is about the same as the length of time for which humans have existed as an identifiably separate species.

A recent report to the Australian government by its principal scientific advisory body, the Australian Science and Technology Council [ASTEC], dealt with the whole issue of Australia's role in the nuclear fuel cycle[14]. The report contained a section on the management of high-level waste. It noted that there was a general view among nuclear authorities that borosilicate glass is a suitable medium for the long-term containment of waste. It also noted that there has been serious criticism of this view in the scientific literature, mainly because of concerns about the possible leaching of radioactive nuclides from the glass by groundwater:

geochemists have shown that borosilicate glass would disintegrate after burial in the earth if it were ever to come into contact with groundwater at quite modest pressures and temperatures. Such disintegration would be accompanied by loss of some radionuclides into groundwater . . . . Advanced waste forms may be better suited to more demanding conditions, but they require further development.

The synroc ("synthetic rock") programme in Australia and the expedited study of twelve alternative waste forms in the U.S.A. were cited as reflections of a desire to find a better material than glass for long-term disposal. While research proceeds on possible forms of containment for the actual waste material, the overall disposal technology is also being studied. Most work has been based on the assumption that an underground repository will be constructed in a suitably stable part of the earth's crust. A recent U.S. government report revealed that four or five sites in a variety of geological environments are to be evaluated before a smaller number are chosen for more intensive study[15]; the hope was then expressed that it might be possible to identify a possible disposal site by 1987. Meanwhile, however, work is proceeding on other possible disposal technologies, of which the following were described as "the more promising":

Very Deep Hole Concept

One alternative to a conventional repository is stacking high-level waste canisters in very deep holes drilled thousands of feet into very hard rock . . . More information in a number of areas is needed . . . the safe emplacement of wastes may present severe engineering problems, and the number of holes (800 to 1300) may be prohibitive.

Geologic Disposal Beneath the Seabed

The feasibility of emplacing wastes beneath the seabed has been studied for the past five years . . . The main technical problem still to be addressed is the response of the geologic medium (red clay) to heat given off by high-level waste in the first five hundred years . . . Since the desirable seabeds are in international waters, legal and political issues will also need resolution before this disposal option can be tested in situ, or implemented.

Rock Melting

The rock-melting concept for geologic disposal calls for direct waste emplacement in a deep underground hold or cavity. Radioactive decay heat causes the surrounding rock to melt, which in turn dissolves the waste.

Eventually, the waste-rock solution solidifies, trapping the radioactive material . . . Major technical uncertainties include the desired properties of the host environment, chemical and physical interactions of the molten zone with the surrounding rock; rock mechanics during the period of the thermal pulse; and mechanics, data, and models for assessing the transport of radionuclides to the biosphere.

Space Disposal

The currently favored [sic] concept is injection into a circular orbit that is about halfway between the Earth and Venus ... The major technology development requirements pertain to safety, the environmental impact of space launches, and the preparation of the waste ... additional study is needed to ensure that it could withstand launch pad and re-entry accidents.".

In summary, it seems fair to say that the technologies for the disposal of high-level radioactive waste from nuclear reactors are still in the relatively early stages of development, with both the waste form and the disposal technology yet to be resolved. U.S. estimates are that actual disposal may commence between 1997 and 2006.

A survey of the estimates by nuclear authorities of the increased radiation dose to civilians from disposal of radioactive waste shows a wide divergence of views. A Swedish study estimated the increased radiation to be between 0.05 per cent and 0.5 per cent of the natural background level, while a U.S. "worst case" gave radiation levels up to fifty times the background. The ASTEC report argued[16] that the worst case doses "could be prevented by the use of advanced waste forms possessing much greater resistance to leaching".

All the evidence of the current state of development of the technology for disposing of radioactive waste was reviewed in detail by the Australian Science and Technology Council, which then concluded[17]:

we consider that any return of radioactivity to the biosphere can be held to safe and acceptable levels over long periods (up to one million years) so that maximum doses to the most exposed individuals would be a small fraction of natural background levels. We consider, moreover, that the technology required to achieve these objectives is available.

That conclusion is, to put it most charitably, a surprising leap of faith from the evidence assembled and set out in great detail in the report. It is quite remarkable to move from a recognition of the need to evaluate a range of possible sites in the hope of finding one which is suitable, and an awareness of the need to develop waste forms which do not suffer from the weaknesses of borosilicate glass, to a confident statement that radioactivity can be contained for up to one million years with technology which is available. There is, of course, no sense in which the confidence that containment will be secure for up to one million years can be based on a measurement. This is simply an assertion of faith, and should properly be associated with similar affirmations of belief in reincarnation or eternal life.

LIFE-TIME OF ENERGY RESOURCES

Since the "oil shocks" of the 1970s, many of the industrialised countries have been reviewing their energy needs and prospects. An integral part of any such assessment is a review of energy resources, known and conjectured, in the hope of determining the scale of resources to meet energy demands. Given the particular importance of oil-derived transport fuels to modern society, most attention has been concentrated on attempts to estimate oil resources. Such estimates are, however, beset by several severe difficulties:

* as the oil is trapped in rock strata far beneath the surface of the earth, determining the physical extent of the oil field is difficult;
* as the oil is located in tiny pores within porous rock, the fraction of the oil which will be recoverable can be as low as 15% or as high as 75%, with the actual fraction only becoming apparent in retrospect;
* the total production from an oil field is partially dependent on the rate of extraction, which may in turn be influenced by economic or political considerations:
* changes in taxation and other government provisions can dramatically alter the financial viability of oil extraction and hence influence the amount of oil obtained from a given field;
* estimates of the extent and nature of oil deposits are usually the commercial property of individual oil companies, who have obvious financial incentives not to make this information public.

Given all those reservations, it is quite remarkable that scientists are ever in a position to make statements about future oil production and the scale of resources. What is more remarkable is that the estimates are sometimes given an appearance of precise and objective measurement. Odell[18] has pointed out that the problems of estimation are so intractable that no estimate can ever be guaranteed within a factor of two, yet statements continue to be made implying that the amount of oil which will be recovered is known to within quite inexplicable precision[19].

VIABILITY OF "ALTERNATIVE" ENERGY SOURCES

It should, in principle, be a straightforward matter to measure the performance of a wind generator or solar hot-water system to determine the economic return, and hence the economic viability, of such an energy source as a replacement for the use of fossil fuels. In practice the "measurements" vary quite significantly, as in consequence do the estimate of viability.

In the federal system of government which Australia suffers as a legacy of its colonial past, the States retain various powers, including responsibility for energy matters. As a consequence, each State makes its own decisions about policies to encourage — or discourage — the use of alternative energy sources. As it happens, the State which does most to encourage the use of solar technology is Victoria, which has the lowest annual solar input of all the mainland States; the State which gives least encouragement to the harnessing of solar energy is Queensland, which has the highest average solar input of any State. Although the State of Tasmania has the electric power system most conducive to the use of wind power, and the attractive "resource" of a wind which regularly batters its west coast, its electricity authority is hostile to proposals to use wind power. These paradoxes can easily be explained in terms of social and political factors, while remaining mysterious within a framework of "objective measurement".

The Queensland government has invested heavily in electrical generating equipment to meet what it foresaw as the needs of a massive expansion in mining and mineral processing. In the absence of that expansion, the level of power capacity is both embarrassing and financially crippling. It is, as a result, in the interests of the electricity industry to encourage the use of electricity even for such thermodynamically-absurd uses as the production of domestic hot water. The official statements of the industry consequently are based on some creative arithmetic to reduce the economic appeal of solar[20], combined with financial inducements to use electricity and misleading claims that imply solar water heaters are likely to damage the roof of your house! By contrast, the Victorian authorities are endeavouring to delay the large capital investment in a new power station, and so have a vested interest in holding down demand, especially in peak periods — winter evenings. It is thus in their interests to promote solar energy, which they do by significant injections of public money[21].

Tasmanian energy policy is dominated by the influence of the Hydro-Electric Commission, long a law unto itself[22]. With a large work-force skilled in the construction of dams and hydro-electric schemes, the authority is naturally hostile to other forms of generation of electricity. The largest problem of harnessing wind power — interfacing the continuously-varying output of wind generators to a thermal system — does not apply to Tasmania, with its large base-load hydroelectric capacity. The social imperative of wishing to maintain its skilled work-force, however, leads the authority to a position of hostility to the use — or even impartial assessment — of the wind power potential of the State.

OTHER CONTROVERSIAL MEASUREMENTS

Lest it be thought that there is anything peculiar to the energy area which predisposes otherwise reputable scientists to allow values to interfere with their normally objective measurements, it is worth mentioning in passing some parallel examples from other fields. The broad question of carcinogenesis is one in which hard data have been difficult to obtain, given the long period between exposure to a carcinogen and detectable symptoms. It is also an area in which controlled experiments would not be ethically acceptable. As a consequence of these problems, the measurement of cancer production has generally been based either on epidemiological studies or extrapolation from other mammals, such as rodents[23].

There are some difficulties in either approach. While there is reason to be suspicious of a substance which induces cancers in rodents, it is not inevitable that the same substance will produce human cancers. The converse is also true; failure to harm rodents does not necessarily mean that a substance will be harmless to humans. Epidemiological studies have been of great value in establishing such links as that between smoking tobacco and cancer of the lung[24]; however, the very nature of such studies means that the link can never be unequivocally established, and so those who choose to do so — in this case, those who profit from the sale of tobacco products — can choose to regard the case as being not proven[25]. While all but a small minority now accept the link between smoking tobacco and lung cancer, a vigorous debate is raging on the question of "passive smoking", or exposure to the tobacco smoke of others. Attempts have been made to measure, for example, differences in lung cancer rates between those married to smokers and those married to non-smokers. At the time of writing, it is probably fair to say that the evidence is still sufficiently inconclusive for those who have a strong view — one way or the other — to see it confirmed[26].

The testing of pharmaceuticals is a similarly vexed area. Bentley[27] describes the general procedures for evaluation and approval of a new drug in the following terms. In the pre-clinical phase, while a manufacturing process and analytical methods are being developed, there must be studies of the fate of the drug in the body. Basic toxicity studies must be conducted on laboratory animals to determine the lethal dose, as well as testing for mutagenicity, teratogenicity and carcinogenicity. The clinical stages of testing involve a further three stages.

In the first of these, the drug is tested for safety and toxicity on small groups of volunteers. In the second phase, the therapeutic efficacy of the drug is tested on a larger group of up to 1000 patients. In this stage the dose range for efficacy is determined, side-effects and adverse reactions are monitored, and the clinical effectiveness of the drug is evaluated. In the third and final stage, there should be a more detailed analysis of possible side-effects and interactions with other drugs, as well as study of the effects of prolonged administration and trials of the drug on special groups, such as children or the elderly. The information and data assembled for an application to market a new drug in the U.S.A. is said by Bentley to average several metres in height.

There are a range of problems associated with this procedure. Some obvious ones are:

possible conflict of interest in the testing programme;
the problem of finding suitable human patients for test purposes;
the ethical problem of deciding what to tell those patients;
the problem of finding a similar control group to receive placebos;
the human problems of separating patients into these two groups;
the ethics of witholding a potentially life-saving treatment for the sake
    of having a controlled experiment;
the possibility of bias in the assessment of the results;
the understandable temptation to take short cuts or fabricate data;
the natural commercial desire to market a drug which has been costly
    to develop.

Braithwaite[28] documents a frightening number of instances of negligence or outright fraud in the testing of drugs, as well as instances of reluctance to withdraw drugs from the market when undesirable effects became apparent. Once again, it is difficult to argue that the science is an objective exercise. As a specific example of the procedures used in testing a medication for widespread use in the community, Diesendorf has recently raised the issue of fluoridation of water supplies[29]. Conventional wisdom attributes to this public health measure most of the credit for the recent dramatic fall in the incidence of dental caries, especially in children. Diesendorf's analysis of the literature reveals two interesting features.

The first is that there have been very few experiments involving a proper control group. Most of the studies reporting the improvement in dental health after fluoridation had no control groups, and so could equally have detected an adventitious improvement due to some other reason, such as improved diet or oral hygiene. Such studies as do involve proper control groups yield much less impressive support for the proposition that fluoride administered through public water supplies is an effective therapeutic agent.

The second point is that studies of non-fluoridated areas in various countries show similar spectacular drops in the incidence of dental caries to the reductions observed in areas where the water is fluoridated. This appears to be at least prima facie evidence that the question is more complex than it seems at first glance. Changes in diet and lifestyle may be playing an important role.

In terms of scientific method, there is nowhere anything approaching the classical double-blind test of clinical practice; because patient behaviour can be influenced by a doctor's attitude, the doctor should not know whether the therapeutic agent or a placebo is being administered to a given patient. This information should be collated later with the clinical results by an independent researcher. In the case of fluoridation, in many cases the evaluation of the efficacy of the technique was being carried out by those authorities which had argued for the measure to be introduced. Diesendorf suggests that these authorities had obvious vested interests in showing that the measure was effective.

This case study is a particularly interesting one, because it examines a measure which has been widely accepted and widely believed to be effective. The examination shows that the evidence for the effectiveness of the treatment is quite scanty. It also calls into question whether there was ever a serious attempt at objective measurement of the value of the therapy.

A final example is the proposed irradiation of various foodstuffs. The treatment is claimed to eliminate pests and increase shelf life of various food items. Unlike chemical pesticides, it does not leave residues which may be harmful — toxic or carcinogenic. On the other hand, irradiation produces a range of radiolytic products, some of which may also be harmful.

Experiments have been conducted on a wide range of foods, including various meats, fish, fruit and vegetables. The use of radioactive isotopes has involved the co-operation of nuclear authorities; not surprisingly, the process has received enthusiatic support from these bodies. International working parties and committees have been established, involving nominees of the World Health Organisation (WHO), the Food and Agriculture Organisation (FAO) and the International Atomic Energy Agency (IAEA).

Those opposed to the process[30] have, equally unsurprisingly, questioned the objectivity of such groups as the IAEA, whose charter includes the promotion of nuclear technology. These groups point to some research which suggests that there are health hazards particular to using irradiated food. Those supporting the process point to the gains from eliminating chemical pesticides or reducing the incidence of illness from such sources as salmonella[31]. The only way of resolving the matter unequivocally would be to feed a large experimental group irradiated food and monitor the results, comparing the experimental group with a carefully-matched control group. In the absence of such a controlled experiment, most of the literature on the subject is written from a committed viewpoint of either support for food irradiation or opposition to it. Once again, the absence of "objectivity" is quite striking.

OBJECTIVE SCIENCE?

Despite difficulties of the style discussed above, many scientists still cherish the notion of objective science. Weinberg[32] argued that there are a class of problems which he labelled as *trans-science*: "though they are, epistemologically speaking, questions of fact and can be stated in the language of science, they are unanswerable by science; they transcend science". Interestingly enough, Weinberg gave as two examples of such questions the biological effects of low levels of radiation and the probability of catastrophic reactor accidents. He argued that such questions cannot be answered *even in principle* by scientific investigation. Weinberg went on to suggest that scientists have a responsibility to recognise when the questions they are asked cannot be answered within the canons of science, and to state this clearly; to do otherwise, he argued, would be to give conjecture the appearance of scientific proof.

It can certainly be argued that many scientists fail to recognise when the questions they are asked are not capable even in principle of scientific analysis. When reviewing the ASTEC report on Australia's role in the nuclear fuel cycle, Ford commented [33] that "the reader is left with the distinct impression that the same facts could easily have been used to reach a totally different conclusion". Indeed, the same facts *were* used to reach a totally different conclusion by another committee with different interests and values [34]. The problem was clearly identified by Kerr [35] with reference to the safety of the nuclear power industry:

there is much more involved than a failure in logical deduction or ignorance of complex systems. Current knowledge remains so incomplete that the gaps have to be filled by assumptions which in turn are based ultimately on values.

The minimal critique of the paradigm of objective science is thus that there are a range of problems which appear to be scientific but defy the scientific method. Where the complexity of the problem, or the impossibility of the experiments needed, makes knowledge so incomplete that the gaps have to be filled by assumptions, which in turn are based ultimately on values, there is no possibility of claiming objectivity. Even this limited critique poses serious questions about science. How many scientists will accept with equanimity that there is no scientific answer to such questions as the health risks from low levels of radiation or the probability of catastrophic reactor accidents? Entire scientific disciplines, such as astrophysics, are potentially threatened; given the impossibility of controlled experiments which chart the evolution of stars on the main sequence of the celebrated Hertzsprung-Russell diagram, what is the status of such explanations for the variety of objects in the observable heavens?

Development of our understanding of the social process of science, however, suggests that the problem is more fundamental. Even within the limited framework of the idealised laboratory science of controlled experiments, peer evaluation and the republic of science, it is now clear that there are several general reservations which should be expressed:

* scientific observations are themselves theory-bound and affected by subjective influences[36];
* scientific knowledge is, as Kuhn showed[37], not a stable body of eternal truths, but a sequence of different world-views, sometimes separated by discontinuities so fundamental that today's good science is tomorrow's irrelevance;
* the acceptance or rejection of scientific paradigms is not entirely a rational process, in the sense of being dictated entirely by the internal logic of the subject[38];
* values play an important role in the choice or acceptance of paradigms[39].

Thus some of the problems, acknowledged by Weinberg to bedevil attempts to move the scientific method into complex phenomena, are arguably endemic to science as a human activity. Far from providing "stable, reliable, certain knowledge of the world", science is itself influenced by human frailties and the structure of the society in which it is practised[40]. The subjectivity of measurement is palpably obvious when the problem is complex or inaccessible and there are obvious political imperatives, such as in the question of the safety of low levels of radiation from the nuclear industry. Few areas of scientific activity, however, can be claimed to be so lacking in complexity that there is only one possible paradigm or theoretical framework for further investigations. The questioning of the notion of objective measurement began as a reaction to the dubious claims being made for political reasons in areas of complexity. Those scientists who made such dubious claims have spurred a more general questioning of the scientific process, particularly when science is translated into the public domain[41]. That more general questioning reveals the entire impressive edifice of "scientific objectivity" to have very shaky foundations. As Albury expresses it[42], "the process of production, evaluation and acceptance of knowledge is already political through and through".

Griffith University, Queensland

REFERENCES

[1] R. K. Merton, *The Sociology of Science: An Episodic Memoir*, Southern Illinois University Press (1977).

[2] W. Broad & N. Wade, *Betrayers of the Truth*, Simon & Schuster, New York (1982).

[3] A. Weinberg, Science and Trans-science, *Minerva* 9, 220—232 (1962).

[4] R. Bertell, *No Immediate Danger*, The Women's Press, London (1985), pp. 42—45.

[5] *Ibid.*, p. 51.

[6] Hinton, 'The Birth of the Breeder', in J. Forrest (ed), *The Breeder Reactor*, Scottish Academic Press (1977), p. 11, quoted in D. Collingridge, *The Social Control of Technology*, France Pinter, London (1980), p. 124.

[7] W. L. Lowrance, *Of Acceptable Risk*, William Kaufmann, Los Altos (1976) p. 73.

[8] U.S. Atomic Energy Commission, *An assessment of accident risks in U.S. commercial nuclear power plants*, AEC WASH—1400 (1974).

[9] H. W. Kendall & S. Moglewer, *Preliminary Review of the AEC Reactor Safety Study*. Sierra Club/Union of Concerned Scientists, Washington D.C. (1974).

[10] W. Patterson, *Nuclear Power*, Penguin, Harmondsworth (1975).

[11] E. Titterton, Energy and Man, in I. Lowe (ed), *Teaching the Interactions of Science, Technology and Society*, Longman Cheshire, Sydney (1987) pp. 39—52.

[12] N. Hawkes *et al. The Worst Accident in the World*, Heinemann/Pan, London (1986) p. 35.

[13] U.S. House of Representatives Committee on Government Operations, *Nuclear Power Costs*, U.S. Congress, Washington (1978).

[14] Australian Science and Technology Council, *Australia's Role in the Nuclear Fuel Cycle*, Australian Government Publishing Service, Canberra (1985).

[15] U.S. Department of Energy, *Energy Technologies and the Environment*, DOE/EP—0026, Washington (1981).

[16] Australian Science & Technology Council, *op. cit.*, p. 257.

[17] *Ibid*, p. 29.

[18] P. Odell, *Oil and World Power*, Penguin, Harmondsworth (1975).

[19] Department of National Development and Energy, *Energy Forecasts for the 1980s*, Australian Government Publishing Service, Canberra (1982).

[20] J. S. Foster, *Policies and the Uneven Diffusion of Solar Domestic Hot Water Systems in Australia: an Interstate Comparison*, Honours thesis, Griffith University, Brisbane (1986).

[21] Victorian Solar Energy Council, Annual Report, Melbourne (1986).

[22] H. Saddler, *Energy in Australia Politics and Economics*, George Allen & Unwin, Sydney (1981).

[23] J. Cairns, *CANCER: Science and Society*, W. H. Freeman, San Francisco (1978).

[24] Royal College of Physicians, 3rd Report, *Smoking or Health*, Pitman Medical Publishing Company, Tunbridge Wells (1977).

[25] Tobacco Institute, 'The Smoking Controversy; a Perspective' (1978), cited in P. Taylor, *The Smoke Ring*, Bodley Head, London (1984).

[26] S. Chapman, *Great Expectorations*, Comedia Publishing Group, London (1986).

[27] P. J. Bentley, *Elements of Pharmacology*, Cambridge University Press, Cambridge (1981).

[28] J. Braithwaite, *Corporate Crime and the Pharmaceutical Industry*, Routledge Kegan Paul, London (1984).

[29] M. Diesendorf, *Community Health Studies* 4, 224—230 (1980); M. Diesendorf, *Nature* 322, 125—129 (1986).

[30] T. Webb, 'Food Irradiation in Britain?', *London Food Commission* (1985).

[31] T. Roberts, 'Microbial Pathogens in Raw Pork, Chicken and Beef: Benefit Estimates for Control Using Irradiation', *Amer. J. Agr. Econ*, Dec. 1985, 957—965.

[32] A. Weinberg, *op. cit.*

[33] J. Ford, 'Hawke gets his nuclear licence', *The Australian*, June 1, 1984.

[34] K. Suter (ed), *Australia and the Nuclear Choice*, Total Environment Centre, Sydney (1984).

[35] C. Kerr, 'Perception of Risk in the Nuclear Debate', in F. W. G. White (ed), *Scientific Advances and Community Risk*, Australian Academy of Science, Canberra (1979) pp. 63—64.

[36] A. Chalmers, *What is this thing called science?*, Queensland University Press, Brisbane (1976).

[37] T. S. Kuhn, *The Structure of Scientific Revolutions*, Chicago University Press, Chicago (1962).

[38] *Ibid.*

[39] B. Easlea, *Liberation and the Aims of Science*, Chatto & Windus, London (1973).

[40] D. R. Biggins & I. D. Henderson, *Australian Science Teachers' Journal* **24** (1), 53—64 (1978).

[41] I. Lowe, 'From Science to Social Policy', *Metascience* **3**, 13—17 (1985).

[42] R. Albury, *The Politics of Objectivity*, Deakin University Press, Geelong (1983).

ROBERT McLAUGHLIN

FREUDIAN FORCES

1. FREUD'S METAPSYCHOLOGY

1.1. *Distinctions*

That psychoanalysis embraces both a psychology and a metapsychology has not always been recognised by its critics (e.g. Popper (1962), Eysenck (1953), (1985)), whose preoccupations have tended to centre on the scientific status of the former, rather than on the metascientific features of the latter. For this, Freud himself may be partly responsible. Thus Grünbaum contends, with some textual support, that Freud regarded his metapsychology as a "speculative superstructure", and took his psychoanalytic corpus to stand or fall with his psychology — that is, his clinical theory or psychopathology, of which the theory of repression is the cornerstone. (Grünbaum (1984), p. 5)

Properly speaking, Freud's *psychology* is a ramified account of human behaviour, with the psychopathology as a more specialised component. The status of Freud's clinical theory has been the topic of vigorous empirical and philosophical inquiry lately. In particular, the probative value of clinical observation, and the efficacy of the technique of psychoanalytic therapy based on free association, have been extensively debated.[1] After decades when psychoanalysis was as unfashionable as astrology in the eyes of most philosophers of science, the present serious philosophical attention is a healthy change.

This recent debate over the clinical theory, fascinating though it may be, is secondary to the present discussion. Despite the indifference of Grünbaum, and the ambivalence of Freud, the latter's *metapsychology* remains the most interesting part of psychoanalysis. This may be construed as a *realist*, causal-explanatory metatheory about human nature, which permeated Freud's detailed "low-level" psychology.[2] While the psychology may be defective in many respects, a scientific realist reconstruction of the metapsychology would allow it to be conceptually connected to modern neurology, thus pointing the way to a causal explanatory neuropsychology — which would be quite different

207

John Forge (Ed.), Measurement, Realism and Objectivity, 207–233.

from present low-level psychoanalytic formulations of the clinical theory. The envisaged reconstruction is incompatible with dualist renderings of the Freudian research programme (*FRP*) (e.g. as a *sui generis* psychology, or the "hermeneutic" interpretations) on the one hand, and with instrumentalist (e.g. behaviourist) dismissals of it, on the other hand.

1.2. *What is Freud's Metapsychology?*

The Freudian metapsychology comprises a set of fundamental hypotheses specifying three aspects of the conceptual framework for the detailed psychology, that is, for the account of the self-regulating control system which is the "mind" (Farrell (1981), p. 22). The *topographic* dimension of this control system distinguishes three categories of process: unconscious, preconscious and conscious. The *dynamic* aspect depicts the interplay of *forces*. The *economic* facet of the system portrays the exchanges and transformations of *energy*.[3]

Well-known detailed concepts within this framework include: the systems of processes labelled 'id', 'ego', 'superego'; bound and unbound forms of energy; primary and secondary process; cathexis; excitation; discharge; instincts; resistance.[4]

Of central interest here are Freud's *mechanical* conceptions — that is, the dynamic and economic components of his metapsychology. Several passages may convey the flavour of his notions of *force* and *energy*. The defining characteristics of his key concept of *instinct* are "its origin in sources of stimulation within the organism and its appearance as a constant *force*." [my emphasis] (1915a, p. 63) His mechanical preoccupation emerges in his characterisation of the *impetus* of an instinct, as "its motor element, the amount of *force* or the measure of the demand upon *energy* which it represents". [my emphases] (*Ibid.*, p. 65) Again, his theory of *repression* is thoroughly dynamic:

... the forgotten memories were not lost ... but hindered from becoming conscious, and forced to remain in the unconscious by some sort of a force. .... One could get an idea of this force ... from the resistance of the patient. [Para.] These same forces, which in the present situation as resistances opposed the emergence of the forgotten ideas into consciousness, must themselves have caused the forgetting, and repressed from consciousness the pathogenic experiences. I called this hypothetical process 'repression', and considered that it was proved by the undeniable existence of resistance. (1910, p. 13f)

He emphasises that continued repression "demands a constant expenditure of *energy*." [my emphasis] ("Repression", 1915c, p. 89) Such quotations could be multiplied indefinitely. Any cross-section of Freud's writings will reveal his regular and natural usage of a range of *mechanical* locutions. The essence of his metapsychology, then, is the *metascientific thesis* that human behaviour is to be explained by means of *causal mechanisms*, in which concepts of *force* and *energy* play central roles.

There has been much debate as to whether Freud's metapsychology is to be understood ultimately as a set of biological, or as a set of *sui generis* (irreducible, causally self-contained) psychological proposals. Certainly Freud originally held a biological construal,[5] which he sought to develop in a thorough-going *neurological* fashion in his *Project for a Scientific Psychology* of 1895.[6] Subsequently, disheartened by the primitive state of 19th century neurology, he seems to have become ambivalent on this (cf. Fine (1963), p. 62) — an ambivalence which has confused many of his commentators. After 1895 his vocabulary became increasingly mentalistic. Nonetheless, Pribram and Gill (1976) have argued that "psychoanalytic metapsychology should be seen to be in all its facets a biological cognitive control theory, based on an explicit neuropsychology." (p. 14) These authors maintain that the *Project* "is in the main a neuropsychologic document" and that "the later metapsychology is ostensibly psychological alone but is in fact neuropsychology, with the neurology rendered implicit in contrast to its explicit statement in the *Project*." (p. 15) The substance of the present paper will complement Pribram and Gill's thesis.

1.3. *History and Philosophy of Psychoanalysis*

Two linked features of the history of psychoanalysis should be mentioned here. First, the metapsychology, in its more familiar psychological — that is, mentalistic — terms, evolved over a *long* period (1900—1923, for the main themes; but it continued into the late 1930s), during which numerous detailed changes occurred — e.g. in the character of the instincts, or of the id-ego-superego structures. It is not surprising that, through these vicissitudes, the underlying continuity of Freud's metapsychological ideas has faded from view. Second, exacerbating this tendency, the *original* metapsychology, written in the biological terms of the *Project* (1895), although it conceptually ante-dated the later mentalistic metapsychology, in fact was almost unknown until it was published posthumously in 1950 in German, and 4 years later in English. A couple more decades passed before its significance began to be properly understood; indeed, it could be argued that Pribram and Gill (1976) were the first to place it in perspective.

Freud was always concerned to put his psychoanalytic theory on a firm scientific basis. His "philosophy of science" was that of the late 19th century, and his own specialised scientific training was, of course, in neurology. Thus it is not surprising that, in his *Project for a Scientific Psychology*, as the name shows, he sought the scientific credentials of his new theory of the mind (i.e. of a cognitive control system) in a *reductive ontological* account, which would reveal the processes of psychology in terms of biology, particularly neurology; and in principle, ultimately, of physics — masses in motion. In his words, ". . . out there, according to the view of our natural science, to which psychology too must be subjected here [in the *Project*], there are only masses in motion and nothing else." (*Project* [7]; Strachey (ed.), *SE 1*, p. 308).

Now Grünbaum has argued that Freud abandoned his attempt to carry out an ontological reduction of psychology to physics; and that after 1895 he sought to base the *scientific* status of psychoanalysis in its *epistemic*, rather than its ontic, character. Hence the controversy over the scientificity of psychoanalysis has been waged almost entirely in epistemic terms — centring on the issue whether the claims of the clinical theory are testable.

The present essay is not primarily concerned with this debate over the *epistemic* status of psychoanalysis — that is, the detailed "low-level" psychology. Grünbaum, Farrell and many others maintain — *contra* Popper — that much of Freud's clinical theory is testable; and that, as a matter of fact, the available data (mainly in the form of clinical observation reports) fail to confirm the theory. But *contra* Grünbaum, it will be argued that Freud's metapsychology, realistically construed as implicitly neurological (Pribram and Gill, *op. cit.*) is very important as an *ontic* thesis. In contrast to the epistemically dubious Freudian psychology, Freud's ontic metapsychology provides a metascientific conceptual framework for a causal explanatory theory of behaviour. The key to this insight is an appropriately *realist* construal of the metapsychology — a construal which, if it could be carried through, would reveal the causal processes involved in molar[7] behaviour as *measurable*, because these would emerge as processes — including holistic or Gestalt ones — of (an enhanced) neurology, and so measurable by the techniques appropriate to neuroscience — EEG's, CAT-scans, chemical tests, etc.

This envisaged construal of the *FRP* admits of a *convergence* rather than a reduction view of the conceptual relations between (molar) psychology and (molecular) neurology, whereby concepts of each can modify and enrich those of the other. As Clark ((1980), pp. 37—38) puts this: "Perhaps the physiological story could come to include the phenomena of moods, feelings, and beliefs, once an adequate psychological theory of those capacities is found." At the molar psychological level, Freud's dynamic concepts — resistances, inhibitions, repressions — and energy concepts — impulses, instincts, drives — admit of nominal and ordinal scaling at best; but it will be argued below that these can be conceptually linked with, and in principle could be measured by means of, quantitative metrical variables of neuroscience. Then repressions could be measured in (fractions or functions of) dynes or newtons; drives in ergs or joules. A necessary condition for such a reconstruction of the *FRP* is a *non-dualist* construal of its causal concepts. These conjectures will be elaborated below but it is worth remarking here that if molar behaviour, as systematically conceptualised in the *FRP*, could be causally explained by objectively measurable neurological processes, then the resulting neuropsychology might properly claim to be scientific — as Freud always envisaged it.

1.4. *Philosophers' Indifference to Freud's Metapsychology*

There are several reasons why Freud's metapsychology has not been studied closely by philosophers of science. First, the distinction between it and the clinical theory is subtle, and has not always been appreciated. Yet its key metascientific thesis that human behaviour is to be explained by means of causal mechanisms is surely clear enough, as is the insight that this metascientific thesis will be unaffected by the disconfirmation of various detailed low-level hypotheses. Second, Freud's psychology has been subject to egregious misunderstanding and misrepresentation — especially by purveyors of an odd dualistic approach to the study of human behaviour, identified by Grünbaum (1984, Introduction) as a "hermeneutic" conception. Grünbaum properly rejects this caricature of Freud's research program. Third, Freud began, in the explicitly neurological version of his metapsychology,

namely the *Project*, with 19th century neurological concepts which —
he came to believe, as did many others — were too primitive to be of
any value. Pribram and Gill have argued that he underestimated
himself, and was underestimated by others, in this respect. Fourth,
Freud's philosophy of science was largely influenced by the Zeitgeist of
the late 19th century, dominated by Newtonian mechanics, Laplacean
determinism, Darwin's theory of evolution, and the empiricism of Mill,
Whewell, T. H. Huxley. He wrote his *Project* against this scientific-
philosophical background, ten years before Einstein published his
Special Theory of Relativity, several decades before the full flowering
of quantum mechanics. In the eyes of many modern philosophers, these
influences on Freud's metapsychology give it a dated, obsolete air. To
these circumstances must be added Freud's own diffidence about
philosophical matters. Although he could not avoid, by the nature of his
enterprise, definite philosophical commitments, yet he was loth to make
these explicit. In fact, it is obvious that he was a thorough-going
materialist, despite a tendency of certain authors to regard his theories
as "mentalistic" in some dualist sense (see Section 2.2 below).

Einstein, influenced by the positivism of Mach, sought to eliminate
the concept of *force* from physics; sought to treat forces instrumen-
talistically, as pure "intervening variables" or "convenient fictions".
Ironically, during the very decades when Einstein was developing his
great theories (Special and General Relativity, and his major contributions to quantum mechanics), Freud was elaborating his psychology in
mentalistic terms — a psychology that seemed filled with "psychic"
forces. Dynamism and determinism characterised Freud's psychoanalysis, just when these notions were becoming outmoded in physics.
(True, Einstein remained a determinist; in this he too was out of
fashion!) Thus it might seem that, even if there was supposed to be an
underlying physical basis to Freud's psychology, that physical basis
itself was obsolete.

There is further irony. Given the suspicion of mind-body dualism
that attached to Freud's psychology, due primarily to his increasing use,
after 1895, of a mentalistic vocabulary, there was a tendency to reject
psychic forces as non-physical and so unscientific. (Freud's common
use of the adjective 'psychic' instead of 'psychological' may have
contributed to this, as suggesting something supernatural.) This style of
objection to Freud could be put: "There are no psychic forces because
there are no psychic anythings!" Hence his whole *psych*-ology was

viewed as ontically illicit. To the contrary, the problem about psychic forces was not that they were "psychic", but rather that they were *forces*. The real objection to Freud's psychic forces should have been put: "There are no psychic forces because there are no forces — psychic *or* physical!" So objections to Freudian forces are just the standard instrumentalist objections to forces in general; and are to be weighed accordingly. Critics of Freud have tended to confuse a straightforward instrumentalist line of attack with a quite different, quite unfair one — namely the charge that Freud was guilty of some kind of illicit ontic dualism, which rendered his enterprise unscientific. Let us proceed to the instrumentalist critique.

2. FORCES: REALIST vs INSTRUMENTALIST CONSTRUALS

2.1. *Varieties of Realism*

Realism has divers faces — for example: scientific, mental, supernatural, Platonic realisms. Scientific realism contrasts with instrumentalism: the issue concerns the semantic status of "theories", and more generally of "theoretical" talk; do "theories" have truth-values, do "theoretical" terms (as opposed to "observational" terms) refer? Scientific realism answers affirmatively, denying any philosophically interesting dualism between "theoretical" and "observational" languages. (Grover Maxwell (1962); Smart (1963); McLaughlin (1967)) Mental realism contrasts with physicalism, and asserts that *both* "the mental" and "the physical" — to use Feigl's (1967) locutions — are real (*cf.* Descartes). Supernatural realism would insist on the reality of "supernatural" entities — gods, demons, souls, infusions of grace — in contrast to the furniture of the "natural" world. Platonic realism contrasts with nominalism, and asserts the reality of abstract entities, universals, Forms, and the like, in contrast to an ontology exclusively of particulars or individuals. Save for scientific realism, these other realisms are *dualistic*, in that they involve *two* "levels" or "realms" of reality. Scientific realism, like its opponent instrumentalism, is not dualistic: both sides accept a materialist ontology. Notice that all the foregoing are versions of *ontic* realism; *epistemic* realism is something else again, having to do with relations between knowers (or perceivers or observers) and what is known (perceived, observed) by them (e.g. direct realism; representative realism).

The issue over Freudian forces is an ontic issue of the first kind: it is the issue of scientific realist versus instrumentalist construals of Freudian force-terms — such terms as 'resistance', 'inhibition', 'repression', 'conflict'. This issue is easily conflated with controversies over mental or other realisms, resulting in injustice to Freud. (There is another, more subtle, conflation, namely of Freudian force-terms with drive-terms, which is addressed below in Section 2.5.)

The key feature of a *scientific* realist view of force-terms is that these terms refer (perhaps very obliquely) to actual *causes* of behaviour; whereas on the opposing, instrumentalist view, force-terms do not refer at all, to causes or to anything else. The problem with viewing Freudian force-terms in a *dualist*-realist manner, as designating mental entities — like "explanations" in terms of supernatural entities ("grace", "demons"), or abstract entities ("the evil" in the agent, say, conflicting with "the good" in her) — is that the manner in which such things could serve as *causes* of behaviour is utterly obscure. Whatever else it may be, behaviour is a matter of physical objects (bodies, limbs, organs) in motion. Any explanatory account of behaviour must be at least consonant with physical causation. Freud saw this quite clearly, as the passage quoted above from his *Project* reveals; and as his scientific background would lead one to expect. It will be argued that it is highly plausible to construe his force-terms in a scientific-realist manner; and highly implausible to construe them in some dualist fashion. Instrumentalists are right to reject all *dualist* species of realism; they are wrong to reject scientific realism, and with it Freud's causal-realist program for psychology.

2.2. *Was Freud a Dualist?*

It is notorious that Freud's psychology and metapsychology, after 1895, increasingly were filled with talk of impulses, desires, egos, ids and divers other psychic entities — seeming at their worst to be "objectively" inaccessible, intangible, immeasurable. But there is no need to construe all this mentalistic talk as disclosing some *ontic* dualism. Certainly the whole tenor of his *Project* is physicalist, in line with his scientific education. There has been much debate about the relation between the neurological conceptual scheme of the *Project* and the subsequent mentalistic conceptualizations from *The Interpretation of Dreams* (1900) onward (cf. Sulloway (1979), p. 118ff.). It is at least widely conceded that the *Project* had a strong *heuristic* influence on the subsequent metapsychology. Without entering this debate too deeply, I want to remark that it seems so far-fetched as to be incredible to suggest that Freud, after his recognition that the neurology available to him in 1895 was inadequate to the task of providing a foundation for his psychology, should totally shift his *ontic* commitments from physicalism to dualistic mentalism. Far more reasonable is it to understand the progression of his thought, from 1895 through 1900 and thereafter, as a matter of setting aside an inadequate conceptual scheme and constructing a different one to serve his ends. These unchanging ends were to develop a *science* of human nature, which would enable the explanation (thence understanding[8]) of the full spectrum of human behaviour, and especially of neurotic or pathological behaviour; for Freud was always a clinician.

On this view, the mentalistic terms with which Freud's writings are so freely peppered do not reveal him as an ontic dualist. Rather, they can be seen as elements of a conceptual system which seeks to depict human beings from the perspective of Freud's "molar" scientific concerns; just as we may think of modern neurology as a conceptual scheme which also seeks to portray human beings (inter alia) from another (more "molecular") perspective. The two schemes may well be complementary, or capable of convergence. It is highly plausible to imagine that Freud considered them so; and the height of implausibility to claim that, after 1895, he abruptly became an ontic dualist and started to believe that the entities and processes with which he dealt were ontically separate from the entities and processes depicted (however crudely) by neurology.

2.3. *Instrumentalism in Physics and Psychology*

According to a modern instrumentalist view with its roots in the concept-empiricism of Berkeley and Hume, and elaborated more recently in the operationism of Mach and Bridgman, force-terms can be eliminated from the conceptual scheme of physical science without loss of explanatory power (although perhaps with some sacrifice of economy). Thus Jammer (1957, p. 248) remarks: "Modern physics recognizes the concept of force in both statics and dynamics, and thereby in every other field of physics as far as motive forces are concerned, as a *methodological intermediate* [my emphasis] that in itself carries no explanatory power whatever." The same author proceeds to insist that not only are force-terms dispensable in general relativity, where they can be replaced by appropriate metrical descriptions,[9] but also the many "nuclear forces" of quantum mechanics[10] are equally superfluous.

This instrumentalist view of forces in physics, namely as "methodological intermediates", concurs with a depiction of the role of various conceptual artefacts in psychology as "intervening variables". This title, introduced by the psychologist Tolman half a century ago,[11] caught on rapidly in behaviourist circles. Tolman conceived the intervening variable as a device for "bridging the gap" between observational variables in an input-output setting. The IV, as I shall abbreviate it, was viewed as an operationally defined[12] symbol, with a purely summarizing role. Introduced by explicit definition (typically as a mathematical function of observational quantities), it was regarded as being in principle eliminable by appropriate substitutions, usually at the cost of increased clumsiness or complexity of expression and computation. Its users (such behaviourists as Hull, Skinner, Lewin) would insist that it had no reference of its own, other than a kind of second-hand reference to the observables whose relationships it summarised. This conception of the IV, as it appeared in various behaviourist systems, was explicated in a seminal paper by MacCorquodale and Meehl (1948):

Such a variable will then be simply a quantity obtained by a specified manipulation of the values of empirical variables; it will involve no hypothesis as to the existence of nonobserved entities or the occurrence of unobserved processes .... Legitimate instances of such 'pure' intervening variables are Skinner's *reserve*, Tolman's *demand*, Hull's *habit strength*, and Lewin's *valence*. These constructs are the behavioral analogue of Carnap's "dispositional concepts" such as solubility, resistance, inflammability, etc.[13] (p. 105)

MacCorquodale and Meehl proceeded to contrast this conception of the IV with that of the "hypothetical construct" or HC — a term not purely operationally defined, and having extra reference; that is, purporting to refer to something other than the variables by which it was introduced and to which it was linked in some syntactic fashion looser than explicit definition (e.g. Carnap's "reduction sentences"). According to the authors, HCs involve

terms which are not wholly reducible to empirical terms; they refer to processes or entities which are not directly observed (although they need not be in principle unobservable); the mathematical expression of them cannot be formed simply by a suitable grouping of terms in a direct empirical equation ... Examples of such constructs are Guthrie's M.P.S.'s, Hull's rg's, Sd's, and *afferent neural interaction*, Allport's *biophysical traits*, Murray's *regnancies* ... and most theoretical constructs in psychoanalytic theory. (p. 106)

The distinction between IVs and HCs proposed by MacCorquodale and Meehl crystallized the issue in psychology over instrumentalist versus realist construals of "non-observational" terms — an issue typified by the contrast between behaviourism and psychoanalysis. With the gradual recognition over the last fifty years, among many philosophers of science if not many psychologists, of the untenability of strict operationist criteria of admissibility for scientific terms (see Hempel (1950), (1951), (1954); McLaughlin and Precians (1969)), the appeal of the early behaviourist vision of the vocabulary of science, including psychology, as limited to "observational" terms and IVs defined operationally thereby has waned. Not only is it the case that various hypothetical constructs — including terms of psychoanalysis — nowadays are acknowledged as potentially scientific, despite their "unoperationality" and possible "extra reference"; but also it turns out that most intervening variables — such as disposition terms — are not "pure" in the manner defined by MacCorquodale and Meehl. Instead, there is a sense — a *realist* sense — in which most interesting IVs do have a degree of "extra reference", pointing to or indicating the existence/presence of "underlying" states or processes, distinct from the observables whereby the IV was introduced.[14]

Take the concept *resistance R* as it appears in Ohm's Law:

$$R = V/I$$

where V is the voltage, and I the current strength, expressed in suitable units, flowing through a wire. One might contrive a tale of wires, for each of which it is observed (under standardised conditions) that the ratio of V to I is a *constant*. Given this observed pattern, one might be led to write the symbol 'R' as a shorthand for this constant ratio; that is, R is explicitly operationally defined by the equation above. Then this R qualifies as an IV, according to the definition. Notice that this shorthand use of R already contains some empirical import — namely, that the ratio V/I is a constant for each particular wire. The plot develops as it is observed that the value of R, although constant for a given wire, is not the same for all wires. Further experiment, indeed, suggests that R differs with different metals, different thicknesses of wire, and so forth.

Each wire has a particular, characteristic R. By now the term 'R' is coming to be taken as indicating, manifesting or measuring something about wires, some intrinsic property of wires; rather than, as it may have seemed to do initially, merely summarising a relation between the variables V and I. At this point, the term 'resistance' is coined for "the property measured by R"; and this recognition of the "extra reference" of the intervening variable R in turn serves heuristically to inspire further investigation of the intrinsic properties of wires — chemical composition, eventually molecular structure — to which 'resistance' refers.

In short, IVs express patterns, correlations, observed relationships; the *fact* that these correlations obtain suggests the presence of underlying intrinsic properties which have a causal role. The IV 'resistance' serves heuristically to *indicate* some (e.g. chemical) properties of wires; these properties, when identified, will be named by some other (e.g. chemical) term, which will be a HC. The direct referent of the HC will causally explain the relationships expressed by the IV; and the IV may be said to obliquely refer to, or indicate, this underlying cause.

Other dispositional properties apart from resistance may be observed to characterise wires — e.g. thermal expandability, elasticity. These too may serve as indicators of the *same* underlying causal state of the wire — say, its molecular structure. So several IVs may indicate the referent of one HC. The several dispositional properties may themselves be inter-correlated — e.g. the higher the resistance, the lower the elasticity — thus reinforcing the inference to a single underlying cause. Notice that, in the case of Ohm's Law, the correlation (under appropriate conditions) of V with I is perfect (= 1). Often the correlation between observed variables, expressed by an IV — or the correlation between IVs — will be less than unity; but *any* significant correlation (i.e. better than chance) by definition expresses a relation of statistical relevance indicating — on this realist view — an underlying cause.

The same point applies to any "black box", including human behaving organisms. The IVs mentioned by MacCorquodale and Meehl — such as Hull's *habit strength*, initially relationally defined as a mathematical function of a conditioning schedule — can be viewed in fact as "low order" constructs, referring obliquely to intrinsic properties or states of organisms. In particular, the psychoanalytic term 'resistance' may well turn out to have a status closely analogous to that of the electrical concept 'resistance' just considered. Within psychoanalysis, 'resistance' and other force-terms may refer obliquely to intrinsic features of human beings, which features also may have more direct names (perhaps in the vocabulary of neurology), and which, further, are the *causes* of the observable behavioural patterns which the psychoanalytic terms serve to express and summarise.

Let me illustrate this. In the passage from "The Origin and Development of Psycho-analysis" (1910), quoted above, Freud speaks of observing *resistance* and inferring to repression. Here 'resistance' is a summarising term, shorthand for a whole pattern of related behaviours — oral, facial, postural, etc.; a pattern Freud has observed regularly in the clinical setting. So 'resistance' is an IV. Now Freud goes on to say that he *infers* from this observed resistance (behavioural pattern) to the existence of underlying impulses or instincts in conflict. These are the *causes* of certain behaviours — both the patterns noted in the consulting room, and larger behavioural complexes manifested in the patient's daily life. In the case cited, these underlying causes consist of a defensive ego-instinct, which *represses* another impulse, namely one which Freud characterises as "a painful memory striving to enter consciousness". This, then, is a case of some enduring pattern of observed behaviour, which is summarised by an IV, 'resistance'; from which is inferred an underlying causal situation, characterised by Freud as a conflict of instincts; which conflict absorbs a steady expenditure of energy (debilitating or exhausting the patient). The more energy involved, the greater the intensity of the conflict — expressed as greater 'force' of *resistance*. A sharper analysis of these concepts of 'force', 'instinct', etc. will be offered in Section 2.5 below.

2.4. *Measurability, Realism and Causality* [15]

What is the connection between measurability and reality? Does the mere fact that an IV can be measured (in the sense, for example, that numerical values can be assigned to it on the basis of some specified procedure) confer reality-status on its purported referent? Does it follow, from what has just been said, that all IV's are "impure", that they all have "extra reference" to entities other than those by means of whose properties the IV's initially were operationally defined?

Evidently not. After all, one can simply *invent* an IV, by defining it as a specified function of any old set of arbitrary variables.[16] Such a capriciously constructed IV would be patently "ad hoc" — it would be useless for heuristic or explanatory purposes. So measurability alone is not sufficient to render an IV interesting, in the sense of "potentially explanatory". That is, measurability alone does not warrant a scientific realist construal of an IV; additionally, one has to ask whether the IV in question has *causal* implications. If it has, then it must refer (however obliquely) to some real process; for it seems fair to insist that reality is a necessary condition of causality.

A mark of those IVs which have causal implications is that they exhibit correlations or systematic connections with other, independently defined and measured, variables — as the tale of electric 'resistance' illustrates. These correlations suggest underlying causal, hence real, processes. In contrast, an IV which lacks such correlations, or one which has merely definitional connections with other variables, is not a good candidate for a realist construal.

This is not to deny that some pure IVs may be of use in science, solely by virtue of their contribution to economy of expression. 'Weight', defined as the product of mass and acceleration (due to gravity, say at Earth's surface), may be an instance of such a pure instrumental IV. But over and above the putative "simplicity-value" of IVs, those here called *interesting* play a heuristic role, suggest underlying causes, and so contribute to the invention and ramification of theories. Such IVs display correlations, systematic connections, and are or become theory-embedded. In particular, Freud's force and energy concepts, although not precisely quantitative, have this sort of connectedness and theoretical embedment, indicative of their realist, causal-explanatory character.

2.5. *Forces and Drives*

Freud's force-talk is readily understood as his way of characterising what he regards as *causal* situations. Here a careful distinction is in order. Strictly, it is not *forces* which resist or repress; rather, resistance and repression are instances of forces. What *does* the resisting and repressing, what "applies the force", is in fact some *energised structure* or "engine" (cf. Maze (1983), p. 152f). When Freud spoke of impulses and instincts — *drives*, as they are often called — he had in mind structures or systems through which energy flows in a controlled manner. (Recall Farrell's ((1981), p. 22) remark that Freud conceived of the mind as a "self-regulating control system".) The distinction between force-terms and drive-terms in Freud's writing is very much in line with his early contrast, in the metapsychology, between *dynamic* and *economic* aspects: the former concerns the interplay of forces; the latter, it will be recalled, treats the transmission and transformation of *energy*. Then we may say that force-terms characterise various causal processes or interactions, where these latter are understood to involve energy propagations and exchanges. Rather than speaking of a conflict of forces, we should speak of a conflict of drives (impulses, instincts), the intensity or strength of which conflict is represented as such-and-such a force.[17]

Freudian force-terms serve to express comparative drive-strengths. The force-terms depict an ordering or ranking of drives; e.g. "Smith's castration-fear *inhibited* his lust for his mother." In this statement, the former drive is ranked as *stronger* than the latter. The drive-terms themselves — 'castration-fear', 'lust for mother' — are molar psychological terms which *refer* to neural complexes, these being describable in the terminology of neuroscience, and being quantitatively measurable by the techniques of that science. Such measures would include the quantities of electrochemical energies in the circuitry and plumbing. A conflict of drives, described initially in the molar psychological terms of the *FRP*, in principle could be quantitatively measured and expressed — on this scientific realist reconstruction of the *FRP* — in terms of electrochemical energies.

Here it should be remarked that, while some progress has been made in connecting broad drive concepts like *libido* (or *lust*) to physiological processes, this is not to suggest that the task of specifying the neural referent of a drive-term linked to a particular object-type ("mother") or individual object ("Smith's mother") will be a simple matter. To the contrary, the latest neuroscientific insights suggest that mapping of such detailed links is a long way down the research track. What presently is being claimed is that this is possible, and eventually achievable — if the right type of psychological conceptual scheme is utilized.

So drives, rather than forces, are the real causes of behaviour, according to this reading of Freud's metapsychology. What can be said about these drives? The conception envisaged here is closely in line with that proposed by Maze (1983), viz.

... there are systems, however complex and anatomically diffuse they may be, sensitive to the internal environment, and controlling the onset and offset of eating [and other] behaviour. That is all that is intended here ... by the concept of instinctual drive, or the expression 'biological engine' . . . To call something an engine does not imply that it has some superior sort of causality, or that it is a *source* of driving energy. An engine is simply an entity that converts one kind of energy into another and applies it to a particular use. (pp. 152—3)

,There is nothing mysterious about any of this. Instincts are to be viewed as energised structures, parts of the architecture of the brain — or, more broadly, of the behaving organism — which determine the patterns and interactions of causal processes. Energy enters the circuitry from two directions: external sources, from the outside environment; and endogenous sources, from the "internal environment". Such energy is then directed and transformed throughout the whole complex system, which may be viewed as a self-regulating control mechanism operating in accord with a homeostatic principle (inter alia). Output energy is expressed in overt behaviour. This sketch is in line with the principal conceptions of Freud's *Project*, and is consonant with a contemporary rendering in neurological terms (cf. Pribram and Gill (1976)).[18] It also concurs with the main features of Freud's mentalistic metapsychology, developed from 1896 onward.

The key feature of this portrayal, as far as the present discussion is concerned, is that it depends on a scientific *realist* view of the drive-terms in Freud's theory. These terms must be understood as referring to real entities — namely, the energised structures or engines which drive the behaving organism. Drives are to be taken, physicalistically, as the same entities which are described in a certain kind of precise detail by means of the conceptual scheme of neurology. Energy, in particular, is to be understood as the same as what is described as electrical, chemical, mechanical energy. The term 'psychic energy' refers to the same energy as do these familiar terms; 'psychic energy' is not to be read dualistically as referring to some mysterious "mental" item, ontically divorced from "physical" energy (and perhaps possessed of some special mysterious causal power, such as that of "agency" or "will"). The *units* of psychic energy carried by drives will be (or be defined in terms of) the familiar energy units of physics: ergs, joules, calories.

An *instrumentalist* approach to drive-talk would forbid any such scientific realist construal, for fear of admitting occult mental entities. Such a black-box instrumentalist line would attempt to construe drives as "that which resists", "that which impels", "those which conflict"; and forces of resistance, impulse and conflict as pure intervening variables summarising observed (somehow quantified) behaviours. But as Maze showed more than thirty years ago, such a *relational* characterisation of drives produces the very occultisms it seeks to avoid: it acknowledges the existence of causes of behaviour, but refuses to allow any description of them in terms of *intrinsic* properties.[19] Nor does that instrumentalist view permit force-terms to serve as heuristic indicators of underlying causes. Nothing could be more inhibiting of inquiry.[20]

On the contrasting *scientific realist* construal of Freud's metapsychology, force-terms may be seen as low-order "constructs" or operationally impure "intervening variables" which serve heuristically as indicators, obliquely referring to underlying drives, which are more directly designated by actual drive-terms. The drives are the causal processes which produce behaviour; and to give an account of them is to causally explain behaviour. Freud's metapsychology, scientific realistically understood, is a causal explanatory theory of behaviour. The causal processes, to repeat, will be those of (possibly enriched) neurology. The resulting neuropsychology will be a genuine scientific theory of behaviour.

Now, how does this story of causal processes and scientific explanation fit prevailing views of these matters in philosophy of science?

3. CAUSAL EXPLANATION: SALMON AND FREUD

Among empiricist philosophers of science, causal explanation has not enjoyed wide popularity. This reflects an empiricist distrust of strong causal notions, springing from Hume's devastating critique of "necessary connections" or "causal glue" purportedly linking events which stand in causal relations. In consequence, empiricists have favoured an *epistemic* rather than *ontic* approach to scientific explanation (Salmon (1984), Chapter 4), which would concentrate on the subsumption of the explanandum-events under regularities, but which would avoid reference to processes or mechanisms which might underlie, or find expression in, the regularities. The most famous version of this epistemic conception of scientific explanation was, of course, Hempel's "covering law" account — introduced in his landmark paper with Oppenheim in 1948, and developed in a series of writings culminating in "Aspects of Scientific Explanation" (1965). If this were the unchallenged contemporary view of scientific explanation among philosophers of science — if there were no feasible *causal* account of explanation at hand or in the offing — then the foregoing sketch of Freud's metapsychology as a causal-explanatory theory would be wanting in philosophical back-up; and empiricists in particular would naturally incline to dismiss it as relying on obscure and discredited causal notions.

Fortunately for Freud, a rigorously developed account of scientific explanation in causal-realist terms — an *ontic* conception of explanation — has recently become available: that of Wesley Salmon. Beginning with a discussion of statistical explanation in 1970, Salmon introduced a causal-realist theme in 1975, and this culminated in his 1984 book, *Scientific Explanation and the Causal Structure of the World*. In this work he gives an account of explanation which is causal but (he insists) compatible with Hume; causal relations are understood in terms of *contingent*, not necessary, connections (furthermore, as *continuous* causal processes or "ropes", rather than as "chains" of discrete events); so Salmon's story fits empiricism. The account is *ontic*, in its focus on the actual processes which find expression in regularities, correlations or patterns in nature; and so it goes hand-in-hand with scientific realism. (Whereas the epistemic conception of scientific explanation is more in harmony with instrumentalism.) Finally, Salmon's account (like Hempel's) is quite incompatible with *dualist* approaches to knowledge, such as those which would divorce "explanation" and "understanding" (the latter being viewed as some peculiar kind of intuition, "verstehen", or whatever), and consequently dichotomize "the natural sciences" and "the social sciences" (as though the subject-matter of these were un-natural!). For Salmon, as for Freud, psychology is straightforwardly a natural science; adequate explanation provides understanding; indeed — as Hempel too emphasized — to explain is to answer a "why?" question, and the questions "Why do people behave thus?", "Why did this person in these circumstances behave as she did?" are perfectly on a par with "Why do the planets move in ellipses?", "Why did this bit of sodium salt burn yellow in the bunsen?". The answers will all be on a par; and, according to Salmon, will all involve revelation of the causal mechanisms (processes, interactions) involved.

Causal processes are central to Salmon's account. In this he departs from the more usual practice of taking events as basic. Causal processes are spatiotemporally larger than events. "In space-time diagrams, events are represented by points, while processes are represented by lines." (Salmon (1984), p. 139) Salmon's causal processes are similar to Russell's causal lines, a key feature of which is a type of *continuity* — the absence of lacunae, of abrupt changes. (Russell (1948), p. 459; cited by Salmon *op. cit.* p. 140) There is a crucial distinction between causal processes and pseudo-processes (*ibid.*, p. 141); the former, but not the latter, are capable of *mark-transmission* (Reichenbach (1956)). This leads to Salmon's contention that "A causal process is one that transmits energy, as well as information and causal influence." (*Ibid.*, p. 146) Salmon proceeds to insist that "causal processes constitute precisely the causal connections that Hume sought, but was unable to find" (p. 147), and argues that the defining characteristic of causal processes, namely ability to transmit a mark, is *not* an example of the "mysterious power" or "necessary connection" which Hume viewed as illicit. On the contrary, "Ability to transmit a mark can be viewed as a particularly important species of constant conjunction — the sort of thing Hume recognized as observable and admissible." (p. 147)

An important element in Salmon's theory is his use of the Principle of the Common Cause (PCC), first enunciated by his teacher Reichenbach (1956). This principle is intuitively very familiar, and states, roughly, "that when apparent coincidences occur that are too improbable to be attributed to chance, they can be explained by reference to a common causal antecedent." (p. 158) Put loosely, when we observe correlations or associations among events which seem unconnected, the PCC bids us postulate an as-yet-unobserved "common cause", of which the correlated occurrences are effects. Reichenbach (1956) explicated this by his conception of the "conjunctive fork", which is a relation among three events (strictly, event-types) A, B, C which can be characterised in terms of probabilities. Given that the joint occurrence of A and B is observed to be more probable than would be expected if they were (statistically, and — it will be argued — causally) independent of each other; i.e. given that

$$P(A \cdot B) > P(A) \times P(B)$$

then the PCC postulates the existence of a "common cause" C such that

$$P(A \cdot B/C) = P(A/C) \times P(B/C)$$

and

$$P(A \cdot B/\overline{C}) = P(A/\overline{C}) \times P(B/\overline{C}) \qquad (ibid., \text{ p. } 159\text{f})$$

C is said to "screen off" A from B (p. 161). The general idea of the PCC and the "conjunctive fork" is that, given an observed correlation between events A and B which seem to have no observable (continuous) connections, then rather than resting content with the mystery of their statistical relevance relation, it is rational to hypothesise the existence of an as-yet-unobserved (or unnoticed) additional event C, which causes both A and B, and so explains their correlation.

Salmon introduced the notions of interactive forks and causal interactions. A *causal interaction* is an intersection of two causal processes such that they undergo correlated modifications that persist after the intersection. (p. 170) When such an interaction occurs, the probability of the joint events (strictly, modifications in the causal processes) ensuing from the interaction is *higher* than that of their joint occurrence had the interaction not occurred. That is,

$$P(A \cdot B/C) > P(A/C) \times P(B/C) \qquad \text{(p. 170)}$$

Notice that here C fails to "screen off" A from B. Thus this situation differs from the conjunctive fork; Salmon has dubbed it an *interactive fork*. (p. 168ff). As he puts it, in many — perhaps all — such cases of causal interactions, "energy and/or momentum transfer occurs, and the correlations between the modifications [in the ensuing causal processes] are direct consequences of the respective conservation laws." (pp. 169–170) These features of causal interactions are stated explicitly by Salmon in his Principle of Causal Interaction [as I term it], or PCI (p. 171). Causal interactions can be thought of as what ordinarily are called "events"; and are described by interactive forks.

With these features of his causal theory in hand Salmon states "three fundamental aspects of causality" (p. 179) which I paraphrase, reversing 2 & 3: 1. *Propagation* of structure and order [and energy] is by causal processes. 2. *Production* of structure and order, i.e. of correlations, is by conjunctive common causes. 3. *Production* of modifications in structure and order, i.e. of *new* patterns, changed processes, is by causal interactions.

There is a great deal more to Salmon's account of causality, which I have here presented so summarily and uncritically. Much of it is currently under lively debate. For example, it has been suggested that interactive and conjunctive forks are not irreducibly distinct, but that the former can be boiled down to the latter.[21] Again, Salmon has insisted throughout on the *probabilistic*, as against deterministic, character of causality; but he also insists on the *continuity* of causal processes. The former feature seems compatible with quantum mechanics, but not the latter, as various critics (e.g. Forge (1982)) have pointed out. Salmon, indeed, has frankly admitted his misgivings about the applicability of his account in the quantum domain. (see, e.g., his p. 278)

However, there can be no doubt that Salmon has proposed an extremely interesting and important set of new ideas on causality, and has lifted this topic out of decades of doldrums. My present concern is not so much critical, as to show how this account, and the theory of scientific explanation that goes along with it, can fit — in its broad philosophical commitments, if not in its details — the requirements for a causal realist psychology, of the type which Freud tried to develop.

Salmon summarises his view of explanation in the following terms.

> To explain a fact is to furnish an explanans. We may think of the explanans as a set of explanatory facts; these general and particular facts account for the explanandum. Among the explanatory facts are particular events that constitute causes of the explanandum, causal processes that connect the causes to their effects, and causal regularities that govern the causal mechanisms involved in the explanans. The explanans is a complex of objective facts. (*Ibid.*, p. 274)

Particularly relevant to the task of psychology — the explanation of behaviours of organisms, notably human ones — is Salmon's idea of "the constitutive aspect of . . . explanation", as captured in the following lines.

> Suppose we want to explain some event E. We may look at E as occupying a finite volume of four-dimensional space-time . . . If we want to show why E occurred, we fill in the causally relevant processes and interactions that occupy the past light-cone of E. This is the etiological aspect of our explanation; it exhibits E as embedded in its causal nexus. If we want to show why E manifests certain characteristics, we place inside the volume occupied by E the internal causal mechanisms that account for E's nature. This is the constitutive aspect of our explanation; it lays bare the causal structure of E. (*Ibid.*, p. 275)

Now E, of course, may be the fact that a given person has certain properties (such and such a "personality") or behaves in a certain way — on one occasion, or with some kind of regularity. These facts about E (viewed as a "large event" or, better, as a complex system of causal processes and interactions) can be explained, Salmon tells us, by disclosing the "internal causal mechanisms that account for E's nature".

But this is precisely what Freud also told us! Look at the identities between the two accounts. Salmon's is an *ontic*, not an *epistemic*, theory of explanation. Freud's metapsychology is an *ontic* theory, counting on this character, rather than on the *epistemic* status of the detailed low-level psychology (clinical theory), for its *scientific* credentials (pace Grünbaum (1984, pp. 1—9)). Both view *causality* as central to explanation. Even more telling, the kind of heuristic inference to a causal process, from an IV expressing a correlation or pattern among observables to a HC designating an underlying structure which is causally related to this pattern, which was developed in Section 2 above, matches precisely the kind of inference expressed in the Principle of the Common Cause. A realist construal of force-terms and drive-terms in Freud fits very neatly the realist injunctions of the PCC. This sort of inference, which I have reconstructed in Freud, and which is explicit in Salmon, effectively would seek to explain — in the case of psychology — a *variety* of overt behavioural patterns, in and across people, by disclosing a relatively small number of underlying causes. (Freud's *The Interpretation of Dreams* (1900) and *The Psychopathology of Everyday Life* (1901) offer numerous examples of such inferences.) We may think of drives or neural structures as providing a framework in which causal processes can transpire, in which information and energy can propagate and undergo change via causal interactions. Drive-terms refer to systems of neural circuitry. Just as there may be many correlated effects of a common cause, so many associated behaviours may be manifestations of the causal activity of a common underlying drive. For example, Freud's *libido* may be conceived as a (very complex) system of neural circuitry through which run a myriad causal processes and interactions, these interactions producing new processes, involving energy transfer and transformation, and eventually ensuing in behaviour.

A detailed working out of a causal psychology would be an enormous undertaking. But without something like the causal metapsychological framework sketched here, based in Freud, such an enterprise would be impossible.

4. THE MEASURE OF FREUD

Measurement of molar psychological variables is a central problem for a scientific psychology. Freud vastly improved on the taxonomies and assumptions of everyday commonsense or "folk" psychology by offering a systematic causal (mechanistic) conceptual scheme, but his concepts were still at a molar level, and did not admit of precise quantification. However, the causal character of Freud's research programme does allow the possibility of its conceptual linkage to a scheme which can provide more exact measurement scales and procedures, namely that of neurology — as it stands today, and promises to develop; not as it was in the late nineteenth century. To achieve this sort of linkage, there must be no conceptual dualism or lacuna between the molar (psychological) and molecular (neurological) systems. It has been argued above that philosophically there need not be, and historically there was not, any such dualism between the *FRP* and the physical sciences, specifically neurology. Then the needed linkage is in fact the connection between some IVs (molar force concepts of the *FRP*) and some HCs (molecular concepts of neurology, designating neurological causal processes and complexes thereof, some of which may be holistically referred to by drive terms in the *FRP* — which thus could conceptually enrich neurology). On this account, it seems plausible to conjecture that the measurement problem in psychology will be solved eventually, not by some utopian refinement of *sui generis* molar psychological taxonomies, but by a convergence between the *FRP* (or a conceptual scheme of similar causal character) and neurology, yielding a causal explanatory metrical neuropsychology.

Of Freud it has been said that the dominant influence on his thought was an implicit faith in the scientific method and the scientific spirit.[22] This guiding commitment is revealed in a scientific realist reconstruction of his research program, by way of a study of his metapsychology, with a focus on its mechanical-causal features. This realist reconstruction of Freud stands opposed to "excessive empiricism" (Grover Maxwell's felicitous term) on the one hand, and to various dualist construals on the other hand. Excessive empiricism is found in divers instrumentalist theses, the common feature of which is a "black box" approach to internal causal mechanisms, these being viewed in terms solely of relational, not intrinsic, properties — and so remaining ineluctably occult and incapable of any explanatory role. Similarly (although this point has not been pursued), various dualist distortions of Freud, and of psychology generally, would exclude causal mechanisms from a role in behaviour and so make impossible a scientific explanation of it.

For psychology to become a science, genuinely capable of explaining behaviour in all its colour and complexity, it must be construed metascientifically along some such lines as I have sought to reconstruct in Freud, and as he himself viewed his work: as a realist, causal-explanatory enterprise within an empiricist epistemology and a non-dualist ontology. By such an enterprise, human beings will be viewed realistically as measurable objects not imponderable subjects; viewed, that is, as parts of the world like everything else.[23]

Macquarie University, N.S.W.

NOTES

[1] Eysenck and Wilson (1973), Fisher and Greenberg (1977), Farrell (1981), Wollheim and Hopkins (eds.) (1982), Grünbaum (1984) comprise a fair sample. Hook (ed.) (1959) is an earlier source.

[2] Farrell (1981) distinguishes a "High Level" and a "Low Level" theory in psychoanalysis, corresponding to the metapsychology and the psychology respectively.

[3] Freud (1915b) "The Unconscious", p. 114; Sulloway (1979), p. 62ff; Fine (1963), p. 45f.

[4] Fine (1963) gives an excellent critical summary of these concepts.

[5] Sulloway (1979).

[6] Freud (1895).

[7] cf. Littman and Rosen (1951).

[8] Some authors, such as the "hermeneutic" figures soundly chastized by Grünbaum (1984) — Habermas, Ricoeur, George S. Klein — try to divorce explanation from understanding (intuition, "verstehen"). So does Eysenck (1953), who should know better. With Hempel, Scriven and Salmon, I contend that an adequate account of explanation must include genuine understanding. There is no separate, occult species of "understanding", detached from scientific explanation. Certainly Freud did not subscribe to this *dualist* fantasy.

[9] ". . . gravitational forces are the outcome of the application of a wrong metric . . . Since the inertial behaviour of a particle, namely, its motion on a geodetic line, is a "force-free" motion, and since in addition all free particles, in general relativity, move only on geodetic lines, providing the appropriate Riemannian metric has been applied, there is no need at all in general relativity for the concept of gravitational "force"." (Jammer (1957), p. 259)

[10] For example: Wigner forces; Heisenberg forces; Majorana forces; Bartlett forces; Serber forces. According to Jammer, all these forces ". . . express merely different methods of introducing the dependence of the forces on the "state" of the particle." (*ibid.*, p. 254)

[11] Tolman (1936).

[12] Bridgman (1927).

[13] Interestingly, MacCorquodale and Meehl seem to have been unaware that Carnap, as early as 1936, had recognised that disposition terms could not admit of *explicit* (exhaustive) operational definition, because part of their meaning was counterfactual. This led him to invent his *bilateral reduction sentences*, which provide a "partial" definition of such terms, leaving them "open" or undefined in some circumstances. This is consonant with my realist line, developed below, to the effect that intervening variables in practice have a degree of "unoperationality" and "extra reference". See Carnap (1936—7).

[14] What follows here is a condensation of a lengthy argument for a realist construal of IVs presented in my (1967), Chapter II.

[15] I am indebted to John Forge for suggestions embodied in this section.

[16] See McLaughlin (1967), p. 44.

[17] Freud's force and drive concepts bear direct comparison with these notions in physics (mechanics), where *work* is done, or *energy* is expended, when a force moves through a distance or sustains an equilibrium (i.e. when a *causal interaction* occurs — see my discussion of Salmon's account of causality, Section 3 below). Freud's contrast between "bound" and "unbound" energy compares with the difference between potential and kinetic energy in physics. In the *Project*, he spoke of "quantity" of "psychic energy", symbolised by 'Q_η', which he conceived as the quantity of excitation in the neurones (which were divided into three systems — ϕ, ψ, ω). Although he never succeeded in elaborating these latter ideas to his satisfaction, his concern with the *physical basis* of psychology, and with the need for *quantitative, measurable* concepts is manifest.

[18] One defect of the 19th Century neurology in which Freud had been trained was its tendency to think of psychopathologies as due to *specific* local brain lesions, by analogy with "physical" illnesses like appendicitis or pulmonary tuberculosis. Nowadays it is clear that the brain and ramified neural systems work in a much more *holistic* manner. On the one hand, this insight reveals the immense difficulty of the task of linking neurology with psychology. On the other hand, it makes much more plausible such an ambitious enterprise, for "mental" properties now can be understood as properties of enormously complex organised wholes (or *Gestalten*), rather than of specific neurological entities such as neurones.

[19] Maze (1954).

[20] See especially my (1967), pp. 84—98.

[21] For instance, by Van Fraassen (1982), p. 205.

[22] Fine (1963), p. 16.

[23] Earlier versions of this paper were read by John Forge, Randall Albury, an anonymous referee, and John Maze, from some of whose comments it has benefited. Responsibility for the final product is solely mine.

BIBLIOGRAPHY

Bridgman, P. W. (1927) *The Logic of Modern Physics* New York: Macmillan.

Carnap, Rudolf (1936—7) 'Testability and Meaning' *Philosophy of Science* 3 (4), 420— 468 (1936); 4 (1), 1—40 (1937).

Clark, A. (1980) *Psychological Models and Neural Mechanisms* Oxford: Clarendon Press.

Eysenck, H. J. (1953) *Uses and Abuses of Psychology* Harmondsworth, Middlesex: Penguin Books.

Eysenck, H. J. (1985) *Decline and Fall of the Freudian Empire* New York: Viking Penguin.

Eysenck, H. J. and Wilson, G. D. (1973) *The Experimental Study of Freudian Theories* London: Methuen.

Farrell, B. A. (1981) *The Standing of Psychoanalysis* Oxford: Oxford University Press.

Feigl, Herbert (1967) *The 'Mental' and the 'Physical'* — The Essay and a Postscript Minneapolis: University of Minnesota Press (First published in *Minnesota Studies in Philosophy of Science* II Feigl, Scriven, Maxwell (eds.): 370—497 (1958)).

Fine, Reuben (1963) *Freud: A Critical Re-Evaluation of His Theories* London: Allen and Unwin.

Fisher, Seymour and Greenberg, Roger P. (1977) *The Scientific Credibility of Freud's Theories and Therapy* New York: Basic Books.

Forge, John (1982) 'Physical Explanation: With Reference to the Theories of Scientific Explanation of Hempel and Salmon' in Robert McLaughlin (ed.) (1982), pp. 211—229.

Freud, Sigmund (1895) *Project for a Scientific Psychology* in James Strachey (ed.) *Standard Edition* 1, 295—391.

Freud, Sigmund (1900) *The Interpretation of Dreams* in James Strachey (ed.) *Standard Edition* 4—5.

Freud, Sigmund (1901) *The Psychopathology of Everyday Life* in James Strachey (ed.) *Standard Edition* 6.

Freud, Sigmund (1910) 'The Origin and Development of Psycho-Analysis' in John Rickman (ed.) (1937), pp. 3—44.

Freud, Sigmund (1915a) 'Instincts and Their Vicissitudes' in Joan Riviere (ed.) (1959) 4, 60—83.

Freud, Sigmund (1915b) 'The Unconscious' in Joan Riviere (ed.) (1959) 4, 98—136.

Freud, Sigmund (1915c) 'Repression' in Joan Riviere (ed.) (1959) 4, 84—97.

Grünbaum, Adolf (1984) *The Foundations of Psychoanalysis* — *A Philosophical Critique* Los Angeles: University of California Press.

Hempel, Carl G. (1950) 'Problems and Changes in the Empiricist Criterion of Meaning' *Revue Internationale de Philosophie* 11, 41—63.

Hempel, Carl G. (1951) 'The Concept of Cognitive Significance: A Reconsideration' *Proc. Amer. Academy of the Arts and Sciences* 80 (1), 61—77.

Hempel, Carl G. (1954) 'A Logical Appraisal of Operationism' *Scientific Monthly* 79 (4), 215—220.

Hempel, Carl G. (1965) 'Aspects of Scientific Explanation' in Hempel, Carl G. *Aspects of Scientific Explanation* New York: Free Press, pp. 331—496.

Hempel, Carl G. and Oppenheim, Paul (1948) 'Studies in the Logic of Explanation' *Philosophy of Science* 15, 135—175.

Hook, Sidney (ed.) (1959) *Psychoanalysis, Scientific Method, and Philosophy* New York: New York University Press.

Jammer, Max (1957) *Concepts of Force* Cambridge, Mass.: Harvard University Press.

Littman, R. A. and Rosen, E. (1951) 'The Molar-Molecular Distinction' in Marx (ed.) (1951), pp. 144—154.

MacCorquodale, K. and Meehl, P. E. (1948). 'On a Distinction Between Hypothetical Constructs and Intervening Variables' *Psychol. Review* 55, 95—107. Abridged version reprinted in Marx, M. H. (ed.) (1951), pp. 103—111; page numbers cited refer to latter.

McLaughlin, Robert (1967) *Theoretical Entities and Philosophical Dualisms* — *A Critique of Instrumentalism* Ann Arbor, Michigan: University Microfilms.

McLaughlin, Robert (ed.) (1982) *What? Where? When? Why?* — *Essays on Induction, Space and Time, Explanation* Dordrecht: D. Reidel.

McLaughlin, Robert and Precians, Robert (1969) 'Educational Psychology: Some Questions of Status' in R. J. W. Selleck (ed.) *Melbourne Studies in Education 1968—1969* Melbourne: University of Melbourne Press, pp. 29—71.

Marx, M. H. (ed.) (1951) *Psychological Theory — Contemporary Readings* New York: Macmillan.

Maxwell, Grover (1962) 'The Ontological Status of Theoretical Entities' in H. Feigl and G. Maxwell (eds.) *Minnesota Studies in Philosophy of Science* III, 3—27, Minneapolis: Univ. of Minnesota Press.

Maze, J. R. (1954) 'Do Intervening Variables Intervene?' *Psychol. Review* 61, 226—34.

Maze, J. R. (1983) *The Meaning of Behaviour* London: Allen and Unwin.

Popper, K. R. (1962) *Conjectures and Refutations* New York: Basic Books.

Pribram, K. H. and Gill, M. M. (1976) *Freud's 'Project' Reassessed* London: Hutchinson.

Reichenbach, Hans (1956) *The Direction of Time* Berkeley and Los Angeles: University of California Press.

Rickman, John (ed.) (1937) *A General Selection from the Works of Sigmund Freud* London: Hogarth Press.

Riviere, Joan (ed.) (1959) *Sigmund Freud: Collected Papers* (4 Vols.) New York: Basic Books.

Russell, Bertrand (1948) *Human Knowledge, Its Scope and Limits* New York: Simon and Schuster.

Salmon, Wesley C. (1959) 'Psychoanalytic Theory and Evidence' in Hook (ed.) (1959), pp. 252—267.

Salmon, Wesley C. (1970) 'Statistical Explanation' in R. G. Colodny (ed.) *The Nature and Function of Scientific Theories* Pittsburgh: University of Pittsburgh Press, pp. 173—231.

Salmon, Wesley C. (1975) 'Theoretical Explanation' in Stephan Körner (ed.) *Explanation* Oxford: Basil Blackwell, pp. 118—145.

Salmon, Wesley C. (1984) *Scientific Explanation and the Causal Structure of the World* Princeton, New Jersey: Princeton University Press.

Scriven, Michael (1962) 'Explanations, Predictions and Laws' in H. Feigl and G. Maxwell (eds.) *Minnesota Studies in Philosophy of Science* III, 170—230, Minneapolis: Univ. of Minnesota Press.

Smart, J. J. C. (1963) *Philosophy and Scientific Realism* London: Routledge and Kegan Paul.

Strachey, James (ed.) (1953—1974) *The Standard Edition of the Complete Psychological Works of Sigmund Freud* London: Hogarth Press and the Institute of Psycho-Analysis (abbreviated '*SE*' in text).

Sulloway, Frank J. (1979) *Freud, Biologist of the Mind* London: Burnett Books, in association with Andre Deutsch.

Tolman, E. C. (1936) 'Operational Behaviorism and Current Trends in Psychology' *Proc. 25th Anniv. Celebr. Inaug. Grad. Stud.* Los Angeles: Univ. of California Press.

Van Fraassen, Bas C. (1982) 'Rational Belief and the Common Cause Principle' in Robert McLaughlin (ed.) (1982), pp. 193—209.

Wollheim, Richard and Hopkins, James (eds.) (1982) *Philosophical Essays on Freud* Cambridge: Cambridge University Press.

CHRIS SWOYER

THE METAPHYSICS OF MEASUREMENT

My thesis is that there are good reasons for a philosophical account of measurement to deal primarily with the properties or magnitudes of objects measured, rather than with the objects themselves. The account I present here embodies both a realism about measurement and a realism about the existence of the properties involved in measurement. It thus provides an alternative to most current treatments of measurement, many of which are operationalistic or conventionalistic, and nearly all of which are nominalistic.[1] This enables the present account to give better explanations of a number of features of measurement and other aspects of science than competing accounts of measurement can, and to be more readily integrated into a realist account of natural laws and causation. It also illustrates a general strategy for combining a familiar and powerful approach to representation with intensional entities like properties, which I think can be useful for dealing with a number of philosophical problems.

In Section I, I explain what I mean by realist theories of measurement and of properties, and defend each. In Section II, I argue that the influential representational approach to measurement provides a useful method for developing the ideas sketched in the first section, and I briefly recall its general features. In Section III, I adapt this approach so that the things represented are properties, for example various magnitudes of length and mass, rather than individual physical objects. I will be more concerned with metaphysical issues than with formal ones, but for purposes of illustration I will sketch a formal language for dealing with properties and formulate a standard set of axioms for extensive measurement in it. For simplicity, my examples will usually involve familiar properties and relations like length, rest mass, and subjective preference, rather than more exotic ones like gravitational-field intensity or the energy-momentum 4-vector, but if the account proposed here is on the right track, it will apply to a fairly wide range of properties and a variety of types of measurement. In Section IV, I examine ways of extending this approach and urge that, since it provides a method for characterizing various aspects of the structure of

235

John Forge (Ed.), Measurement, Realism and Objectivity, 235–290.

the world, it should be useful for dealing with a number of issues in metaphysics and other areas of philosophy. Because my concerns are with metaphysical issues, I will focus on the more abstract aspects of measurement, but this is not intended to suggest that other aspects are less important.

I. MEASUREMENT, PROPERTIES, AND REALISM

What realism in general amounts to is notoriously obscure, but I think that many of the issues concerning measurement and properties that divide the realist and anti-realist can be made clear enough for present purposes. Three sorts of realism will be important here: *realism with respect to measurement, realism with respect to properties, and realism with respect to theoretical entities in science (scientific realism)*. I do not want realism about measurement to be formulated in a way that presupposes realism about either properties or theoretical entities, and so I will first defend it using more-or-less observational magnitudes like length and without assuming that magnitudes are properties in any full-blooded sense.[2] I will then argue that the best way to develop a realist theory of measurement is to employ a realist account of properties. I do not regard my arguments here as demonstrative — I don't think that demonstrative arguments are to be found in many areas of philosophy — but I believe that their cumulative force is sufficient to motivate the development of a realist account of measurement in terms of properties.[3]

Anti-realism with respect to measurement can assume a variety of forms. The simplest is an austere operationalism of the sort expressed by the refrain that intelligence just is whatever intelligence tests measure. The idea here is that terms like 'intelligence', 'length', and 'temperature' do not stand for objective things in the world, but derive their meaning entirely from our measurement practices. According to an extreme version of the doctrine, the only objective facts about length, for example, are grounded in the ways in which we use rulers or other instruments to order objects and assign numbers to them. This outlook is a species of a more general and widespread view, according to which the fundamental facts about measurement are grounded in *conventions* about the way scientists use words like 'length', what they are willing to count as a measurement of length, and so on, rather than being rooted in objective facts about length itself (as in [43], 130ff.).

A much more sophisticated conventionalism has been developed by Ellis, who regards the ideas that there are objective properties like length, or true and objective orderings of objects according to their lengths, as serious errors proceeding from a "concealed realist standpoint" ([10], 2, 49). A notorious difficulty with classical operationalism was its inability to account for the fact that we often employ quite disparate operations to measure what is almost always viewed as a *single* magnitude. Ellis goes a good distance toward solving this problem with his carefully qualified development of the idea that measurement operations can be *said* to measure the same thing if they give rise to the same orderings of objects under the same conditions, and he usefully notes the sorts of considerations he believes guide our choices among alternative conventions that we might select ([10], 48). For current purposes, however, the important point is that Ellis too concludes that measurement involves a large dose of convention.

By contrast, I take realism with respect to measurement to be the view that in many cases measurement can give us information about objective features of phenomena that is tinged with few interesting elements of convention. This claim will be developed more carefully below, but the general idea can be conveyed with a simple example like length. The realist's thesis is that there are objective facts about what the length of something is, facts that are — within precisely specifiable limits — independent of our linguistic and scientific conventions, the particular theories we happen to accept, and the beliefs we happen to hold. Length can be measured on a ratio scale, and this means that once a unit (e.g., the meter) is conventionally selected, there will be an objective fact as to how many meters long any given object is (since this will just be a fact about the ratio of its length to that of the meter bar). Moreover, any other satisfactory scale for length will be related to this one in quite specific ways. Indeed, one way to view representational accounts of measurement is as attempts to isolate sufficient and (where possible) necessary conditions for there to be such objective facts about magnitudes like length.

The realism—anti-realism dispute over measurement is not usually cast in terms of semantic issues, but it is important to realize that they are there just beneath the surface. Classical presentations of operationalism, for example, often boil down to claims that the meanings of terms like 'length' are given by the operations used to measure length. Ellis' view is more subtle, for he maintains that concepts like length are cluster concepts and that no one feature in the cluster — e.g., any specific procedure for determining length — necessarily defines the concept. Other forms of anti-realism incorporate the view that the meanings of at least the more theoretical terms like 'charge' and 'temperature' are "implicitly defined" or "partially interpreted" by the theories in which they occur. Like Ellis', such accounts are versions of the so-called "cluster-theory" of meaning, according to which the reference or extension of a term is given by a cluster or disjunction of descriptions — in this case the descriptions involve general claims embodied in the theory. These views about meaning play a more important role in the various accounts of measurement, and hence in the dispute over realism with respect to measurement, than a first glance at the debate would suggest. Consequently, an adequate realist account of measurement must include some treatment of the semantics of magnitude terms.

I believe that there are a number of reasons to be a realist about measurement. They are most easily brought out by noting several important facts about measurement that are not easily explained by the anti-realist. First, simpler versions of anti-realism cannot account for the fact that we often use quite different operations to measure what is commonly taken to be the *same* magnitude. Quite different procedures are used to measure lengths and distances in astronomy, geology, histology, and atomic physics, for example, and we may measure small lengths using X-ray diffraction, altitude with an altimeter, larger distances with radar, huge ones by determining the red shift, and so on. Indeed, it was not so long ago that scientists could not measure astronomical distances or the diameters of atoms at all. The assumption that there are nevertheless facts about such things seems to have been frequently borne out, however, and it has often been an important stimulus to scientific progress.

The use of different procedures to measure what we take to be the same magnitude is not a practical contrivance to avoid more magnitudes — ruler length, sonar length, etc. — than we can keep straight. It plays an important methodological and epistemological role in science, enabling us to obtain a general and unified picture of nature that minimizes the number of unrelated, brute facts by displaying the same laws and causal mechanisms at work behind apparently unrelated phenomena. And realism about measurement offers a simple explanation for both the *legitimacy* and the *success* of this practice of measuring the same magnitude in quite different ways. Even a fairly small group of objects can be ordered in a good many ways, and it is difficult to see why vastly different operations would so frequently yield quite similar orderings if they weren't reflecting the same facts about the world. According to the realist, there *are* objective facts about length that transcend particular methods of measuring it, and the reason why different measurement procedures so often yield similar results is that they are sensitive to the same facts. This also helps explain why things often work so well when we devise new procedures for measuring things on the basis of complicated theory, for if the theory is a good one, it too should be sensitive to facts about lengths, masses, and the like.

Second, measurement error is a fact of scientific life. In many sciences it is expected that estimates about the magnitude of error be reported along with results, and the theory of errors is now a well-developed field. Yet it is difficult to see what it is to be in error — to get things wrong — if there is no such thing as getting them right. To be sure, there are various accounts the anti-realist might offer about measurement error. For example "correct" measurements might be identified with measurements made in various idealized situations, in the limit of inquiry, or the like. But aside from their unclarity, such notions as ideal circumstances would strip the anti-realist position of its empiricist flavor, which provides its chief motivation in the first place. Of course the anti-realist could bite the bullet and maintain that our traditional picture of measurement error involves a realistic prejudice that should be abandoned ([10], 49ff.). But in light of the other attractions of realism about measurement, it seems to me worth looking for an account that makes sense of it.

Third, alternative procedures of measuring a given magnitude often do not agree completely. As technology and theory progress, we typically obtain slightly different measurements of (what is taken to be) the same magnitude, with later results serving to correct earlier ones; experimenters, for example, are still trying to make more accurate determinations of the values of such fundamental constants as the universal gravitational constant. This is easily explained if there are objective facts about magnitudes like length that we learn more about in the course of time, facts to which our successive claims approximate more closely.

Finally, anti-realism about measurement will lead to anti-realism about many aspects of science, since objective facts about even such familiar things as length or mass won't be countenanced. Those who think that there are good reasons for accepting at least a modest realism about laws, causation, or the like will find this a strike against anti-realist accounts of measurement.

Such considerations show that realism about measurement has much to recommend it; I will now argue that the most satisfactory way to develop a realist account of measurement requires properties and relations. On the view I will defend, properties and relations are genuine entities in their own right. Elsewhere I have argued that we have good reasons to believe in their existence because of their ex-planatory value, and that we learn what they are like by seeing what something would *have* to be like to fill the *explanatory roles* they are invoked to fill ([38], [39], [41]). This means that properties are the-oretical, in the sense that our belief in them is justified by the fruitful-ness of (philosophical) theories that require them, but it doesn't mean they must be unobservable — though some of them might be — as theoretical entities are sometimes said to be. To call a given length or color an observational *property* simply means that the magnitude can be theoretically redescribed *as a property*, much as the perfectly observ-able sound of a bell can be theoretically redescribed as a conditioned stimulus.

On this conception of properties and relations, they are altogether different from intensions, that is, from functions that pair possible worlds with sets of n-tuples of objects in them. Intensions can often provide a useful representation of properties in clear, set-theoretical terms, but this should not lead us to conclude that properties and relations *are* intensions, as some philosophers do. We can, for example, recognize previously unencountered and unimagined instances of nu-merous properties *as* instances of those properties. Yet we don't — we couldn't — learn what things are in some large set like the set of all human beings, much less what is in the set of human beings in every world. The natural explanation of our ability to recognize novel instances of a property is that we are attuned to it, perhaps in simple cases because the property produces certain sorts of perceptual ex-periences in us. But whatever the details, we couldn't be attuned to intensions.

We frequently specify units of measurement for magnitudes like length and time in a general way; for example, rather than appealing to an object like the standard meter bar, we now think of the meter in terms of something that can be instantiated in distant laboratories, say as *the length* equal to a certain number of wavelengths (in vacuum) of a particular color of light emitted by krypton 86 atoms. Moreover, many measurements do not involve anything like an ordering of different objects but, intuitively, measurement of some magnitude (like temperature) of a single object at different times. These intuitions and observations suggest that we often think of measurement as involving properties, and they can be backed by a number of arguments to this effect. In particular, there are a number of specific features of science that suggest the need for properties and relations, and given the central role of measurement in science, nearly all of them support the view that measurement involves them too.

Quantification over properties and relations is common and important in science, and in many cases it does not look like it can be paraphrased away or otherwise dismissed as a figure of speech. As we will see, a central claim in many discussions of measurement is that a property is meaningful or objective just in case it is preserved under a certain group of transformations. Indeed, talk of transformations is a common way of defining large classes of properties; for example, a geometrical property is projective just in case it is invariant under all transformations of a given sort.

Many important properties like those of *being a belief, being a pain*, or *being a gene* are quite plausibly thought to be functional properties. They are especially important in fields like cognitive psychology and artificial intelligence, where we want to be able to investigate properties like belief without first having to identify them with specific physical properties. In such cases, to say that P is a functional property is to say that x exemplifies P just in case *there are* physical properties $Q_1, \ldots,$ Q_n that x exemplifies and that these play a certain functional or causal role (e.g., they realize some particular theory). The quantification over properties is especially important here, since we want to allow the possibility that widely different — and perhaps infinitely many — constellations of properties could play the same functional role.[4]

Elliot Sober argues that well-confirmed theories in biology contain claims that in particular types of environments given *sorts* of properties have selective value, and he makes a strong case for thinking that such quantifications over properties are ineliminable ([33]). Perhaps one reason for this is that such claims involve functional properties; at all

events, like many functionalistic claims, they involve causation, and
there is strong reason to suppose that causation and related notions
involve properties. We often think of an object's or event's having some
property as causing (or being causally relevant for) some other object's
or event's existing or having some particular property: The liquid in the
glass caused the litmus paper to turn blue because the liquid is an
alkaline (not because the liquid also happens to be blue), a force of this
magnitude causes an acceleration of that amount, smoking tends to
cause cancer. Explanations also frequently advert to properties, often
because they cite causes; the liquid's being an alkaline, for instance,
explains why it turned the litmus paper blue.

We frequently hear claims to the effect that one thing, say temperature, is reducible to another, say mean molecular kinetic energy or that
things of one sort, e.g., mental states, are supervenient on things of a
different sort, e.g., physical states. The most natural way to interpret
such remarks is as claims about the relationship (perhaps identity,
perhaps something weaker like correlation) between various properties
(perhaps, in the case of supervenience, infinitely complex properties).
There are also a number of reasons for thinking that the only way to
account for various features that natural laws are commonly acknowledged to have, e.g., being confirmed by their instances or supporting
counterfactuals, is to regard them as involving properties ([2], [3], [8],
[38], [42]).

Finally, it is difficult to find a semantics that is compatible with realism concerning measurement — or with any sort of realism in science
— unless we avail ourselves of properties. Writers like Feyerabend and
Kuhn maintain that theoretical terms draw their meaning from the theories in which they occur, so that a change in a theory causes a shift in
the meanings of all its constituent terms — with the result that different
theories simply talk about different things and cannot be rationally
compared. The common, and I believe correct, realist response to such
versions of semantic holism is to argue that the denotations of terms
can remain constant through some changes in a theory and, indeed, that
different theories may contain terms that denote the same thing. Since
denotation is what determines truth value, one theory may make a
claim about a given sort of thing, e.g., rest mass, which a competing
theory denies. Insofar as epistemological holism is supported by semantic holism, as it is in philosophers like Quine, this response also opens
the way for less holistic accounts of the testing and confirmation of

theories. But this response cuts little ice unless *there are* things that the terms of a theory can have as denotations, and since most terms in theories are predicates and function symbols, the only very obvious candidates for their denotations are things like properties.

Nearly all of these considerations bear on measurement in one way or another, since we may want to measure functional magnitudes (degrees of belief, subjective utilities), magnitudes involved in causal relations (the amount of force is related to the amount of acceleration), and so on. Less centrally, various modal notions play a role in measurement. For a set of bathroom scales to be working properly, it is not enough for them to just happen to register the right weight when you step on them. It is also important that if your weight had been different (within limits), the scales would have given a different, but still correct, reading. We may not want to build such facts about counterfactual sensitivity into a formal account of measurement (though the account below would allow it), but we shouldn't develop an account that makes it impossible to explain them. One explanation is that there is a lawful dependence between the various magnitudes of weight and readings of the scales.

All of this suggests that we view measurement as involving properties of the sort that W. E. Johnson called *determinates*, specific lengths, like the property of *being two meters long*, rest masses, and so on. We also need higher-order properties and relations that can be exemplified by such determinate properties, for example the relation *being longer than* that holds between two determinate lengths just in case the first is longer than the second. This is not intended in any way to deny that our measurements involve individual objects, but only to assert that the facts discovered in such measurements involve properties.

So far we have found much to recommend realism with respect to measurement and with respect to observable properties, but what about *un*observable properties like determinate charges? Debates over scientific realism typically involve general and elusive issues, and one way to come to grips with them might be to ask what realist formulations of specific theories like classical mechanics or thermodynamics or Gibson's ecological psychology would look like. There are doubtless competing criteria for success in constructing realist interpretations of a theory, e.g., historical faithfulness and philosophical illumination, but in many cases the problem is not to choose among alternative realist interpretations, but to devise even one. I am inclined to think, for example, that much of the difficulty in saying anything very satisfying about measurement in quantum mechanics stems from the fact that we lack a satisfactory realist conception of quantum phenomena, rather than from any problems peculiar to measurement.[5]

In the case of at least many theories the standard arguments for realism seem quite powerful. Realism unifies, allowing us to view a wealth of apparently diverse phenomena as instances of the same sort of thing obeying the same laws (as bodies with mass being acted on by gravitational forces, as forms of electromagnetic energy, as techniques for increasing fitness). It fills in causal chains between observable phenomena with connecting micro-events, provides some explanation why science works as well as it does, offers a non-anthropocentric ontology, and holds out the hope of a reasonably straightforward semantics that takes our sentences pretty much at face value. The drawback is that it is epistemologically less secure than empiricism and conventionalism. But the epistemological problems are diminished if we accept a naturalistic epistemology — which has much to recommend it anyway — that acknowledges that an unshakable foundation for knowledge simply isn't in the cards and encourages us to use empirical methods to discover how we come to know as much as we do. Such an epistemology *can* make room for properties of the sort I have defended, ones that exist here in the actual world and whose causal powers are responsible for our perceptual experiences, experimental effects, and other phenomena around us. We can learn about such properties in empirical, naturalistically respectable ways — by observing their instances and making inferences on the basis of such observations — rather than through some mysterious faculty of intuition or the like ([38], [39]).

This approach to epistemology, together with the virtues of realism noted above, supports a modest, *discriminating* scientific realism — one in which we don't accept global claims about the commitments of science in general, but consider each theory on its own terms. Within this framework, there is no *a priori* reason to banish unobservable properties, and hence no reason to think that the following account of measurement could not apply to them.

I think that the arguments in this section show the need for *some* properties in an account of measurement, but philosophers who agree with this general conclusion might hope to avoid determinate properties like *being 3 meters long*, straying from nominalism only to admit a few qualitative, *relational* properties like *being less than in length* that hold between individual objects. One could then regard determinate magnitudes as equivalence classes under relations like *being the same length as*. Russell called such approaches relational accounts of magnitude ([30], Ch. 19), and although he rejected them, various philosophers since have found them appealing (e.g., [10], 25; [23] — though these philosophers are nominalists who would not regard *being less than in length* as a genuine relation). An object's position or velocity is not an intrinsic, monadic property of it — not something that it would have even if there were no other objects in the world — and the formal similarities that one can find, or concoct, between relational accounts of position and relational accounts of extensive magnitude are sometimes taken to show that we shouldn't think of determinate lengths or rest masses as genuine monadic properties either. Instead facts about length and mass are supposed to be somehow grounded in facts about relations like *being longer than* or *being a greater mass than*.

However it seems to me that there is some intuitive sense in which the spatial locations of objects can be changed without altering any of their intrinsic features, whereas one cannot change the relative lengths of two objects without altering something in them. Put another way, the relation *being longer than* does not seem capable of relating two objects whatever their monadic properties might be; it relates them *because* of monadic properties they have, namely their lengths. I don't want to rest much on such intuitions, though, for it would be difficult to spell out such notions as an intrinsic feature of an object, and besides these claims will probably strike the relationalist as question begging. But together with the following four considerations, I think that they motivate an attempt to work out a non-relational account of extensive magnitudes and measurement.

First, there are important differences between relational accounts of space (or space-time), on the one hand, and relational accounts of extensive magnitudes, on the other. One of the main attractions of relational accounts of space is that spatial points play no causal role in the world and it is difficult to see how we could ever know anything about them. Exemplified properties, by contrast, do have causal powers, including ones to influence our perceptions. Second, on a simple relational account of length, the claim that everything doubled in length overnight would seem to lack empirical content, since ratios of length would remain the same. Yet there are ways in which we could tell that everything had doubled in length, since various laws tie lengths in with other magnitudes. No doubt the relationalist can complicate his account to deal with such matters, but they show that a detailed relational account would lack much of the intuitive appeal that simpler sketches of the idea may have.

Third, as we will see, many theories of measurement include closure and Archimedean axioms that postulate the existence of objects that may not actually exist. Theories incorporating determinate properties can accomodate this fact; it is not clear how qualitative relational theories can. Fourth, although relational accounts require fewer properties and relations than non-relational accounts, the edge in parsimony is slight, for philosophical economy is achieved by having as few *kinds* of things as possible, not by scrimping on things of a given kind. After all, properties and relations have rarely been maligned on the grounds that there are *too many* of them; the charges have instead been that it would be difficult to gain knowledge about them, that their identity conditions are obscure, or the like. And if these objections are telling, they will not be countered by cutting back on properties.

II. MEASUREMENT AS REPRESENTATION

It will be useful to begin by recalling a few simple, but quite central, types of measurement. Many magnitudes in physics, including length, mass, time, and charge, are said to be *extensive*. Roughly speaking, this means that the things being measured can be ordered with respect to the property being measured and that there is an addition-like, "summation" operation on them, so that, for example, it makes sense to say that the combined masses of two objects is equal to the mass of a third. As we will see, when the ordering and the operation satisfy certain fairly straightforward conditions, the objects can be measured using a *ratio scale*. In such a situation, ratios of a magnitude are the same regardless of unit, and a ratio scale is said to be *unique* up to a choice of unit. It is a matter of convention which object or property we select as the unit, but because there is an objective fact about the ratios of the lengths of objects — including that of the length of each to the unit — once we select a unit the scale values of all the other objects fall into place. Since the work of S. S. Stevens, it has been common to define scale types in terms of the group of their "admissible transformations"; for extensive measurement, these are similarity transformations, those of the form $\phi^* = \alpha\phi$ ($\alpha > 0$). This means that if ϕ is an acceptable scale, then ϕ^* will be too. The two scales will convey the same objective information, since ϕ^* simply represents a change of unit (if we are converting feet to meters, for example, $\alpha = 0.30480$).

Some properties do not admit any natural summation operation, but we may still be able to order differences among them; for example, the difference between the temperature yesterday morning and yesterday evening might be greater than the difference between the temperature this morning and this evening. This applies to temperature (as it is measured in everyday life), calendar time, and is sometimes thought to be true of judgements of preference or scores on intelligence tests. Such properties can often be measured on *interval scales*. Interval scales involve two conventional decisions — choice of a unit and choice of a zero point — and are unique up to positive linear transformations (those of the form $\phi^* = \alpha\phi + \beta$, $\alpha > 0$; in converting from Celsius to Fahrenheit, for example, $\alpha = 9/5$ and $\beta = 32$). For other properties, the best we can do is an *ordinal scale*, one that is unique merely up to order. A frequently cited example is the Mohs scale of hardness, which assigns numbers to minerals in accordance with the rule that if one is harder than another, as determined by the scratch test, it is assigned a higher number than the other.

The chief role of measurement in science is to enable us to *classify* things in such a way that we can fruitfully apply our mathematical concepts, theories, and techniques in reasoning about them. Consequently a central question for a philosophical theory of measurement is a traditional one from the philosophy of mathematics, and, indeed, from metaphysics: how can an abstract body of mathematical theory apply to concrete things in the world?

I think that the *best explanation* why a mathematical theory applies to certain concrete phenomena — when it does — is that the theory involves some of the same structural features as the phenomena. On this account, when we establish a measurement scale, for example one for measuring lengths in meters, we set up a *correspondence* between the lengths of objects and the positive real numbers. Each length is paired with a corresponding *numerical counterpart* or *representative* (e.g., the standard meter bar, or its length, is paired with 1), and fundamental, qualitative relational properties of lengths like *being longer than* are correlated with *corresponding numerical relations* like $>$. Roughly speaking, we can say that the representing numerical system has the same structure as a set of objects of varying lengths if certain concrete objects stand in a given relation just in case their numerical counterparts stand in the corresponding numerical relation; for example, it might be that the number representing the object a is less than the number representing b just in case a is shorter than b. This picture is a very influential one; it was developed by Helmholtz, Hölder, Campbell, Stevens, and has been elaborated by Suppes and his collaborators (e.g., [16]; [36], pt. I). It is sometimes known as the *representational approach to measurement*.

Contemporary workers in the representational tradition use what I will call *extensional relational structures* to make these ideas precise. Such a structure is an ordered set, $\langle C, (R_i) \rangle_{i \in I}$, where C is a non-empty set, I is an index set $\{1, \ldots, m\}$, and the R_i are n_i-ary relations-in-extension on C, i.e., they are subsets of $C^{n(i)}$. Such structures can provide useful models or representations — *in the intuitive senses of these words* — of actual and possible facts, situations, or states of affairs, representing atomic facts by telling us which objects in C stand in which relations to each other. Models typically represent phenomena in a selective, condensed, and systematic way that highlights some things while ignoring others. Different sorts of models have different virtues; for instance some represent phenomena in geometrical or even visualizable form; others facilitate calculation. Relational structures can provide useful models because they are set-theoretic entities, and when we use them we can avail ourselves of the concepts, techniques, and results of a rich and powerful theory in which a great deal of mathematics can be developed. Relational structures are also rich enough that additional notions like isomorphism, substructure, and direct product also become available, and many interesting situations can be modeled by well-known sorts of structures like linear orderings or vector spaces over the reals. Finally, if we interpret a formal language in an extensional relational structure, we can use the tools and results of model theory.

When we use an extensional relational structure to model an actual situation, we abstract from many of the situation's features, concentrating on those aspects that we represent by relations in our structure. All this bears on our present concerns because science typically deals with various types of situations by abstracting from all but a few of their properties and relations. When we are concerned with measuring length, for example, an object's smell and taste are usually irrelevant. Similarly, we abstract from all but the projective properties of objects when we do projective geometry, or from all but mechanical properties like position and momentum when we do classical mechanics. Indeed, intellectual progress often goes hand in hand with such abstraction, prescinding properties like momentum or energy from the welter of observed phenomena. This is important because it allows us to view apparently diverse phenomena as being similar in theoretically relevant ways, as manifesting the same projectible properties and obeying the same laws. A relational structure allows us to represent such features of objects as relations, and laws can often be represented by axioms constraining their behavior.

Representational accounts of measurement associate different types of measurement with relational structures of specific similarity types. For extensive measurement we have an ordering relation, $>$, and a three-place, summation relation, $\circ$, which relates three objects just in case the "sum" of the first two is equal to the third. For difference measurement, we can get by with a four-place relation that holds among four things just in case the difference between the first two is less than the difference between the second two. With ordinal measurement, we will have only a binary ordering relation. On the mathematical side, we will want numerical relational structures of the same similarity type as the concrete extensional relational structures that they are intended to represent. There is typically some room for choice here, but by way of illustration the mathematical relational structure for extensive measurement might be $\langle R^+, >, + \rangle$, where R^+ is the set of positive real numbers and $>$ and $+$ are the obvious relations on it (the latter relation being a function). With an interval scale, we might map to $\langle R, D^4 \rangle$ (where $D^4 wxyz$ iff $w - x > y - z$), and with an ordinal scale to $\langle R^+, > \rangle$. In each case, a measurement *scale* would be thought of as a mapping from objects in the domain of the concrete relational structure to their numerical counterparts in such a way that familiar relations between numbers mirror relations between objects back in the concrete structure.[6]

Two relational structures C and R have the *same structure* in the strongest possible sense just in case they are *isomorphic*, that is, just in case there is a $1-1$ function ϕ (an isomorphism) from C onto R that preserves structure. ϕ preserves structure just in case, for every relation R_i^C of C, *corresponding relation R_i^R of R*, and n-truple $\langle a_1, \ldots, a_n \rangle$ of elements of C,

$$R_i^C(a_1, \ldots, a_n) \text{ if and only if } R_i^R(\phi(a_1), \ldots, \phi(a_n)).$$

The requirement that ϕ be onto turns out to be rather stringent, and we will typically settle for C's having an *isomorphic embedding* in R. This means that C is isomorphic to a substructure of R; it has an isomorphic copy inside R.[7]

The function ϕ, which correlates objects of C with their numerical counterparts in R, is a scale of measurement. We can now see why the application of the mathematical theory of R to the objects of C is justified when the above biconditional is true. For if it is true, then whenever an atomic fact, $R^C(a_1, \ldots, a_n)$, holds in C the corresponding atomic fact, $R^R(\phi(a_1), \ldots, \phi(a_n))$, will hold in R (and conversely for atomic facts of R whose constituents are all counterparts of elements of C). Thus we can begin with empirical facts about concrete objects in C, use ϕ as a bridge to their counterparts in R, mobilize logic and the mathematical theory associated with R to infer that further facts obtain in R, and finally travel back along the inverse of ϕ to a conclusion about our original concrete objects.

Within this framework, the central task is to devise qualitative (i.e., non-numerical) axioms which will ensure that a given type of measurement is legitimate. The axioms should be sufficient (and, if possible) necessary for two things. First, they should allow us to prove a *representation theorem*, which guarantees the existence of an isomorphic embedding of each model of the axioms in an appropriate numerical structure R. Second, the axioms should allow us to prove a *uniqueness theorem*, which tells us "how unique" the representation is; we will consider its significance in a moment, but first it will be useful to step back and consider the representational approach in a larger perspective.

If a representational account of a given sort of measurement is to be of any but mathematical interest, its primitive terms like '$>$' and '∘' must be interpretable as non-contrived (e.g., non-Goodmanesque), real-life relations of genuine scientific interest. Moreover, the axioms containing them should (perhaps with some exceptions discussed below) be thought of as empirical, testable, lawlike claims about these relations. Many exponents of representationalism go a step further, holding that terms like '∘' must be given a very simple and direct empirical interpretation via operational definitions, or at least that such operationalizations provide a partial interpretation for them or related terms (e.g., [13]. 68). For example, in the case of the measurement of mass using an equal-arm balance, we might interpret '$x \circ y$' to mean that x and y are placed together in one pan of the balance and '$x > y$' to mean that when they are placed in different pans, y's pan rises and x's pan falls.

The central ideas of representationalism are quite rich and general, however, and it seems to me that the representational method can be employed to serve at least three ends, only one of which need involve operationalism or any strict empiricism. First, the representational method can be used by those with an interest in practical methodological problems concerning measurement. Second, the method can be used to clarify the evidential or operational basis of a given sort of measurement. Third, the method can be used to help characterize the structural features of the world that underlie various sorts of measurement.

My interest here is primarily with the characterization of structures. This use of the representational approach may itself have different motivations. It may be found in Suppes' own work, particularly when he extends the representational approach from measurement to a variety of substantive scientific theories ([37], [36] pt. iv), Field adopts it in an effort to show that physics does not really make any ontological commitment to numbers ([12]), and Mundy uses it primarily in a critical study of the foundations of current physical science ([21], [22]). My interests are different from Field's, and differ from Mundy's and Suppes' at least in emphasis, for I am more concerned with metaphysical questions than with questions about the details of current theories. My hope is to use the representational method to provide a direct characterization of certain aspects of the structure of reality, those that underlie — and so account for the success of — successful measurement. This would also explain the applicability of mathematics to reality, for the key here is that certain aspects of the world have the same structure as some mathematical system, not that operational tests reveal that they have the same structure.

More generally, I believe that the use of relational structures to model or represent or map a given sort of phemonena can be an important one in metaphysics. In many ways it is reminiscent of the logical atomism of Russell and Wittgenstein. Writing almost twenty years before the advent of model theory, it is not surprising that they spoke of isomorphisms between reality, on the one hand, and language (rather than a relational structure), on the other. But their goal of devising a clear and rigorous mapping of various aspects of the structure of the world is a venerable metaphysical one, and in Section IV, I will suggest several ways of applying an intensional version of the representational apparatus to achieve this end.[8]

If our goal is a rigorous characterization of certain aspects of the world, the ideal is a categorical axiomatization — one that characterizes the sorts of relational structures involved up to isomorphism — for this is the most complete formal characterization possible. Failing this — perhaps because we don't know enough to be sure about the cardinality of the sorts of structures involved, perhaps because different cardinalities are possible — we would still like to come as close to capturing the intended interpretations as possible. This may well require the use of a logic considerably more powerful and less well-behaved than first-order logic, but given our purposes, this is not a serious defect. For our interest is *not* in anything linguistic — *not* in any particular axioms we might adopt — but in the class of models that the axioms characterize.

In stressing the metaphysical use of representationalism, I do not mean to suggest that other uses of the approach are less interesting or less important or that they are incompatible with it. Many of the original practitioners of the representational approach were psychologists with practical questions about which statistics could legitimately be used in connection with various types of scales or about how measurement might work when no summation operation was available. I believe that such concerns were often needlessly mixed with concerns involving operationalism, perhaps because of the prominence of behaviorism and positivism in psychology at the time when the formal elaboration of the representational approach began, and that the method's use in providing characterizations can play an important role in discussions of methodological and practical problems in measurement.

To note a simple example, a characterization of the structure of a family of magnitudes would be quite relevant to questions about the meaningfulness of the uses of different statistics in different cases. Again, we may recall the numerous attempts that have been made to provide a qualitative axiomatization of subjective utility and probability at the same time. One way to interpret some accounts of this sort involves offering a subject a number of gambles, two at a time, and letting a qualitative, observable, preference ordering for these be determined by his choices among them. On the basis of this and a bit more, e.g., the assumption that he acts so as to maximize his subjective expected utility, a utility function and a subjective probability distribution may be constructed. The details don't matter here; the point is just that in many such approaches the subject's beliefs and desires are constructed from the choices he makes, so that beliefs and desires are indissoluably linked with observable, qualitative, behavioral relations. As we move away from positivism and behaviorism, it appears more natural to think of beliefs and desires as genuine states that are causally responsible for our actions — including the gambles we choose — but not characterizable in any simple way in terms of behavior or measurement operations. On this picture, representationalists can still hope to tell a metaphysical story about the structural features of an agent's beliefs and desires that is sufficient for the existence of the appropriate sorts of numerical representations of his subjective utilities and probabilities. The axioms in such stories would not be interpreted in terms of simple observable relations, but that doesn't mean that they would not be important when it came time to test hypotheses about the agent's beliefs and desires. For such axioms place quite stringent constraints on the relations between his mental states, and so together with other assumptions could generate a wide variety of testable predictions.

With the shift of emphasis from operationalism to characterization, such traditional matters as the distinction between fundamental and derived measurement fade in importance, but this does not mean that questions about the testability of measurement theories become unimportant. Theories of measurement are still empirical theories, though this may be obscured by the fact that some of their axioms, e.g., the claim that the summation operation is associative, seem like truisms when applied to familiar properties like length and mass. But even such truisms are not knowable *a priori*; indeed conceptual revolutions in science, for example those in relativistic and quantum mechanics, can overthrow very central beliefs about quite familiar magnitudes. Furthermore, outside the physical sciences it often is a pressing empirical question whether a given family of properties satisfies a particular axiom; for example, it has been debated whether preference is transitive or can be aggregated in an addition-like way.

From the current perspective, there is little point to talk about *the* empirical or observational content of measurement axioms, taken one-by-one. Once we move away from operationalism and other exuberant empiricist doctrines like the partial interpretation picture of theories, few sentences of interest have such content. Consequently, although there can be quite genuine philosophical problems about the testing of various measurement axioms, these problems do not appear to be ones about measurement as such. They are instead instances of more general problems about the nature of evidence, the description and reduction of data, the testing and confirmation of theories, the foundations of statistical inference, the theory of error, and so on. Such problems are obviously important, but since they are not uniquely ones about measurement, I will not discuss them here.

We are now in a position to deal with an apparent problem for my earlier claim that the best way to secure a realist account of measurement is to adopt a realist account of properties. The problem is that standard presentations of representationalism make absolutely no mention of properties, yet they appear to be quite compatible with realism. Matters are not so simple, however, as the case of extensive measurement will illustrate. The simplest accounts of the extensive measurement of, say mass, assume that the summation operation is closed, so that for *any* two objects there exists a third — their "sum" — that is equal in mass to the combined masses of the original objects. But as Mundy has stressed, it is most unlikely that the required sums, construed as normal physical objects, will always exist [21]. Many representationalists agree that the relevant sums don't exist in the case of objects that are too big to put together in any pan balance, too long to place end to end in any lab, and so on, and they abandon the closure requirement for such cases. For example Hempel suggests recasting axioms for extensive measurement as conditionals, asserting, for example, that *if* the sum of two objects exists, it satisfies such conditions as commutativity and associativity ([13], 86).

As it stands Hempel's suggestion is too unsystematic to yield representation and uniqueness theorems, and Krantz et al. adopt a different approach ([16], Ch. 3.4). Their goal is to capture the intuitive idea that the summation operation should be closed for all objects that will "fit together in the lab"; for example, if the sum of a and b exists and c is less massive than b, then their axioms require that the sum of a and c exist ([16], Ch. 3.4). They realize that even here the sum of two objects may not *actually* exist, but they attempt to finesse the point, taking their theories as idealizations whose intended domains are sets of "all realizable entities" of some sort (e.g., rigid rods) and concentrating on "potential observations" of pairs of objects that "can be concatenated." ([16], 26—27, 81—87; [26], 223). So in cases where no *actual* object is the sum of two small, concatenatable objects, the requirement that pairs of small objects have a sum seems to mean that there *could* have been an object equal in mass to the combined actual mass of the two objects.

If we are concerned with the evidential basis of actual measurement practices — with measurements of mass using an equal arm balance, to take an extremely simple example — then when a and b are small enough to put in the same pan, it may well seem an irrelevant accident that no object c exists which is their sum. In this case, the idealization that such an object exists is harmless. By contrast, it will not seem a mere accident that some objects are too big to be placed together in any balance we could construct. *Given* the concern with actual measurement procedures using balances, this is an essential limitation that should not be idealized away by assuming the existence of potential sums of huge objects.

Idealizations involving possible objects and potential observations are much less appealing when we shift attention to the metaphysical foundations of measurement, to questions about the *truth conditions* for sentences ascribing scale values to things in the world. There seem to be two ways to make metaphysical sense of nominalistic theories that require closure of ○ in cases where no natural physical objects correspond to the sums of various pairs of actual objects. The first is to axiomatize a theory of measurement in Leonard and Goodman's calculus of individuals and take the "concatenation sum" of two objects to be their Goodmanian sum. Even if we learn to live with the unnaturalness of scattered objects, however, this means that two objects cannot be placed in one pan while their concatenation sum is placed in the other. And even if we are not so concerned with actual measurement practices, in many cases it is important that the "concatenation sum" of an object with itself exists and that it be greater than (e.g., longer than or heavier than) the object itself. But the Goodmanian sum of b with itself is just b.

The second idea is to take the representationalists' talk about possible objects as a more serious piece of metaphysics than they probably intended it to be and introduce possibilist quantifiers that include merely possible individuals in their range. The problem with this is that it is unclear what sort of epistemological, semantical, and causal contact we have with merely possible individuals. How can we ever have knowledge — or even justified beliefs — about them; how can terms of our language come to denote them; how can facts about them make our sentences in this world — for example sentences about actual measurements of actual objects — true or false? No doubt various response can be given to these questions, but the fact remains that merely possible individuals are no more attractive than properties and relations; they are no easier to individuate or to acquire knowledge about, for example. Moreover, we have already found good reasons to think that properties and relations exist and that they play a role in measurement. And since, as I will argue, they can do the work that the standard representational account reserves for merely possible objects, it is better to base a realist account of measurement on them.

Before working properties into the picture, let us return to the uniqueness theorem. The claim that Kay's age is 40 is true if we measure age in years but false if we measure it in days, and so the number 40 does not signify anything about objective, scale-independent facts involving Kay's age. But if Kay is twice as many years old as Susan, then Kay is twice as old as Susan *period* — this is an objective fact about their ages that is completely independent of the units we use for measuring age; as it is sometimes put, this ratio is invariant under any legitimate transformation of scale. By contrast, the temperature yesterday was 20°F and today it is 40°F, but if we had instead used the celsius scale, which is equally informative, we would have said that the two temperatures were −6 2/3°C and 4 4/9°C. So the claim that today it is twice as warm as yesterday does not reflect an objective, scale independent fact about the weather; it is an artifact of our choice of scales and is only true relative to the Farenheit scale. On the other hand, ratios of differences between temperatures would be the same whichever scale we use.

The moral is that *not all* of the numerical properties and relations of a number system need be numerical counterparts of objective properties and relations of the phenomena that the numbers are used to represent. One point of a uniqueness theorem is to give us a handle on which numerical properties represent something objective in the phenomena (like the fact that Kay is twice as old as Susan) and which do not (in our story, which does not involve an absolute scale for temperature, there is no objective, scale-independent fact that it's twice as hot today as yesterday).

In his seminal work on these topics S. S. Stevens classified scales in terms of their invariance properties. His idea was that the "admissible transformations" of a scale are those that won't "lose any empirical information" (e.g., [35]). Spelling these ideas out in a general way turns out to be difficult and is still a matter of some controversy. According to an early attempt, a statement involving scale values conveys objective, scale-independent information — or, as it's often put in the literature, *is meaningful* — just in case it has the same truth value under any admissible transformation of scale ([36], Ch. 5). The notion of a statement's involving numerical scales is not as clear as it could be nor is that of an admissible transformation. Much recent discussion has focused on the latter notion, and it has been argued that if ϕ is an isomorphic embedding of C in N and @ is an automorphism of N (i.e., an isomorphism of N to itself), then the functional composition, @ $\circ$ ϕ, of @ and ϕ is an admissible transformation of ϕ — and that the only admissible transformations are obtained in this way (cf. [19]).

The intuitive idea behind such claims is that since ϕ is an isomorphic embedding, its scale values in N will reflect the properties and relations among individuals in C, and since @ $\circ$ ϕ will also be an isomorphic embedding of C in N, it too will reflect these properties and relations. Automorphisms respect structure, and the set of all automorphisms exhausts those transformations that respect structure in the strongest sense. So by considering the automorphism group, we erase any particular facts that depend on the specific measurement scale we happen to be using and thereby get at the objective, scale-independent features of reality that are invariant across things like arbitrary differences in units among ratio scales for length.[9]

We can now rephrase our earlier claims about realism and conventionalism with respect to measurement in a more definite way. We can think of an axiomatization as affording a perspective on the objective features of reality that underlie a given type of measurement. If the axioms are true when they are interpreted as being about a given magnitude like length, then there are objective facts about things like the ratios of the lengths of objects, and the scope of such objective facts is specified by the uniqueness theorem.

In the case of extensive measurement there are three elements of convention. First, the choice of unit is clearly a matter for decision. Second, there will be some choice about which numerical system to use for the representation; for example, we could use the additive, positive reals, the additive reals, some extension of these like the non-standard reals, and so on. Which numerical system we use to represent facts about a given magnitude will depend, in part, on tradition and what systems we are familiar with, but we can at least say that the system we select must have enough structure to represent what we believe to be the fundamental objective facts about the magnitude that we want to measure. For example $\langle R^+, > \rangle$ does not have enough structure to capture various central features of extensive magnitudes. Finally, it is a convention to use the *additive* reals. It is frequently noted that we can change a ratio scale to the additive reals into a scale to the multiplicative reals by taking exponents, and that the new scale will convey the same information as the original, additive one (define a scale $\phi' = \exp \phi$, and associate $\circ$ with multiplication rather than addition). Like any acceptable scale for extensive magnitudes, however, ϕ' has but one degree of freedom, and the groups of admissible transformations for all such scales are isomorphic. Such scales simply represent the same objective facts in slightly different ways ([16], 13, 99—101, 152; cf. [10], 81—86).

III. PROPERTIES REVISITED

In Section I, I argued that there are good reasons to accept a realist account of measurement and that the best way to explain how such an account could be true requires the existence of properties. In Section II we saw good reasons to suppose that measurement is possible when the structure of the things being measured is the same as that of some numerical system (or substructure thereof) used to represent them. To put these two lines of thought together, we need to adapt our picture of a relational structure so that it incorporates genuine properties and relations, rather than their set-theoretic surrogates.[10]

Extensional relational structures furnish useful models or representations for many purposes, but sometimes it is better to use properties and relations themselves in our structures. One reason for this is that sets are extensional, and thus ill equipped to represent distinct properties and relations that are exemplified by exactly the same things. Such failures of extensionality are less common in the physical sciences than in psychology, but there is some reason to think that they infect laws, causation, and counterfactual sensitivity. Moreover, some properties and relations are probably structured; for example, there may be conjunctive properties of the form *being P & Q*, which are somehow composed of P and Q. But sets are too undifferentiated to represent such structural features of properties.

The idea of putting properties and relations themselves in what I will call *genuine relational structures* is a very natural one that has occured to a number of philosophers. So far as I know, it was first worked out in detail by Zalta ([45]) and Bealer ([5]), though others have developed similar ideas (e.g., [20], [21]).[11] A genuine relational structure contains a non-empty set of individual objects, some properties or relations or both, and an *extension assignment function* that assigns extensions of the appropriate sort to each property and relation (it may assign the same extension to different properties or relations of the same level and rank). As a simple example, consider $A^* = \langle A_0^*, R^1, R^2, R^3, \text{ext}^{A^*} \rangle$. A_0^* is a set of individuals, the next three elements are properties or relations, and the extension function ext^{A^*} assigns a subset of A_0^* to R^1 (giving us, for each object in A_0^*, a *yes* or a *no* answer to the question whether or not it exemplifies the property R^1), a set of ordered pairs of members of A_0^* to R^2 and a set of ordered triples to R^3.

A precise account of these notions would take us too far afield, but I will try to convey the intuitive idea informally. On the present view, properties and relations exist, and so it is useful to explicitly locate them in *domains* over which high-order quantifiers may range. We also want to allow properties of properties, and so on up; the most elegant way to handle this is with a type-theoretic hierarchy, but for current purposes I will adopt the cruder but simpler requirement that properties and relations at one level can only have extensions involving entities one level down. Finally, we want our apparatus to be sensitive to the fact that there may be relations of differing ranks (i.e., with various numbers of argument places) at various levels.

Such considerations suggest that we think of a genuine relational structure as an ordered set $A = \langle A_0, \, ^{q(j)}A_j, \, \text{ext}^A \rangle_{j \in J(\subset I+), \, q \in P(I+)} J$, where A_0 is a non-empty set and ext^A is a (possibly partial) function that, for each j in J, assigns a subset of $(A_{j-1})^n$ as an extension to each n-ary relation in A_j (if $j - 1$ is in J; otherwise ext^* is undefined for arguments from A_j). J is an index set and q is a function from J to the powerset of I^+; we will require that J be non-empty and that $q(j)$ be non-empty for each element of J. Intuitively, A_0 is the set of individual objects and the A_j are sets of level j properties and relations, i.e., properties and relations that can be exemplified by entities in A_{j-1}. At each level there may be relations of different ranks; we may form the set of all level-j monadic properties, that of level-j 2-place relations, and so on, and let A_j be the union of these sets. Every entity in A_i is of rank n for some n in $q(i)$; one might think of these as the filled *cells* of A_j and of $q(j)$ as indicating which cells of level j are filled. Although this machinery makes some use of set theory, it requires only such things as the power set of a given set, and nothing like the full iterative hierarchy.

As an illustration, A^* above has only level-one properties, so

$J = \{1\}$, and there are only properties and relations here of rank 1, 2, and 3, so $q(1) = \{1, 2, 3\}$. Again, we might have a structure whose only properties and relations were level-one, 1-place properties and level-two, 2-place properties (so that $J = \{1, 2\}$, $q(1) = \{1\}$, and $q(2) = \{2\}$). The domain of individuals here might be the set of all colored objects, that of first-level properties the set of determinate colors, and that of second-level place properties the relation that one color bears to another just in case it is darker than the first.

For current purposes this rough sketch will suffice, but it is worth noting that it is easily enriched in various ways. For example, we could also allow A_0 to be empty, move up to a full type-theoretic hierarchy [45], or allow relations with infinitely many argument places. Or we could add a domain of times (with an ordering relation on it) and a domain of worlds (with accessibility and similarity relations on it) and extend ext to assign extensions at times in worlds.[12] We could also add operations that mapped properties onto properties, for example ones that guaranteed the existence of conjunctive or negative properties (as in [5], [20], and [45]).

Two structures are of the same similarity type just in case for each property or relation in one there is a corresponding property or relation in the other (at the same level and of the same rank; one way to spell this would be to order the contents of each cell). And the structures A and B are isomorphic just in case they have the same similarity type and there is a one-one mapping i from the union of domains of A to the union of the domains of B that carries the individuals of A to individuals of B and properties and relations of A to the corresponding properties and relations of B. In addition the mapping must preserve structure; of each n-tuple of things $\langle x_1, \ldots, x_n \rangle$ in each cell at each level of A, each relation R^A of A, and each corresponding relation R^B of B, $\langle x_1, \ldots, x_n \rangle \in \mathrm{ext}^A(R^A)$ just in case $\langle i(x_1), \ldots, i(x_n) \rangle \in \mathrm{ext}^B(R^B)$. We can often compare genuine and extensional relational structures that have corresponding properties and relations by adding an "extensional extension function" to be extensional structure that simply carries each relation-in-extension to itself.

Theories of measurement are rudimentary scientific theories. A good deal of work in the philosophy of science suggests that theories are usefully viewed as — or at least represented by — classes of relational structures or related set-theoretic entities (e.g. [36], Chs. 1, 4; [37], [34], sec. 1). On such accounts, theories of extensive measurement, as well as much more complex theories like classical mechanics, are treated as classes of structures that satisfy axioms for extensive measurement, axioms for classical mechanics, and the like. Such an approach often provides a fruitful representation of the *structure* of a theory, and there is a good deal to be said for the view that theories characterize types of systems (which can be viewed as types of relational structures). For example, we might want to find out whether a given concrete system like the solar system or the pendulum in the livingroom clock is a specimen or realization or a model of a classical mechanical system.

As Suppes and others have stressed, the notion of a model in such cases has a good deal in common with the logician's concept of a model or, in our terms, with that of a relational structure. But I believe that there is one important disanalogy. We don't count something as an instance of a classical mechanical system simply because it can be described by a set of sentences that satisfies some consistent axiomatization of Newton's theory. It is too easy to find things that do this. What we are more often interested in are the models that give the non-logical words in the axioms their intended interpretation — in which mass is mass and length is length. We want to keep this fixed, while allowing the individual objects and initial conditions to vary from model to model. Using genuine relational structures, we can say that we are interested in intended interpretations for the domain of properties like mass and force — we want to keep this fixed — while allowing the domain of individuals to vary from one model to another.

One reason for adopting such a view is that if we suppose the meanings of terms like 'rest mass' are completely determined by specific versions of a theory, it will be quite difficult to explain how successive theories can make claims *about* rest mass, as it seems they can. But on the present account, this is no more mysterious than the fact that different witnesses in a courtroom can tell conflicting stories *about* one and the same person. In short, theories are not mere formal devices; typically they are linked to the world, and one way that they can be linked to it is by a relationship of denotation between their predicates and genuine properties and relations in the world.[13]

The claims that a statement of a theory makes about various families of properties need not be literally true for the theory to be useful. Theories often involve idealizations, dealing with only some of the properties that are relevant to explaining the behavior of actual, concrete systems, or only with properties that are sufficient for explaining it in some special case like that where a system is closed or, indeed, only sufficient for explaining the workings of some idealized systems like frictionless surfaces or perfectly fair coins that some actual system approximates. But a theory needn't make all or only true claims about a property to make claims *about* that property.

These ideas can be applied to extensive measurement in the following way. Our earlier discussion suggested that there are first-level determinate, extensive properties like *being 5 meters long* which hold of individual objects. There are also second-level relational properties, like the binary relation *being a greater length than* ($>$) and the trinary relation *being the sum* (with respect to mass) *of* ($\circ$), which hold among pairs and triples of such determinate first-level properties.[14]

An *extensive property structure* is a genuine relational structure of similarity type $\langle E_0, E_1, E_2, \text{ext} \rangle$, where $q(1) = \{1\}$ and $q(2) = \{2, 3\}$. E_1 will be thought of as containing first-level, determinate properties like lengths; E_2 contains only $>$ and $\circ$, and it will often be convenient to make this explicit by writing it as $\langle E_0, E_1, >, \circ, \text{ext} \rangle$. When we are only concerned with properties, we may simply drop E_0, and the resulting structure is of the same type as the extensional extensive structures studied by representationalists (once we add an extensional ext function to the latter). This means that we can simply appropriate some satisfactory axiomatization of extensive measurement — perhaps making small changes to reflect the fact that we are now thinking of it in terms of properties rather than objects — and this will get us our representation and uniqueness theorems. Slightly more work is needed if we wish to do justice to the fact that magnitudes are exemplified by objects, and we still need to say something about closure conditions and existence claims (for the moment I will provisionally assume that properties can exist unexemplified).

Most representationalist theories are presented in some technical fragment of English rather than in a formalized language. I think that the philosophical significance of formalization is often greatly over-rated, but there are several reasons why a formal language for a theory of measurement might be of value. First, it should make it easier to give an account of the notion of a sentence involving a scale, and hence help us explicate that notion of meaningfulness which takes it to be a property of sentences. For example, we could explore such options as associating a classical model with each particular scale in the class of isomorphic embeddings of A in R and regard a sentence as making a meaningful claim about A just in case it is true or false in the super-valuation defined over this set of models. Second, we could more readily avail ourselves of tools and results of model theory like diagrams and preservation theorems. Third, modeling and representation can sometimes be thought of in more linguistic terms than we have so far; sometimes, for instance, we may be interested in structures that are indistinguishable with repect to some language (elementarily equivalent structures, for example, if the language is first-order). Fourth, setting out a language will allow us to give a semantics for predicate terms, which is needed for a complete development of a realism with respect to measurement. Finally, a number of extensions and applications of the present approach to measurement will be easier if we have a formal language at our disposal.

Most writers who have been interested in things like our genuine relational structures have adopted a higher-order syntax ([20], [21], [45]), though it is possible to use a first-order theory of the exemplification or predication relation instead (cf. [5]). The use of a higher-order syntax allows us to capture the fact that some natural language predicate terms, e.g., 'red', can occur both as grammatical subjects and as grammatical predicates of sentences. On the other hand, first-order theories of the exemplification relation have been defended on the grounds that predicative terms like 'wise' cannot occur as grammatical subjects, though correlated singular terms, their nominalizations, do (such terms are often represented formally by lambda abstracts; on this account, 'red' would presumably be ambiguous). Moreover, it is sometimes added, grammatical subjects and grammatical predicates play quite different roles in sentences; subjects are used to pick out or *refer* to things, and predicates are then used to say something about them, to *ascribe* properties to them.

It does not seem to me that considerations of either sort are decisive. It is true that grammatical subjects typically refer to things and that predicates then ascribe properties to them, but this asymmetry can be formally represented in various ways. With a higher-order syntax of the sort sketched below, it will not be reflected in a *semantic* difference between singular terms and predicates, for both will *denote* their semantic values. A grammatical predicate of a sentence, 'F' in 'Fa' to take a simple example, indeed ascribes a property to something, but this will be reflected in the fact that the thing denoted by the grammatical subject 'a' *is a member of the extension* of the property denoted by the grammatical predicate 'F'. In the end it doesn't seem to make a great deal of difference which sort of syntax we select, since the semantic consequence relation will work the same way in either case, and it will not be difficult to go back and forth between our higher-order syntax and a many-sorted first-order one and thence to a standard first-order one. I will adopt the higher-order syntax here simply because is familiar, does not require many special axioms or notational conventions for an exemplification relation, and there is some evidence that many English predicates do have a denotation ([18]).

All I aim to do here is to give an informal picture of enough of the logic to show how the formulation of a theory of extensive measurement would go, and this means that we will only need first and second level properties. The logical terms of the language are the parentheses, the connective '&' and ' $\sim$ ', the quantifier '$\forall$', the logical constant '=', and variables. The set of variables is the union of a countably infinite set of level-zero (individual) variables v_i, countably infinite sets of n-place, level-one variables V_i^n, and countably infinite sets of n-place level-two variables V_i^n, for each positive interger n. We have a similar set of individual constants, made up of sets c_i, C_i^n, and $\boldsymbol{C}_i^n$ for each positive i and n. The variables and non-logical constants are *terms*, and the *type* of a term is given by its level and number of argument places (all individual variables and constants will be regarded as 0-place). If t_1 and t_2 are terms of the same type, $t_1 = t_2$ is an (*atomic*) *formula* and if t^n is a term of level two (one) and $t_1, \ldots, t_n$ are terms of level one (zero), then $t^n(t_1, \ldots, t_n)$ is an (*atomic*) *formula*. If ψ and χ are formulas and μ is a variable, than $\sim \psi$, (ψ & χ), and ($\forall\mu$)ψ are formulas. We define additional connectives, quantifiers, and such notions as open and closed formulas in the usual way. Finally, any standard proof techniques for first-order logic with identity may be appended, with the provision that the axioms or rules for identity and the universal quantifier be recast to apply to all types of terms (and that we substitute one term for another only if they are of the same type, avoid clashes of variables, and so on).

One important feature of such a language is that it can easily be enriched in philosophically interesting ways. For example, we may add variable-binding term operators like descriptions, or intensional operators and connectives, along with axioms and rules to govern their behavior. The language could also be emended and elaborated to get a richer, more flexible hierarchy of types (as in [45]) or to get something type-free (as in [20]). Finally various non-logical axioms can be added to yield various substantive theories of properties. Those who relish a rich ontology might add a comprehension schema, perhaps along with λ-terms and a principle of λ-conversion, enriching the relational structures with various operations on properties to guarantee that the class of properties is closed under such operations as conjuction and projection. Those who are more frugal might prefer an axiom schema of instantiation, requiring that all properties be exemplified, perhaps adding a free logic that doesn't require every predicate to denote a property (though the logic might still allow predicates to be true of sets of things of the appropriate sort; this would be a natural way to develop a line of thought in [2] and [39]).

We could give a semantics for our full language by requiring that each cell in level two and three be non-empty, but things will be more manageable if we cut the syntax back to a language appropriate for a less elaborate similarity type. In general this can be done by requiring that for each non-empty cell at any level there be infinitely many variables of the associated type and allowing (but not requiring) constants of that type. Now an extensive genuine relational structure has level-one, 1-place properties and one level-two 2-place relation and one level-two, 3-place relation, and so the "language of extensive measurement theory" will have variables of the associated types and two constants for the level-two relations (I will abbreviate as '$>$' and '$\circ$', use infix notation and other obvious abbreviations of variables, and drop parentheses where ambiguity won't result). We can now define semantic notions for this language in the obvious way.

A *model E* for the language of extensive measurement is an relational structure E of the sort just described, together with an *interpretation function I* that assigns elements of the appropriate cell of the appropriate domain to constants; for example, it will assign a 2-place, level-two relation to '$>$'. A *value assignment v* is a function that assigns to each variable a member of the appropriate cell of the appropriate domain. And a denotation function with respect to a model E and a value assignment v, $den_{E, v}$, agrees with I with respect to the constants and with v with respect to the variables. We can now provide a precise realist semantics for predicates by defining the notion of the *satisfaction* of a formula ψ by a value assignment v with respect to a model E (for readability I'll drop subscripts and reference to E). If ψ is of the form $t_1 = t_2$, v satisfies ψ just in case $den(t_1) = den(t_2)$, and if ψ is of the form $t^n(t_1, \ldots, t_n)$, v satisfies ψ just in case $\langle \text{den}(t_1), \ldots, \text{den}(t_2) \rangle \in ext(\text{den}(t^n))$. Clauses for the connectives are as expected, and if ψ is of the form $(\forall \mu)\chi$, v satisfies ψ just in case every assignment v' exactly like v, save perhaps in what it assigns to μ, satisfies χ. A formula is true in a model just in case it is satisfied by every value assignment with respect to that model, and satisfiability, validity, and semantic consequence are defined in the usual way.

This system has all of the familiar metatheoretic properties of first-order logic, including completeness and compactness.[15] This is possible because its models need not be as rich as those required in *standard* higher-order logic — whose semantic consequence relation is not recursively axiomatizable — where the n-place, level-one predicate variables range over the *entire* power set of $(E_0)^n$, and so on up. By contrast, our domains of properties are like those in Henkin's general models, which can be as small as you like, so long as the relevant cells are not empty.

The only problem with this language is that we cannot formulate Archimedean axioms in it. There are various ways to formulate Archimedean properties of number systems, but the basic idea is that no two numbers x and y are infinitely far apart; we can get from one to the other by some finite number of steps of multiplication, so that for all numbers x and y are infinitely far apart; we can get from one to the obvious numerical relation and n is some standard, positive integer; for the moment I will ignore complications needed to also rule out infinitely close numbers). Related ideas arise in the case of extensive measurement, for we can define the notion of an *integral multiple property, nX*, of the property X inductively; letting '$X \circ Y$' denote the sum of X and Y, set $1X = X$ and $(n + 1)X = (nX) \circ X$. Now a simple Archimedean axiom in measurement theory would tell us that for any two properties X and Y, there were some positive integer n, such that $nX > Y$. This means that X and Y are commensurable; hence, if X and Y are lengths and X is the property of *being one foot long*, Y can be compared to X in terms of some finite number of sums of X (or else of Y) with itself.

The Archimedean property cannot be expressed when our logic is compact. Of course we can write sentences in our higher-order language that would capture the Archimedean property *if* we interpreted our quantifiers as they are interpreted in the standard semantics for higher-order logic — with $\forall X$ meaning *all* subsets of E_0 for example —

but in our semantics our quantifiers need not do this, and sentences that look like Archimedean axioms will still allow non-Archimedean models. In the last few years several authors have challenged the need for Archimedean axioms, proposing non-Archimedean theories of utility ([31]), (weak) extensive measurement ([24], [22]), and the like. Archimedean properties are easily seen to be necessary for the representation of structures in Archimedean numerical systems like the reals, but it may be that the common use of the reals in measurement is a historical accident resulting from the fact that such things as non-standard analysis were not available until very recently.

There seems to be no reason to suppose some determinate lengths are infinitely greater than others, and it is plausible to hold that a full characterization of the family of lengths should reflect this fact. The question here, though, is whether a theory *of measurement* should reflect it. Typically, non-Archimedean theories aim for a more limited representation theorem, say one to the effect that $x \geqslant y$ only if $f(x) \geqslant f(y)$ (or $x > y$ only if $f(x) > f(y)$; perhaps with corresponding changes for sums as well). Some arguments for this approach have a good deal of force, but I am not convinced we should abandon Archimedean axioms when we are concerned with the metaphysical foundations of a realist account of measurement. The realist assumes that each property has a true scale value, though our inexact measurements are incapable of determining it precisely or even of disclosing small differences in magnitudes below some threshold of sensitivity. In such a case, we may still be interested in necessary and sufficient conditions for the existence of true scale values, preferably for each determinate length, and assignments like $f(x)$ will be viewed as involving an objective state of nature — say a true ratio of some magnitude to a unit property — rather than as involving assignments that we actually make using our imperfect instruments and methods. In such a case, it seems reasonable to suppose that a realist would hold that $x > y$ just in case $f(x) > f(y)$. Such matters are controversial, however, and a second reason for including an Archimedean axiom is simply to show *how* an account of measurement in terms of properties would handle it.

One way to capture the Archimedean property is by modifying the syntax and semantics of our logic to allow infinite conjunctions and disjunctions (the atomic formulas and facts remain unchanged; for details see [15]). On the syntactic side we add the connective '$\wedge$' to represent infinite conjunction, require that whenever $\{\psi_1, \ldots, \psi_n, \ldots\}$

is an at most countably infinite set of formulas, '$\bigwedge(\psi_1, \ldots, \psi_n, \ldots)$' is a formula, and define a symbol, '$\bigvee$' for infinte disjunction using DeMorgan's law. On the semantic side we add that a value assignment satisfies an infinite conjunction just in case it satisfies each of the conjuncts. If we add a standard infinitary rule of inference to enable us to derive such things as the infinite conjunction of infinitely many premises, the logic is sound and complete in the sense that exactly the valid sentences are provable (though it now becomes possible for proofs to grow infinitely long).

This infinitary logic is much more manageable than alternatives, like second-order logic, in which Archimedean properties are often characterized. When our interest is a metaphysical one of characterizing a given structure, rather than epistemological or proof theoretical, infinite conjunctions and disjunctions seem reasonable, even though we cannot manipulate them. After all, we frequently use first-order logic to mimic infinite conjunctions, for first-order theories are just infinite sets of sentences. But what we need to capture the Archimedean property is an infinite disjunction, and first-order logic cannot mimic this.

Finally we must ask about the existence claims embodied in the Archimedean axiom and the assumption that the summation operation is closed (this assumption can be weakened in various ways, but not enough to avoid the following problem). If we adopt the epistemologically attractive view that the only properties that exist are those that are actually exemplified (as suggested in [2] and [39]), we will encounter difficulties exactly analogous to those which drove traditional representationalists to merely possible objects. The general problem here is not confined to measurement. Some statements of vacuous laws — laws involving properties that are never exemplified, e.g., being a body free of impressed forces — are thought to be true, and it is also natural to suppose that they involve unexemplified properties in some way. Indeed, claims about sums of properties that are not exemplified are really just a species of vacuous laws. There are two strategies for dealing with this problem. One might argue that properties and relations can exist even if they are never exemplified, or one might deny this but argue that merely possible properties and relations avoid the difficulties raised by merely possible individuals.

The view that properties and relations can exist without being exemplified is often said to be *the* traditional view, though historical documentation is rarely provided to substantiate the claim. Unexemplified properties are highly theoretical creatures whose connection to actual, concrete things is quite tenuous. This sort of view may come in degrees, however, and is it compatible with a tempered empiricism according to which it is an empirical question just what *families* of properties exist. If a given extensive family like length exists, however, our account of measurement would require that a good many of its members exist as well, whether they are exemplified or not.

On the other hand we might hold that only exemplified properties and relations exist, but introduce higher-order possibilist quantifiers that include merely possible properties and relations in their range. Uninstantiated and merely possible properties are similar in the sense that with each it is unclear how our linguistic expressions could come to denote them, how they could play a substantive role in the causal order, or how we could gain knowledge about them. At least this is so for merely possible properties if we think of them as primitive things. But if we could show that facts about them were somehow grounded in facts in the actual world, some of these difficulties would be overcome. One strategy for doing this suggests itself if we reflect on the problem of naming merely possible individuals. We cannot interact causally with such things, and so the only way to fix the reference of names for them seems to be to use definite descriptions that they satisfy. But as Kripke and others have argued, in most cases there simply aren't suitable descriptions of merely possible individuals that are rigid, descriptions that are satisfied by the same thing in every world (in which they are satisfied at all). If some merely possible person performs the feats attributed to Sherlock Holmes in one world, different people do so in others, and so none of them uniquely satisfies our description.

When we turn to determinate properties like lengths, the situation is different. It is at least plausible to suppose that the place which a given length occupies in the $>$ ordering of its family is necessarily occupied by it and it alone, and similarly for facts about $\circ$. It is difficult to imagine, for example, that the property *being* 2 *inches long* might have been greater than the property *being* 4 *inches long*. If this is right, then there is some prospect of finding rigid descriptions of merely possible lengths in terms of the $>$ and $\circ$ relations they would have borne to actual, exemplified lengths; for example, we could speak of the property that, had it existed, would have been the sum of the actual properties X and Y. One might hope to use such facts to fix the reference of enough terms (or at least to assign values to free variables) to set up a scale of measurement.

If essential features of merely possible determinate lengths are determined by the place they would have occupied in the > ordering and the like, then we can learn a good deal about them by studying exemplified properties like > and ○ and other members of their family. Moreover, it would be facts here in the actual world involving these exemplified properties that would ground the truth of claims about merely possible properties. I think that these ideas may point the way to an actualistic account of merely possible properties that would make talk about them unobjectionable — that would highlight the fact that talk about possible properties is just talk about what properties there might have been.[16]

Fortunately, for many purposes the differences between merely possible properties and uninstantiated actual properties are less striking than one might suppose. As we have seen, they face many similar problems. Moreover, many claims of either approach can be translated into those of the other; where the first speaks of merely possible properties, for example, the second speaks of actual but unexemplified properties. Although I prefer possible properties, the higher-order quantifiers in what follows can, with quite minor changes, be read either as ranging over uninstantiated properties all of which are actual *or* as ranging over actual and merely possible properties (though individual quantifiers are still confined to actual things); where the two accounts diverge, I will for definiteness speak of unexemplified properties.

A set of axioms for extensive measurement couched in a higher-order syntax has been presented in Mundy ([21]). I shall present a slightly different set here, since his incorporates several novel features and dispenses with the Archimedean axiom, and this makes it harder to compare with more standard axiomatizations. The transliteration of existing axioms into a higher-order language is quite straightforward, but I shall go ahead and do it in order to illustrate the sorts of axioms involved in extensive measurement (for readers unfamiliar with them) and to have the axioms before us so that we can see whether any of them raise special difficulties when we take them to be about properties and relations. A variety of axioms and, indeed, of primitive terms could be used here, but to facilitate comparisons with standard representationalist accounts I shall consider a higher-order version of a (trivial variation of) the well-known axioms in Krantz *et al.* ([16], 73).

We will be dealing with only one extensive family at a time, and for definiteness we will continue to think of it as length.[17] This means that our axioms well ensure the existence of a unique sum of any two family members, and it will be convenient to treat '$X \circ Y$' as a singular term. We could bring our notation more closely into line with Krantz *et al* by introducing special notation for operations (like $\circ$, these being functional relations), but to emphasize the fact that we are dealing with relations, we do not. We will say that a genuine relational structure of the appropriate similarity type is an *extensive property structure* just in case it is a model of the following axioms (formulas with free variables are equivalent to their universal closures).

A1a, b, c $(X > Y \ \& \ Y > Z) \rightarrow X > Z, \ X > Y \rightarrow \sim Y > X,$
   $X \neq Y \rightarrow (X > Y \lor Y > X),$

A2a, b $(\exists Z)(\circ(X, Y, Z)), \ ((\circ(X, Y, Z) \ \& \ \circ(X, Y, W)) \rightarrow W = Z,$

A3 $X \circ (Y \circ Z) = (X \circ Y) \circ Z,$

A4 $X > Y \leftrightarrow (X \circ Z > Y \circ Z) \leftrightarrow (Z \circ X > Z \circ Y),$

A5 $X \circ Y > X,$

A6 $X > Y \rightarrow \lor ((1X \circ Z) > (1Y \circ W), (2X \circ Z)$
   $> (2Y \circ W), \ldots, (nX \circ Z) > (nY \circ W), \ldots).$

A1 tells us that $>$ is a strict simple order. A2 tells us that $\circ$ is a function, which A3 says is associative. A4 is a monotonicity axiom telling us how $>$ and $\circ$ interlock. A1, A3, and A4 are necessary axioms for the existence of a representation in the positive reals. A5 is a positivity axiom, and is necessary for a representation in the *positive* reals; for some purposes it might be dropped. The infinite disjunction A6 is our Archimedean axiom. The numerals in it are defined symbols, and it is important to keep in mind that nX is a property. The axiom is slightly more complicated than the one discussed above, since it captures the Archimedean property for differences, not only ruling out infinitely large properties but infinitesimal ones as well ([16], 74; 132 ex4.ii).

We are interested in functions from an extensive property structure $E = \langle E_1, >, \circ, \text{ext}^E \rangle$ to $R^+ = \langle R^+, >, +, \text{ext}^{R+} \rangle$ (where ext^{R+} simply maps $>$ and $+$ to themselves; I shall use '$>$' to stand for the corresponding orderings on both structures). A1—A6 suffice for a proof of a representation theorem telling us that E is an extensive property structure just in case there exists an isomorphic embedding ϕ of it in R^+ (this shows, incidentally, that A1—A6 are consistent). That is, for all X

and Y in E, $\langle X, Y \rangle \in \text{ext}^E(>)$ just in case $\langle \phi(X), \phi(Y) \rangle \in \text{ext}^{R+}(>)$ and similarly for triples of objects in E, $\circ$ and $+$. The axioms also allow proof of a uniqueness theorem, telling us that ϕ' is an isomorphic embedding of E in R^+ just in case there is some $\alpha > 0$ such that $\phi' = \alpha\phi$ ([16], Ch. 2).

We can now see that *if* a particular family of properties satisfies these axioms — and it is always an empirical question whether a given family does — we are justified in measuring them on a ratio scale and applying the mathematics that goes along with this to them. For the axioms ensure that the family has much — though by no means all — of the structure of the positive real numbers; indeed, they even tell us what that structure is, namely that of an Archimedean ordered commutative semi-group.

On the present account measurement involves the assignment of numbers to properties, but we obviously make such assignments by observing and manipulating the objects that exemplify them, and so it is important to work individuals into our picture. To this end we can add domains of individuals to our extensive structures, as well as additional properties and terms as the need arises.

When W. E. Johnson introduced the notion of a determinate, he stressed that an object could exemplify only one determinate of a given family at a time, and it seems quite true that an object cannot have more than one length at the same moment. As long as we are dealing with one family of properties at a time, we could simply require that each member of the domain of individuals exemplify exactly one determinate property. We will have more flexibility, however, if we introduce Johnson's notion of a *determinable* property, saying that two determinate lengths fall under the same level-two, 1-place determinable *being a length* (L) just in case they are identical or either bears the *being longer than relation* to the other (and similarly for other extensive determinates). Then we could add '$((Xx \ \& \ LX) \rightarrow (\forall Y)((LY \ \& \ Yx) \rightarrow X = Y)$' as an axiom to our theory for the extensive measurement of length.

We sometimes speak of one length's being greater than a second, but we also speak of one object's being longer than a second, and so it will be useful to add level-one constants, '$>_i$' to stand for a two-place relation $>_i$ that holds between two *individual* objects just in case the first is longer than the second, and '$\circ_i$' to stand for the three-place summation relation on individuals. In simple cases $>_i$ and $\circ_i$ may be observable, but they aren't always. However our account does not require any crisp division between what is observable and what isn't, as many standard representational accounts seem to.

We can now express *tracking principles* for corresponding pairs of relations like $>_i$ and $>$, according to which they covary or track each other in the expected way; for example, $x >_i y$ just in case x and y exemplify determinate lengths and x's is longer than y's. We would also expect that scale values could be assigned to objects that would track the scale values of their relevant properties. Mundy discusses these topics and gives first-order axioms for relations much like '$>_i$' and '$\circ_i$' which are analogues of the above axioms except for A2 and A6 and the fact that the axioms involving summation are conditional on the existence of the relevant sums ([22]). So instead of pursuing the matter here, I want to ask about the evidential and explanatory relationships between claims about the first-level relations $>_i$ and $\circ_i$, on the one hand, and corresponding claims about their second-level counterparts $>$ and $\circ$, on the other.

Mundy views the tracking principles as bridge laws which allow us to derive claims about the behavior of $>_i$, and $\circ_i$ from measurement axioms involving properties. To take a very simple example, we can use the principle linking $>$ and $>_i$, together with A1a, to deduce the claim or prediction that $>_i$ is transitive. We can then check this prediction by examining triples of individuals, thereby testing the axiom A1a which entailed it. Indeed, Mundy argues, properties should be treated just like any other theoretical entities in science, and an account of measurement in terms of them generates successful predictions that yield empirical confirmation for the claim that properties exist.

The picture suggested by our discussion in earlier sections of this paper is somewhat different. According to it, we begin with observations of individual objects and make inferences about regularities involving $>_i$ and $\circ_i$. For example, we find good reason to suppose that $>_i$ is transitive (it is also asymmetric, but since different physical objects may have the same length, it is not likely to be connected as $>$ is) and that whenever sums of objects can be found, summation is commutative and associative. Indeed, we find regularities involving individuals that are analogues of all the axioms for extensive measurement except for the closure and Archimedeans axioms (A2 and A6), though some of these regularites are conditional on the existence of the relevant sums.

Then, as discussed in Section I, in order to *explain* these regularities involving individuals, $>_i$, and $\circ_i$, we make an inference to the existence of level-one determinate lengths and of the second-level relations $>$ and $\circ$. This inference is what yields the tracking principles, which might at this stage be expressed in Ramsey-style claims to the effect that $x >_i y$ just in case $(\exists X)(\exists Y)(\exists Z^2)(LX \ \& \ LY \ \& \ Xx \ \& \ Yy \ \& \ Z^2(X, Y))$ (and similarly for $\circ_i$ and $\circ$; Z^2 of course turns out to be satisfied by $>$). In tandem with the regularities involving $>_i$ and $\circ_i$, these principles place tight constraints on the behavior of $>$ and $\circ$ in cases where they relate exemplified determinate lengths. Indeed, it follows that in such cases $>$ and $\circ$ obey analogues (in some cases ones conditional on the relevant properties being exemplified) of all the axioms but A2 and A6. Finally one last inference takes us to uninstantiated (or merely possible) determinate properties, and, indeed, to enough to satisfy the unconditional axioms A1—A6. This inference is again justified because it helps us to explain facts, in particular facts about scales and measurement.

Now the tracking principles do indeed tell us that many corresponding claims about $>$ and $>_i$ (and $\circ$ and $\circ_i$) have the same truth value. So if we come to believe that $>_i$ is not transitive, this will falsify the claim that $>$ is. But this test of the claim that $>$ is transitive proceeds through a philosophical principle in a way that most scientific tests do not. On the present account, the tracking principles are not really empirical laws that might be found to be false. Although they are devised to help explain empirical phenomena, they result from more-or-less *a priori* philosophical reasoning and are not likely to be abandoned in the light of new evidence about the details of the behavior of $>_i$ or $\circ_i$, for example the discovery that $>_i$ isn't transitive. In fact, the tracking principles tell us that *whatever* features $>_i$ and $\circ_i$ are thought to have will automatically be passed on to $>$ and $\circ$ (insofar as they apply to exemplified properties); *whatever* we learn about $>_i$ and $\circ_i$ will be transfered directly to $>$ and $\circ$. In short it is claims about $>$ and $\circ$ that are deduced from a principle based on philosophical reasoning and our empirical conclusions about $>_i$ or $\circ_i$, rather than the other way around, and so at this stage these empirical claims about $>_i$ and $\circ_i$ are not really used to test claims about properties like A1—A6.

It may be argued that our axioms involving properties are really only tested once we add unexemplified properties and move to unconditional claims like A1—A6. However, weaker versions of these axioms

— for example that when three properties are instantiated and one is the sum of the other two then sums are commutative — are all that are needed to deduce the claim that individual sums commute when they exist and the like. On the present account, the claim about individuals and their sums is used to obtain the former, and so does not really provide a test for it in any normal sense. To be sure, the full set of axioms has consequences that truncated versions in which summation isn't closed do not have; indeed, the full set is needed to show that scales having some of the features discussed in Section II exist. But even here support for the claim that properties exist is more philosophical than empirical. There can, for example, be nominalistic versions of axioms like A1—A6 that quantify over merely possible objects and together with a few background assumptions these would have the same empirical consequences concerning actual individuals that our axioms involving properties have. I have argued that such accounts are not as good as an account in terms of properties and relations, but the reasons for thinking this are almost purely philosophical.

Certainly the success of our measurement practices, and more generally of our scientific practices, supports the claim that there are properties of the sort required by A1—A6, but it does so more because of that claim's explanatory power than because it entails predictions checked by scientists. Of course things are complicated by the fact that the boundary between science and philosophy is probably not sharp. Moreover, deductive, explanatory, and evidential relations can run in various directions at the same time, and we may focus on one direction when thinking about discovery and another when thinking about justification. Still, the moral I would urge is that our reasons for believing in properties of the sort required by A1—A6 are based on philosophical argument to a much larger extent than most conclusions in science are.

A2 does seem to commit us to uninstantiated properties. On Quine's criterion of ontological commitment, the fact that a theory is highly confirmed is itself a good reason to believe in the existence of *all* of the things over which circumspect formulations of it quantify. In many cases this view is difficult to reconcile with a plausible, naturalistic epistemology, for we often have no causal contact, however indirect, with some of the things various well-confirmed theories mention, e.g., point particles, numbers, or space-time points, and so it is difficult to see how we could ever acquire knowledge about such things. But in practice it is implausible, even for a realist, to take all of a theory's claims literally. Well-confirmed theories often incorporate a variety of assumptions that almost everyone would allow are not to be taken as literal descriptions of anything but that are simply included to make life easier for users of the theory. Such assumptions may involve idealizations about objects (point particles), simplifying assumptions that make it easier to apply mathematics (some claims about continuity seem to be of this sort), and so on.

A discriminating realism can make room for a distinction between the literal claims of a theory and those that merely grease the wheels, but this isn't of great help unless we can find a principled way of drawing the distinction in specific cases. One rough guide here is that putative entities exist only if they play a causal role in the world. By this test mass, length, and spin fare well; numbers, space-time points, *and unexemplified properties* do not. The most convincing demonstration that talk about a given sort of thing need not be taken literally is to show that it need not be engaged in at all in order to say the things that we want to say (as, for example, Field tries to show for talk about numbers in classical gravitational theory in [12]). The question is whether A2 (and A6) can somehow be interpreted as convenient fictions, rather than claims that need to be taken literally. In part this involves mathematical questions about the minimal conditions required for various sorts of representation and uniqueness theorems, but it also involves philosophical questions about the interpretation of theories. At this stage it is unclear what the answers to some of these questions are, but whatever the answers, the theory of extensive measurement sketched above is an attractive one that can be integrated with a realist account of measurement, science, and semantics.

IV. EXTENSIONS AND CONCLUSIONS

Extensions to ordinal, difference, and additive conjoint measurement are straightforward, for existing axiomatizations of these can be easily accommodated to properties. We have viewed theories of measurement as rudimentary scientific theories, and the question naturally arises whether we can recast axiomatizations of more substantive scientific theories like classical particle mechanics or statistical learning theory in terms of properties. Many theories relate several families or dimensions of properties, often by equations relating products of powers of various families (e.g., $F = Gm_1m_2/d^2$), and a first step toward handling this would be to introduce relational structures containing several different families of extensive properties and to add second-level relations capable of relating members of different families. Many properties of interest in science like vectors and tensors are more complicated and abstract than properties like length, and so the numerical relational structures used to represent theories may be more complicated than R^+. That is all right, however, so long as the empirical properties have a structure — conferred by higher-order relations on them — in common with the appropriate sort of mathematical structure or substructure.

To get this approach off the ground, we would need qualitative or synthetic axiomatizations of various theories; these now exist for a number of theories, including at least parts of classical particle mechanics, rigid body mechanics, classical gravitational field theory, thermodynamics, special relativity, statistical learning theory, and various parts of economics (e.g. [12], [32], [34], [36], [37]). The goal would be to find a way to reinterpret their axioms in terms of properties, so that theories could be regarded as claims about the structure of various families of properties in the world.

Synthetic axiomatizations have not been found for some central theories like quantum mechanics, and even where we have a fairly detailed axiomatization of a given theory, adapting it to properties may not be nearly as easy as it was in the case of measurement. Moreover, on the current account it is an empirical question just what properties exist, and it is not yet clear precisely what the required second-level properties and relations would be like; indeed, there is still uncertainty about the proper way to handle physical dimensions and laws within the representationalist framework ([16], Ch. 10; [26], Sec. 5). But by way of a simple illustration, we might imagine that a product of powers of dimensions represents a second-level multiplicative relation, *, that can hold within families (length squared is area) and across families (mass times velocity is momentum) (cf. [16], Ch. 10).

To the extent that synthetic axiomatizations of interesting theories can be recast in terms of properties, they would provide a natural framework for the recent view that the laws of nature involve higher-order relations among properties. Indeed, they would tell us something about *what* these higher-order relations were and *how* they behave. In simple cases the relations might be ones like > and ∘ and, perhaps *, which hold among the determinate extensive magnitudes of various families, and the exact form of the relationships, when known, could be specified by sentences telling us how the relations interlock. Laws restrict the (physically) possible configurations and sequences of properties, and these restrictions can be captured, with varying degrees of accuracy, by axiomatizations of theories which restrict configurations of the scale values of the magnitudes involved.

The current framework is also well-suited for developing the idea, suggested by various writers, that the identity conditions of properties are determined by their nomological role (their "causal powers" or the like). At least in the case of an extensive property, this role could in principle be specified by a functional characterization in terms of the laws in which the property figures, i.e., in terms of relations like $>$, $\circ$, and perhaps $*$, that it bears to other properties. This leaves open the identity conditions for second-level relational properties like $>$, though the most obvious suggestion is that their identity conditions are determined by their formal features (separately and together with those involving other relations, e.g., transitivity, monotonicity) and the fact that they relate the families of extensive properties that they do. If so, the identity conditions of at least many of the properties involved in science are given by the algebraic structure of the various families of properties taken together.

Questions about the identity conditions for things within a given world should not be run together with questions about identity across worlds, but the latter can also be of interest. Given our semantics, simple lawlike statements of the form '$X = Y \circ Z$' and perhaps '$X = Y * Z$' will be necessarily true if the singular terms '$Y \circ Z$' and '$Y * X$' are rigid. Whether they are or not depends on whether the relations $\circ$ and $*$ are rigid, i.e., on whether, when they relate three objects in the actual world, they relate them in all worlds in which they exist. I think that there is something to be said for the view that all such relations are rigid, but it would be natural to draw a line between the essential and the contingent higher-order relations among extensive properties in terms of those that hold within families (like $>$ and $\circ$), on the one hand, and those that can cut across families (perhaps like $*$), on the other. It is difficult to see how 1 meter might have been greater than 3 meters, but it may be easier to imagine that multiplicative relations across families could have been different.

Our account of the applicability of mathematics to reality is compatible with platonistic views of mathematics, but once we encounter synthetic axiomatizations — ones that abjure mention of numbers — that are sufficient for the existence of extensive measurement, it is natural to wonder whether numbers are really required in any sort of measurement or, more generally, in science. Of course numbers are extremely convenient, but perhaps instrumentalism is a bad philosophy of science but a good philosophy of mathematics. Hartry Field has developed this sort of idea in ([12]). Like traditional representationalists, he begins with a qualitative axiomatization of a physical theory and a representation theorem that allows us to move from claims about concrete phenomena to claims about their numerical counterparts, reason about the latter, and transfer our conclusions back to the former. He goes a step further, however, arguing that this procedure can be legitimate even if the numerical claims are not literally true. All that is required is that the mathematical theory be conservative over the physical theory. This means that by using the mathematics and our theory of the concrete phenomena, we can't draw any conclusions about the concrete phenomena that we couldn't have drawn (with a lot more effort) without using the mathematics. It takes a great deal of work to establish such a claim about any particular body of theory. Field tries to show it for classical gravitational-field theory by developing a synthetic axiomatization of it in a language that allows quantification only over space-time points and their Goodmanian sums.

One might ask whether similar conservativity results could be obtained for theories using a language like ours, with quantifiers ranging over properties rather than over space-time points and their Goodmanian sums. Before becoming too sanguine about this, though, it should be noted that Field's approach is not free of difficulties. The general tendency of his program is to scrimp on ontology by expanding logic, and even then it is not clear that he can say everything that he would want to about the case he deals with or just how widely the approach can be extended. What is more important here, Field gets the effect of standard second-order quantification over sets of sets of individuals by quantifying over Goodmanian sums of points. By contrast there is no obviously appealing way to achieve this effect with properties without bringing in more set theory, which would not be an auspicious beginning for an attempt to show that science does not need numbers or similar abstract entities.

On our account, families of extensive properties have a good deal of structure in common with the real numbers, and a second way to think of mathematics in the current framework would be to read even more mathematical structure into properties — in effect viewing extensive families of properties as isomorphic copies of the reals. One could then do mathematics right on the properties themselves, defining such notions as absolute value, limit, continuity, and so on on them (cf. [44]). Hölder's original axiomatization of the notion of an extensive property structure was closer to this view than more recent ones are, for he assumed that families of extensive properties enjoy continuity, have no minimal element, and so on (cf. [16], 54). Once we introduce either unexemplified properties or merely possible properties, it is fairly natural to suppose that each extensive family has the cardinality of the continuum. This suspicion is stregthened by the fact that many theories (at least if they are to be formulated in a manageable way) require such things as continuity and differentiability. We have said nothing to preclude requiring continuum many lengths and the like, though it would take a logic stronger than ours to do it. But the suggestion that numbers mirror the reals requires even more than this. There would have to be some fact, for example, about which property played the role of the number one, and our uniqueness theorem tells us that there is no such fact.

Earlier we thought of scientific theories in terms of classes of intended models, ones in which the domain of individuals could vary but the domains of properties were fixed. It would be in keeping with this picture of scientific theories to think of ' > ' and '+' as denoting genuine, autonomous mathematical relations that can have extensions in different domains like the sets of integers, rationals, and reals. Once we start thinking along these lines, mathematical truths seem to have more to do with such numerical properties and relations than with the specific things that exemplify them; the important thing is not the particular objects that are involved, but the overall structure of the relations in which they stand. This structure might itself be thought of as a complicated, rather abstract property somehow composed out of simpler properties like >, *successor*, and +.

This picture makes the epistemology of mathematics a little less mysterious than it often seems, for structural properties can be instantiated in the concrete things we perceive. Most of us easily recognize the same tune when played in different keys at different speeds, for example, and some people can recognize even more abstract patterns like a sonata form or blues progression. In such cases, we are recognizing structural relationships that are invariant throughout a variety of concrete instances. Similarly, a family of extensive properties might be seen as instantiating a complicated mathematical structural property.

On this picture, we could think of individual numbers in terms of roles, and roles could be specified by sentences having the logical form of functional definitions. As with all functional definitions, they specify an entire set of roles in one fell swoop. For example, something realizes the role of *zero* just in case it has certain properties (not being the successor of any number, being the identity element for addition, and so on). Of course there are many open questions about the details of such a picture of mathematics, but a good deal of work on these matters has already been done by Resnik and others (see [27] and references therein). What the present account offers is a way to simplify this "structuralist" account of mathematics through the use of properties.

Our marriage of representationalism and properties has taken properties and relations as the things being represented, but there is no reason why properties and relations themselves — rather than numbers or sets — could not also be used to represent other things. To illustrate the scope of the possibilities here, I will conclude with a rather speculative example.

The claim that Kay is five feet tall is often expressed in familar formal languages by a sentence like '$h(k) = 5$ ft'. This has sometimes been taken to show that Kay stands in some *two-place* (functional) relation like "having a height in feet of . . ." to an abstract entity, the number 5. By constrast, our account of measurement tells us that Kay exemplifies a *one-place* property that has nothing particularly to do with numbers, though given certain qualitative facts about lengths, we can establish a scale and use numbers to classify such properties as her height. In a similar spirit, several philosophers have recently suggested that belief, which also seems to involve a relation between a believer and an abstract entity (a proposition or the like) really isn't a relation either. Instead, beliefs are monadic, functional properties (often called 'states') of believers, though propositions can be used to classify them.[18] I think that the present framework would be a very natural one in which to develop this idea.

The first step would be to enrich our machinery so that we could deal with propositions, as well as properties and relations, where propositions are to be thought of as properties or relations that have all of their argument places filled with objects or other properties or by quantification. This can be done in a precise way by adding axioms, operations on properties, and an "algebraic semantics" to our machinery (as in [5], [20], or [45]). The operations ensure the existence of more complex properties given simpler ones. For example, we might have a conjunction operation that guarantees the existence of a conjunctive property given two other properties, much as ∘ guarantees the existence of the sum of two lengths. We might also have an operation that yields converses of 2-place relations, one that "plugs" an object into a monadic property to yield a proposition, one that provides existential generalizations of other properties and relations, and so on. We found earlier that families of *extensive* properties have the algebraic structure of a certain sort of semigroup. The current approach finds that properties also have the structure of an intensional analogue of a Boolean algebra with some extra operations tossed in to deal with relations, identity, and quantification — in short, they have a familiar *logical* structure.

The claim that there is some structure to the family of belief states that is mirrored by logical relations among propositions is suggested by the fact that it is so often useful to apply our theory about the structure of propositions, namely logic, in our thinking about believers and their beliefs. So the question is whether belief states, construed as monadic, functional *properties*, are related by higher-order relations that can be matched with abstract counterparts in the domain of higher-order relations on propositions. Beliefs and propositions are of a different logical type, the former being monadic properties and the later being in effect 0-place properties, but so long as they share enough structure, we can extend notions like that of an isomorphism to deal with this.

We may call a mapping from belief states to propositions that preserves the structure of the family of belief states an *interpretation function*. The most obvious sorts of qualitative relations on beliefs would be ones whose abstract counterparts were logical relations and operations like connectives, quantifiers, and entailment. For example, there might be a 3-place relation that triples of belief states stand in just in case the interpretation function maps them to propositions in such a way that the image of the third is the conjunction of the images of the first two.

Taken alone, such relations among belief properties probably would not suggest many interesting explanations or predictions, any more than > and ∘ by themselves do. The latter acquire real interest when embedded in theories relating lengths to other magnitudes, and similarly this picture of belief would be more interesting if it was embedded in a more general theory about the acquisition or change of belief, the role of belief in guiding behavior, or the like. One form such a theory might take may be illustrated with a simple and familiar example. Let us think of a psychological theory as including a two-dimensional mapping of belief states to propositions (representing their content) and to real numbers in the interval [0, 1] (representing degree of belief), along with a similar mapping of desire states to pairs of propositions and real numbers (representing utility). We might then devise a theory, perhaps one including the hypothesis that agents maximize expected subjective utility, that would tell us about what intentions would be formed in various circumstances. Of course current work in rational decision theory, epistemic logic, and related fields is highly idealized, but our account of theories allows that theories may make idealized claims about real properties, properties that might eventually be more accurately described by better theories.

If we could identify qualitative properties of belief states and find axioms governing their behavior, we could even ask about representation and uniqueness theorems. If we could find a representation theorem, we would have an account of why it works so well to classify belief states in terms of propositions. And if we could find a uniqueness theorem, we would see to what degree such classifications involved convention — *indeterminacy* — by seeing which transformations of the set of propositions were admissible, i.e., what its group of automorphisms was. We might then add various constraints on interpretation functions, some of which involved more than formal or structural features, to preclude some transformations and reduce indeterminacy; for example, it appears that Ronald Reagan can have a belief *about* Nancy, and there may be good reasons to require that she be a constituent of the proposition used to represent it. Those who are not fond of propositions might even hope to excise them by showing that the theory of propositions is conservative over the psychological theory. It is too early to say whether such projects would succeed, but they are the sorts of things that are needed to back up a number of recent claims about belief, and the current framework would be a good one in which to develop them.

A more common claim about knowledge and belief also finds a natural expression in our framework. A variety of thinkers have found it enlightening to view thoughts, at least true beliefs, as models or maps or representations of various aspects of reality. For example, Heinrich Hertz urged that the agreement or correspondence between mind and reality is quite analogous to that between two systems that are models of each other, and this stimulated Wittgenstein, who made it a central theme of his *Tractatus* that a thought is a logical picture — where this is construed in a very wide sense rather like our intuitive idea of a model — of a fact. Ramsey likened thoughts to maps, Tolman and other psychologists offered detailed accounts of "cognitive maps" in rats and men, and more recently the conception of thought as a map has been developed in detail in Armstrong's work on epistemology ([2]).

Kenneth Craik's treatment of this idea is of special interest here, for it incorporated the key elements of representationalism. He urged that models have the "same relation-structure" as the thing that they model and stressed three steps in the mind's modeling of reality: first, we "translate" facts into corresponding mental representatives; second we manipulate these representations or mental counterparts in thought; and, finally, we translate the result back into some conclusion or prediction about the world ([7], esp Ch. 5). These three steps correspond closely to the steps of "translating" facts about lengths into their numerical counterparts, engaging in mathematical reasoning with these, and then translating the result back into some prediction or conclusion about length. In the latter case, an isomorphism of an appropriate sort explains the applicability of mathematics to reality. In the former, a hypothesis about mental models being somehow isomorphic or homomorphic to aspects of the world could be seen as an attempt to explain *the applicability of thought to reality*. The representational framework is obviously well-suited for developing such ideas, but because its traditional reliance on *extensional* relational structures, it was bound to encounter difficulties in modeling thoughts, which are intensional in a variety of ways. What I want to stress here is that the insertion of properties, relations, and perhaps propositions into our relational structures and the picture of beliefs and other mental states as themselves structured properties, opens the door to a more fruitful use of this approach in both psychological and the philosophical investigations of the mind.

Craik's themes are now in vogue, as the currency of talk of mental representations, images, and mental models attests (cf. [14]). The transformations or manipulations of mental models that figure so centrally in thought — which perhaps *are* thought — may be viewed in a variety of ways: as applications of some standard logic, of a non-monotonic logic, of probablistic canons, of various heuristics, or of still more exotic principles for the processing and transformation of information. Our general framework is neutral on this, as it is on the precise nature of the elements of models. They might be language-like, mental images, some much more abstract models, or some combination of these. All that our approach requires is that the elements and their properties and relations can be treated as a genuine relational structure that is the isomorphic (or perhaps some weaker sort of) image of a genuine relational structure that models certain aspects of reality, and similar sorts of relationships could even hold between the a computer simulation and the working of a mind it simulates.

For one of my beliefs to be *about* something, say Ronald Reagan, it is not enough for me to get lucky and have certain aspects of my thoughts mirror the structure of certain features of the president. To be an actual representation *of him*, my belief must be related to him in some more intimate way. A similar situation arises in measurement.[19] As we saw in Section I, it is not enough for the bathroom scales to get lucky and register your actual weight; in order for them to be *measuring* it, their reading must be related to your weight in some non-fortuitous way. Similarly, a belief must be related in some non-fortuitous way to a thing in order to be a belief *about that thing* and a belief that accurately represents some fact (i.e., a true belief about it) must be related in some non-fortuitous way to that fact in order to be knowledge. Such, at least, is the epistemological doctrine of externalism, and its partisans have devoted much work to finding an enlightening characterization of non-fortuity in terms of appropriate causal connection, counterfactual sensitivity, subjunctive probabilities, nomic connections, or the like (e.g., [2], [9]).

A person is unlikely to have beliefs about all of the facts in any given corner of the world, and so his mental models are likely to be incomplete. Still, when his beliefs are sensitive to the facts, if a proposition isn't true, he won't believe it. If, in light of our discussion of counterfactual sensitivity, we upgrade this conditional to a counterfactual, we arrive at a condition that many counterfactual accounts of knowledge tell us is necessary for knowledge. Similar counterfactuals about measurement, e.g., that if a hadn't been warmer than b the thermometer wouldn't have said that it was, combined with its converse, is, in effect, a subjunctive version of a representation theorem.

These ruminations suggest that mental models and at least many cases of measurement are instances of a more general sort of phenomenon in which the properties in one family are related to those in other families in such a way that, as a matter of natural law, they co-vary within a certain range of circumstances. We can view such cases in terms of two genuine relational structures containing properties that are, because of their nomic relationships, counterfactually sensitive to each other, so that in a given range of circumstances, facts in one track carry information about facts in the other. A simple example of this is the way in which — when temperature, air pressure, and so on fall within a normal range — the *length* of a column of mercury varies systematically with the *temperature* of the surrounding air.[20]

University of Oklahoma, U.S.A.

NOTES

[1] In particular it differs from Ellis' account [10], which is anti-realist about measure-ment and about properties, and from the representational approach developed by Suppes and his collaborators insofar as this involves operationalism and nominalism, as it often does (e.g., [16]; [36], pt. 1). Few realist accounts of measurement have been de-veloped, but I have benefited from those in [6] and especially [21].

[2] Observability is a vague notion, but in many circumstances such things as colors and length are more directly observable than charge or spin.

[3] When we take demonstrability as a standard for evaluating philosophical arguments, nearly all of them fail, many quite transparently. I think that much philosophical argument, including most of the following, is better viewed as a species of inference to the best explanation ([39]).

[4] In addition to the notion of a functional property discussed here, we will sometimes speak of an $n + 1$-place relation's being functional when it is an n-ary function or operation, i.e., for any n things, there is a unique thing that stands in the $n + 1$-th place of the relation. Context will disambiguate.

[5] At all events, I shall have nothing to say about measurement in quantum mechanics or about properties of quantum phenomena.

[6] These examples are intended as a simple illustration of how extensional relational structures can be used to model quite familiar features of measurement involving scale types. However, a general treatment of scales turns out to be complex (cf. [25]).

[7] Even this may be too strong for some purposes. If we were assigning numbers to physical objects, rather than to their properties, we would want to allow for the possibility that two objects have the same property, e.g., the same length, and so would also drop the requirement that ϕ be $1-1$. But such variations do not affect my central thesis about the role of properties in measurement.

[8] Not all facets of the atomists' programs — their fondness for extensionality, their reductive yearnings for absolutely simple bits of reality, Russell's ardent empiricism, or

even the atomism itself — need have counterparts in our use of models. As one moves toward a Montagovian treatment of languages (viewed syntactically) as set-theoretic objects that are homomorphic to their semantics, the difference between a linguistic representation like the atomists' picture theory of language and representation by a relational structure diminishes. A number of interesting applications of related ideas — thoroughly laced with relativism and nominalism — may be found in much of Nelson Goodman's work. There are of course many important differences between representations, models, maps, pictures, replicas, copies, analogies, and so on, but I will not try to sort them out here.

[9] The discussion here oversimplifies a number of points, but I do not think they will be relevant here. Endomorphisms are sometimes needed rather than automorphisms, and so on. More importantly, substantive disagreement remains about uniqueness, admissibility, definability, and even about whether the attribute of meaningfulness attaches to sentences, relations, or families of relations. For a recent and relatively non-technical discussion of meaningfulness, together with numerous references to the relevant literature, see ([29]).

[10] I suggested this line of thought in ([38], 206), but didn't develop any of the details. An earlier suggestion along related lines may be found in ([12], 55). The first sustained development of these sorts of ideas is Mundy's [21].

[11] These versions differ from each other in a number of ways, and the first three incorporate elements, e.g., comprehension schemata, that I do not endorse. Nevertheless, the underlying ideas are quite similar. Related ideas may be found in early work on situation semantics where a fact is represented by a set $\langle R^n, a_1, \ldots, a_n, i \rangle$, where i is yes or no according as the objects $a_1, \ldots, a_n$ (in that order) stand in, or exemplify, the relation R^n or not; situations are then represented by sets of such sets ([4]).

[12] The introduction of times seems a natural way of treating the evolution of systems through time, and for some purposes it might be useful to represent the fact that laws determine the physically possible configurations of properties in terms of constraints on the extensions of various combinations of properties in all worlds.

[13] This is not to deny that for some purposes it is useful to concentrate on a theory's formal features. It is a pressing question in the philosophy and psychology of language how predicates come to denote or express properties and relations. There need be no single way this happens, but part of the answer seems to be that we are causally connected in a variety of ways to many properties through their observational and experimental effects and that these connections can often be exploited to help fix the reference (as opposed to giving the meaning) of many predicates. Such considerations would also help explain how we can get some handle on certain properties, independently of what a given theory says about them, which is clearly important when we want to test that theory's claims about them. This general approach to semantics is not free of problems, but there are a number of compelling arguments that something like it — and quite unlike cluster theories and stories about partial interpretation — is on the right track (e.g., [17]).

[14] If we imagine 'o' added to English, the intuitive idea is that 5 $meters$ o 7 $meters$ will be the monadic, determinate, first-level property of $being$ 12 $meters$ $long$. One could think of scales as systematically providing a set of $names$, like '2.5 meters', for determinate extensive properties, which are not themselves intrinsically numerical. An

interesting alternative to separate relations like *being longer than, being a greater mass than*, etc., is Russell's use of a generic, all purpose ordering relation that can hold within different families of determinates ([30], 165). Mundy has extended this idea to ∘, and offered an interesting axiomatization of all extensive properties in one fell swoop ([21]). One might even go a step further and view > as a relation that can also hold between first-level two-place relations such *being the distance between*, and use it in an account of difference measurement, but I shall not make any of these extensions here. The fact that relations like > and ∘ hold among determinate lengths may well be constitutive of these properties being members of the same family ([38], 219). Armstrong has made the intriguing suggestion that two specific magnitudes are members of the same family of determinates just in case one is part of the other ([2], 121ff). It is not clear to me that there is a unique, natural order — and hence any natural part-whole relation — for such families of determinates as shape. Nevertheless > has the right formal features to be some sort of part-whole relation, and so the framework should be a natural one for developing Armstrong's suggestion,

[15] A Henkin-style completeness proof for a slight variant of this logic is given in [28], Ch. 6.

[16] The idea sketched here is an instance of a more general approach of trying to account for claims about merely possible things of a given sort in terms of actually existing things of a higher-order (*cf.* e.g., [38], 218ff; [40], 621). For it to bear much philosophical weight — which it won't be asked to here — it would need to be backed by a formal semantics; I hope to provide one in a future paper.

[17] There are separate orderings of each family of extensive properties, *being longer than, being a greater mass than*, and so on. We might label these '$>_l$', '$>_m$', etc. (and similarly for $\circ_l$, etc.), but since we are only considering one family at a time, I will continue to supress such subscripts. In what follows we may think of > and ∘ as $>_l$ and $\circ_l$, but the axioms for any other extensive family would look the same.

[18] Several people appear to have noticed this analogy, though no one seems to have worked it out at all. An early instance is Field ([11], 114), who credits it to David Lewis. Beliefs are functional properties in the sense, discussed in Section I, of interest in the philosophy of mind, rather than in the mathematician's sense of functions or operations. It should be stressed that a monadic property need not be unstructured — the conjunctive property *being P and Q* is structured, but is still monadic. Although beliefs are monadic properties, they may have the same sort of structure as atomic propositions, conjunctions, existential quantifications, and so on.

[19] The general difficulty here involves what Kim calls the *problem of fortuitous satisfaction*; for a general discussion see [40, Section III]. The similarity here between measurement and the present picture of thought — the "thermometer view" — is no accident, for this view of thought and knowledge draws inspiration from simple cases of measurement (e.g., [2] 166ff; [9] *passim*).

[20] I want to thank Monte Cook, John Etchemendy, John Forge, Michael Hand, Chris Menzel, Brent Mundy, and Ed Zalta for helpful discussions on the topics of this paper, and the Center for the Study of Language and Information at Stanford University, where it was written.

BIBLIOGRAPHY

[1]   Armstrong, David. *Belief, Truth, and Knowledge*. Cambridge: Cambridge University Press, 1973.

[2]   Armstrong, David. *Universals and Scientific Realism, Vol II: A Theory of Universals*. Cambridge: Cambridge University Press, 1978.

[3]   Armstrong, David. *What is a Law of Nature?* Cambridge: Cambridge University Press, 1983.

[4]   Barwise, Jon & Perry, John. *Situations and Attitudes*. Cambridge, Ma.: MIT Press, 1983.

[5]   Bealer, George. *Quality and Concept*. Oxford: Clarendon Press, 1981.

[6]   Byerly, H. and Lazara, V. 'Realist Foundations of Measurement.' *Philosophy of Science* **40** (1973): 10—23.

[7]   Craik, Kenneth. *The Nature of Explanation*. Cambridge: Cambridge University Press, 1943.

[8]   Dretske, Fred. 'Laws of Nature.' *Philosophy of Science* **44** (1977): 248—268.

[9]   Dretske, Fred. *Knowledge and the Flow of Information*. Cambridge, Ma.: MIT Press, 1981.

[10] Ellis, Brian. *Basic Concepts of Measurement*. Cambridge: Cambridge University Press, 1966.

[11] Field, Hartry. 'Mental Representation.' Reprinted with a postscript in *Readings in the Philosophy of Psychology*: Vol. I, pp. 78—114. Edited by N. Block. Cambridge: Harvard University Press, 1981.

[12] Field, Hartry. *Science without Numbers*. Princeton: Princeton University Press, 1980.

[13] Hempel, Carl. *Fundamentals of Concept Formation in Empirical Science*. Chicago: University of Chicago Press, 1952.

[14] Johnson-Laird, Philip. *Mental Models*. Cambridge Ma.: Harvard University Press, 1983.

[15] Keisler, H. J. *Model Theory for Infinitary Logic*. Amsterdam: North-Holland, 1971.

[16] Krantz, D., Luce, R., Suppes, P., and Tversky, A. *Foundations of Measurement*, Vol. I. New York: Academic Press, 1971.

[17] Kripke, Saul. *Naming and Necessity*. Cambridge: Harvard University Press, 1980.

[18] Linsky, Bernard. 'General Terms as Designators.' *Pacific Philosophical Quarterly* **65** (1984): 259—276.

[19] Luce, R. Duncan. 'Dimensionally Invariant Numerical Laws Correspond to Meaningful Qualitative Relations.' *Philosophy of Sciences* **45** (1978): 1—16.

[20] Menzel, Chris. 'A Complete, Type-free Second-Order Logic and its Philosophical Foundations.' Report #CSLI-86-40, Center for the Study of Language and Information, Stanford, University.

[21] Mundy, Brent. 'The Metaphysics of Quantity.' *Philosophical Studies* **51** (1987): 29—54.

[22] Mundy, Brent. 'Faithful Representation, Physical Extensive Measurement, and Archimedean Axioms.' *Synthese* **70** (1987).

[23] Nagel, Ernest, 'Measurement.' *Erkenntnis* **2** (1932): 313—333.

[24]   Narens, Louis. 'Measurement without Archimedean Axioms.' *Philosophy of Science* **41** (1974): 374—393.

[25]   Narens, Louis. 'On the Scales of Measurement.' *Journal of Mathematical Psychology* **24** (1981): 249—275.

[26]   Narens, Louis & Luce, Duncan. 'The Algebra of Measurement.' *Journal of Pure and Applied Algebra* **8** (1976): 197—233.

[27]   Resnik, Michael. 'Mathematics as a Science of Patterns: Ontology and Reference.' *Noûs* **15** (1981): 529—550.

[28]   Robbin, Joel. *Mathematical Logic.* Amsterdam: W. A. Benjamin, 1969.

[29]   Roberts, Fred. 'Applications of the Theory of Meaningfulness to Psychology.' *Journal of Mathematical Psychology* **29** (1975): 311—332.

[30]   Russell, Bertrand. *Principles of Mathematics.* Cambridge: Cambridge University Press, 1903.

[31]   Skala, H. J. *Non-Archimedean Utility Theory.* Dordrecht: D. Reidel, 1975.

[32]   Sneed, Joseph. *The Logical Structure of Mathematical Physics.* Dordrecht: D. Reidel, 1971.

[33]   Sober, Elliot. 'Evolutionary Theory and the Ontological Status of Properties.' *Philosophical Studies* **40** (1980): 147—176.

[34]   Stegmüller, W., *et al. Philosophy of Economics.* Berlin: Springer-Verlag, 1982.

[35]   Stevens, S. S. 'Quantifying the Sensory Experience.' In *Mind, Matter, and Method*, pp. 213—233. Ed. P. Feyerabend & G. Maxwell. Minneapolis: University of Minnesota Press, 1966.

[36]   Suppes, Patrick. *Studies in the Methodology and Philosophy of Science.* Dordrecht: D. Reidel, 1969.

[37]   Suppes, Patrick. *Set Theoretic Structures in Science.* Forthcoming.

[38]   Swoyer, Chris. 'The Nature of Natural Laws.' *Australasian Journal of Philosophy* **60** (1982): 203—223.

[39]   Swoyer, Chris. 'Realism and Explanation.' *Philosophical Inquiry* **5** (1983): 14—28.

[40]   Swoyer, Chris. 'Causation and Identity.' *Midwest Studies in Philosophy. Causation and Causal Theories* **9** (1984): 593—622.

[41]   Swoyer, Chris. 'Belief and Predication.' *Noûs* **15** (1982): 197—220.

[42]   Tooley, Michael. 'The Nature of Laws.' *Canadian Journal of Philosophy* **7** (1977): 667—698.

[43]   Toulmin, S. *Philosophy of Science.* New York: Harper Torchbooks, 1960.

[44]   Whitney, H. 'The Mathematics of Physical Quantities, II: Quantity Structures and Dimension Analysis.' *American Mathematical Monthly* **75** (1968): 227—256.

[45]   Zalta, Edward. *Abstract Objects: An Introduction to Axiomatic Metaphysics.* Dordrecht: D. Reidel, 1983.

JOHN FORGE

ON ELLIS' THEORY OF QUANTITIES*

A statement of the account of quantities to be discussed and defended
in this essay can be found in the second chapter of Brian Ellis' book
Basic Concepts of Measurement (Ellis 1966). I refer to this account as
"Ellis' theory of quantities", although I should say at the outset that
Ellis himself does not fully and unequivocally embrace the account.
This has caused some confusion about precisely what his position is
with regard to quantities, and has led two of his critics to write "It is not
clear in what sense he [Ellis] does allow that quantities exist" (Byerley
and Lazara 1972, p. 14)[1]. In the second part of this essay — there are
three parts in all — I shall state what I believe to be Ellis' theory and in
so doing indicate in what sense he allows that quantities exist. In the
third part I shall attempt to defend the theory against five objections,
one of which is due to Ellis. It will be necessary to modify the theory in
two respects in the light of these objections.

Ellis presents his theory of quantities in Chapter 2 of his book after
making some critical remarks about the realist view of quantities. And
in the introduction to *Basics Concepts* he writes

My main thesis is that certain metaphysical presuppositions, made by positivists and
non-positivists alike, have played havoc with our understanding of the basic concepts of
measurement, and concealed the existence of more or less arbitrary conventions. (Ellis
1966, p. 3)

The non-positivists referred to in this passage are those who incline to
realism about quantities. I assume that a realist account of quantities is
one that takes a quantity to be a certain kind of *property*. This sort of
account is the main rival to the Ellis theory, operationism having
dropped by the wayside long ago[2]. There are, however, several varieties
of realism, distinguished from one another in terms of the way in which
properties are construed. For example, David Armstrong's realist account of quantities is based on his theory of universals (see Armstrong
1978, pp. 120–125 and Armstrong 1983, pp. 114–116), and it is a
'sophisticated' version of realism. Ellis' critical remarks are not so
telling against Armstrong as they are against 'naive' viewpoints. While it

291

John Forge (Ed.), Measurement, Realism and Objectivity, 291–310.

is not my intention here to try to uncover a form of realism about quantities that is not subject to the kind of criticism that Ellis gives, it is nevertheless worthwhile to see what Ellis has against realism. This will set the scene, so to speak, for an exposition of his own theory, and will serve to sharpen the issue between partisans of the Ellis position, such as myself, and the realists.

1. QUANTITIES-AS-PROPERTIES: THE REALIST ACCOUNT

We tend to use the language of realism when we talk about measurement. Thus we say that in measurement, values (numbers or numerals depending on whether one sides with Suppes or Campbell) are assigned to objects as measures of the quantities they possess. This suggests that quantities are entities quite distinct from the physical objects[3] that 'have' them. But it does not follow that we must therefore be committed to realism. If some non-realist account were to be adopted, then no doubt suitable paraphrases could be found for the realist language. However, any account of quantities must accord with measurement as it is practised in science. So for instance an account of quantities that took what are in fact "more or less arbitrary conventions" to represent actual features of the world would not accord with our measurement practices. This is the charge which Ellis levels against the realist. I shall therefore begin by giving a very simple example of measurement in which several conventions are apparent, and then I shall go on to consider realism about quantities and Ellis' criticism thereof in the light of this example. I shall also make use of this example in the next two parts of the essay.

The assignment of values to objects by means of measurement is either direct or indirect, but all measurement presupposes the existence of standards. What I mean here by direct measurement is simply the *direct* comparison of an object with a standard[4]. Consider the measurement of mass by means of the equal arm balance. Suppose we want to determine the mass of an object *a*. We place *a* on one of the balance pans and then we add standard objects to the other pan until there is no deflection of the balance. The sum of the values of the standard masses is the mass value of *a* on the scale in question. This procedure presupposes what Ellis calls an extended system of standards on a scale for mass. Such a system can be constructed in the following way: Let us refer to the state of affairs where there is no deflection of the apparatus as an instance of balancing or as an instance of the balancing relationship B. If there is no deflection when x and y are placed on the pans, we write $B(x, y)$. Now, to construct an extended system of standards, it is necessary to begin by choosing an object s to serve as the basis of the system and which is assigned by convention a value that is normally, but not necessarily, 1. Replicas of s, s', s'', etc., are then manufactured so that $B(s, s')$, $B(s', s'')$, etc. Each of these replicas of s is assigned the value 1. Next we make an object s_{10} such that $B(s_{10}, 10s)$ — i.e. s_{10} balances 10 s, s — and then assign a value 10 to s_{10}. Replicas of this standard are also made. In similar fashion we manufacture s_{100}, s_{1000}, etc, together with their replicas. Finer divisions on the s-scale are possible when we construct $s_{1/10}$, $s_{1/100}$, etc, where, for instance, $B(10s_{1/10}, s)$, and hence 1/10 is assigned to $s_{1/10}$. There is, in principle, no limit to the range of standards that we can construct in this way and so in principle we can find some combination of standard objects that stand in B to any object whatsoever. There are, of course, practical limitations to extending the s-system indefinitely.

A direct measurement thus involves the manipulation of standard objects which are juxtaposed in some appropriate way with the object under investigation. But direct measurement is not always convenient and in some cases it is not possible. The spring balance, for example, is in many respects more convenient for the measurement of mass than is the equal arm balance. However, a spring balance must be calibrated with reference to standards. The divisions on the dial attached to a spring balance are marked in accordance with the extensions produced by standards. So if the pointer indicates one s-unit when a is placed on the pan, then this means that a produces the same extension as s. In this case let us say that a and s stand to one another in the relationship "produces the same extension as" and symbolise this by $SG(a, s)$. Measurement here is indirect because objects like a are not directly compared with the standards; rather the comparison is mediated by the calibration of the instrument.

In some instances direct measurement is not possible because the objects in question are too large or too remote to be subject to any measurement operation. A planet cannot be loaded onto a balance. However, the mass of a large remote body such as a planet can be inferred in the following manner:

A probe of known mass is fired off towards the planet and the acceleration that it experiences as it approaches the planet is determined. The mass of the planet can then be inferred, provided that we know the mass of the probe. On the assumption that the mass of the probe has been determined by comparison with standards of the s-system, then the experiment just described effects a comparison of the planet with the s-system. This comparison between our planet and the standard objects may seem rather more indirect than was the case with the spring balance, nevertheless I do not think we should deny that it constitutes a measurement. Furthermore, the fact that we use law statements that relate mass to acceleration does not represent a significant point of difference with the previous cases. Use of the spring balance and the equal arm balance as means to measure is warranted by laws — the principle of moments, Hooke's law, and so forth. Indeed, all methods of measurement are embedded in some framework of laws that establish relations between quantities.

If it is true that all measurement involves some comparison with standards, be this direct or indirect, then any conventional choices made in setting up the system of standards will, as it were, be embodied in assignments of values on that scale. There are three conventional choices made in setting up the s-scale. Two of these are obvious, while the other is perhaps not obvious. The two 'obvious' conventions are the choice of s as the basis of the system and the assignment of 1 to s. These choices are conventional in that we could have chosen some object r rather than s, where it is not the case that $B(r, s)$, and we could have assigned a value different from 1, say 2, to r, and our subsequent assignments of mass values would have been just as correct as they were on the s-system. The third convention is the decision to *add* mass values of standards that stand in B to the object whose mass is being determined. We decided that if $B(a, 3s)$, i.e. that if three replicas of s stand in B to c, then 3 is assigned to c. But suppose we chose the r-scale and decide to *multiply* rather than add values of the standards. Then if three copies of r stand in B to object d we assign $2^3 = 8$, not 6, to d. In general, if n replicas of r stand in B to an object, then 2^n will be assigned to the object on this multiplicative scale (*cf.* Ellis 1966, p. 79). Additive scales are normally preferred to multiplicative scales since the latter are apparently less simple than the former. But both kinds of scale are just as correct as one another.

In the passage from Ellis quoted above, it is claimed that certain metaphysical presuppositions have led certain non-positivists (whom I identify as realists) to overlook the existence of conventions such as those just mentioned. We now need to see why Ellis makes this claim.

The realist believes that a quantity is a sort of *property*, namely one that admits of degrees or *magnitudes*, as I shall call them. Mass is a quantity, or quantitative property, because an object has more or less mass; it has, in the realist idiom, a particular degree of mass. Quantities are therefore to be distinguished from qualities, or qualitative properties, such as "being on the board of Exxon". Either one is or one is not on the board of Exxon; this is not a matter of degree. Now we expect the realist to tell us what a property is — realists do not take properties to be reducible to classes — for the idea of a property is by no means absolutely clear. How it is that properties are construed will determine the outlines of the realist account of properties to be given. But it seems that Ellis will not hold with any variety of realism about quantities:

A quantity is usually conceived to be a kind of property . . . According to this conception, objects possess quantities in much the same way as they possess other properties, usually called 'qualities'. Quantities, like qualities, are supposed to be inherent in the objects that possess them. Just how this is to be understood is not clear, but at least it is supposed to make sense to imagine a universe consisting of a single object possessing a variety of different quantities and qualities. It may, for example, be imagined to be round, three feet in diameter, made of steel, coloured yellow, and being five grams in mass. All this exists, so to speak, before measurement begins. The process of measurement is then conceived to be that of assigning numbers to represent the magnitudes of these pre-existing quantities; and ideally it is thought that the numbers thus assigned should be proportional to these magnitudes. (Ellis 1966, p. 24).

Ellis believes that the realist misses the essentially *relational* character of quantities. He believes that objects possess quantities, but that when we say that object x has quantity q, we are not making a statement about x in isolation, rather we are referring to the relations in which it stands to other objects in q.

A quantity is relational in the sense that if x and y have q, then necessarily x and y stand to one another in one of the following mutually exclusive relations

1.1a. $x >_q y$

1.1b. $x =_q y$

1.1c. $x <_q y$

where "$x >_q y$" is read "x is greater in q than y", etc. The formal properties of $>_q$ and $=_q$ are as follows: $>_q$ is transitive, asymmetrical and irreflexive, while $=_q$ is transitive, symmetrical and reflexive. (In

Part 3 I shall introduce binary relations $\geqslant_q$ and $\leqslant_q$ that are equivalent to 1.1). Suppose we give Ellis' yellow ball the name c. Then it would appear that c cannot possess quantities because there are no other objects in its universe with which it can stand in a 1.1 relationship. There is no doubt that Ellis is right in that if x and y have q, then either they have the 'same amount' of q or one has 'more' q. Hence, if more than one object has q, then necessarily they stand in relations of form 1.1. But does it follow that c cannot have any quantities?

As a matter of fact no values can be assigned to c because there are no resources with which to set up and implement measurement procedures in c's universe. So, as a matter of fact, the value 5 grams could not be assigned to c. But there are many objects in our universe which are inaccessible, and hence are such that no values can be assigned to them. These objects possess quantities; it just happens that they cannot be measured. The realist would say the same about c. c has quantities but these cannot be assigned values. If c were in our universe, and were accessible to us, then values such as 5 grams could be assigned to it. The realist may invite us to consider a thought experiment in which c was originally a member of a universe much like ours. In this case c has quantities. Now the objects in c's universe are extinguished one by one until only c is left. Given that there is still a universe for c to be in — and I take it that Ellis would accept this, as he is willing to talk about c's universe even though he denies that c has any quantities — could not c still have its original quantities? If not, when does c lose them? When c has just one partner left, it can still stand as a term in 1.1, and hence still has its quantities. Does c lose its quantities at the instant this partner is extinguished? I find that suggestion implausible. Certainly it makes *sense* to say that c retains its quantities even when it has no partners left.

The relevance of these thought experiments — the one just described and that given by Ellis — for a theory about quantities should not be overestimated. What our theory tells about whether object c has or has not quantities is not particulary important, as the theory is supposed to be about quantities in *our* universe subject to our measurement practices. In our universe, as far as we know, all quantities have more than one instance, and hence objects possessing quantities stand in relations of form 1.1. The realist must acknowledge this and be prepared to give an account of it in terms of inherent magnitudes. Since the realist postulates the existence of properties that have magnitudes in his or her theory of quantities, then these must be invoked to explain those characteristics that objects have in virtue of their having quantities. If this explanation is such that values are assumed to be proportional to inherent magnitudes, then the realist is guilty of mistaking arbitrary conventions for facts about the world.

Realism about quantities need not, however, incorporate the idea that values are assigned to objects so as to proportional to inherent magnitudes. But, as I have said, the realist must give as an account of the nature of magnitudes which shows why it is that objects having those magnitudes stand in 1.1. As was said in the introduction, this is not the place to go into details about particular versions or realism. Notice, however, that the existence of magnitudes needs to be inferred from measurement, in that measurement is carried out on objects and not on magnitudes. We place object a on the balance, not a's mass. If it were possible, then, to come up with a theory of quantities that took quantities to be relations between objects without introducing magnitudes, then we would have a more economical theory from the point of view of our ontology. Magnitudes would be seen to be unnecessary and talk about magnitudes would be translated into talk about objects. Ellis theory, when suitably modified, does, I believe, show that we do not need to concern ourselves with properties that have magnitudes in our analysis of quantities.

2. QUANTITIES AS ORDERS OF OBJECTS: ELLIS' THEORY

Let us now turn our attention to objects to which measurement can be applied and leave to one side the case of object c. When the mass of an object a is determined by means of the equal arm balance, the sum of the values of the standards that are in B to a is assigned to a in accordance with the condition

2.1. $B(x, y)$ only if $x =_m y$

that is to say, x balances y only on condition that x is equal in mass to y. $B(x, y)$ is not a necessary condition for $x =_m y$ because it may happen that x and y are never placed on the pans of a balance, and it is not true that objects only have mass if they are subject to measurement. Furthermore, values need not be assigned to a by the equal arm balance operation. The spring balance can be used instead. In which case we have the condition

2.2. $SG(x, y)$ only if $x =_m y$

These two relationships, $B(x, y)$ and $SG(x, y)$, are the results of performing certain operations on x and y — placing them on the pans of a properly constructed balance — but they themselves are physical states of affairs. In fact, $B(x, y)$ represents an interaction between x and y in accordance with the principle of moments. We can also interpret $SG(x, y)$ as an interaction involving x and y together. Suppose we place x and y together on the spring balance pan. If the extension produced was then twice that produced by either x or y alone, then we can think of $SG(x, y)$ as representing an interaction of *both* objects with the balance. This interaction is governed by Hooke's law. It is helpful to think of relationships like B and SG as physical interactions, as this enables us to distinguish them from a state of affairs such as $x =_m y$ which does not entail the existence of any actual interaction between x and y.

The two other possible relations between x and y given by 1.1, besides $x =_q y$, are $x >_q y$ and $x <_q y$. In the example of mass, the latter states of affairs may also be represented on the beam balance and the spring balance. If x's pan goes down and y's pan goes up, let us write $D(x, y)$ and assume

2.3. $D(x, y)$ only if $x >_m y$

If x produces greater extension on the spring of a spring balance than y, we write $G(x, y)$ and likewise assume

2.4. $G(x, y)$ only if $x >_m y$

Now if M' is a finite class of objects on which measurement operations can be performed — the members of M' are not too large or too far away — then the members of M' can be arranged in *linear order* by means of either of our two measuring operations. Using the beam balance, for example, we can find the minimal element (or elements) $min \in M'$, such that all $x \in M'$, $D(x, min)$, and similarly the maximal element (or elements) $max \in M'$ for which $D(max, x)$ for every $x \in M'$ can be found. Between these end points, the other members of M' can be arranged depending on how they stand in B and D. These objects can be physically arranged in order, e.g. by lining them up on a bench top. Or — and this may be easier — numbers can be assigned to the objects so that min is given the smallest number and max the largest, with the order of the other members of M' being recorded by the order of the assigned numbers — in which case we have an ordinal scale. Provided that 2.1 and 2.3 are true, this order is precisely the order of mass of the objects.

We are now in a position to present Ellis' theory of quantities. In the first place, he believes that the existence of a quantity q entails, and is entailed by, the existence of a class of linear ordering relationships (Ellis 1966, p. 3). We know what linear ordering relationships are. They are certain modes of physical interaction that we can make use of in our measurement practices so that, at least in principle, the members of the class Q of objects that have q can be arranged in order. B, D and GE, SE are linear ordering relationships for mass. Ellis maintains that the existence of linear ordering relationships entails the existence of quantities, but he does not assert that quantities *are* linear ordering relationships nor does he assert that the latter provide criteria for the identity of quantities. Operationism has the consequence that there is a $1:1$ correspondence between quantities and linear ordering relationships, from which it follows that there are at least two mass quantities: mass according to the equal arm balance and mass according to the spring balance. Ellis is not an operationist. He believes that it is the *order* generated, not the ordering relationships, which give the criteria for the identity of quantities. If B, D and SE, GE give the same linear order in the same class of objects, then they pertain to one and the same quantity. This answers Byerley and Lazara, who wondered in what sense Ellis allows quantities to exist.

The provision of criteria for the existence and identity of a certain kind of entity does not necessarily amount to a *theory* about the nature of that entity. Indeed, Ellis seems more concerned to explicate our *concept* of quantity than to investigate the nature of quantities themselves. Of course, an analysis of the concept of quantity cannot be conducted without saying something about what a quantity is taken to be. Nevertheless, the analysis may focus on the conditions under which it is correct to apply the concept. For instance, Ellis states that our concept of quantity is *cluster* concept; which is to say it is a concept that has no one set of necessary conditions for its application. For mass, the fact that B, D and SE, GE both generate linear orders means that the applicability of either of the respective measuring operations constitutes a sufficient condition for applying the concept of mass. But knowing when to apply the concept of mass need not amount to having a theory about mass. What Ellis has to say about how we apply quantity concepts is consistent with the view that quantities are properties with associated magnitudes.

In the light of Ellis' critical remarks about realism, it would be most surprising if he were to endorse some version of the quantities-as -properties viewpoint. The theory of quantities which I believe is strongly suggested in *Basic Concepts* is that a quantity is a particular linear order of objects, or equivalently that it is a linearly ordered class. I have already said that Ellis is not willing to assent unreservedly to this account, but I would maintain that insofar as he does have a theory of quantities, as opposed to an analysis of the concept of quantity, then it is the theory just mentioned.

3. A DEVELOPMENT OF ELLIS' THEORY

I have taken the theory to state that a quantity is an ordered class. Since 1.1 are equivalent to

3.1a. $x \geqslant_q y$

3.1b. $x \leqslant_q y$

where $\geqslant_q$ is transitive, reflexive and antisymmetrical, we can represent these ordered classes by the ordered pair $q = \langle Q, \geqslant_q \rangle$. As we shall see, it is a necessary but not sufficient condition for something to be quantity that if is an ordered class of physical objects. The Ellis' theory is in fact too weak. In this part I shall try to defend the theory by answering five objections to it. It turns out that we need to strengthen the theory in two ways to answer these objections.

The first objection comes from the realist camp. It is that we are unable to distinguish one quantity from another because we unable to define the classes Q and $\geqslant_q$ — unless we appeal to properties. Obviously we cannot list the members of these classes, but we can define the classes another way, namely by appeal to operations of measurement and laws of nature. This is, I believe, what is done in the context of science.

The memberhsip of Q can be specified as follows. Certain measurement operations apply in actual practice to a subclass Q' of Q. These operations apply because the elements of Q' are accessible to us and hence can be placed on balance pans, connected up to galvanometers, aligned with measuring rods, etc. Strictly speaking, several subclasses of

Q should be distinguished here, as it is unlikely that the measuring operations for q apply to exactly the same objects, although we expect these subclasses to have common members. Different operations of measurement specify the domain Q' of the same quantity q because they generate the same order $\geqslant_q$ in Q'^*. Hence we are in no danger of the operationist proliferation of quantities. There are, however, members of Q to which our measurement operations do not apply, and there are also all manner of objects *not* in Q to which our measurements do not apply. How are the former distinguished from the latter?

This can be done with reference to laws of nature. Laws, as these are conceived in physical science, are relations or *correspondences* between quantities. If quantities are taken to be ordered classes, then a law of nature will be a correspondence between ordered classes. For example, if g_1 and g_2 are mole samples of ideal gas, then if $g_1 \geqslant_v g_2$ and $g_1 \geqslant_p g_2$, then $g_1 \geqslant_t g_2$, where the orders in question are associated with volume, pressure and temperature respectively. Provided that q stands in a law L to some other quantity, say p, then the members of Q are just those objects that are in L to the members of P. This in itself is not satisfactory as a way of specifying Q since it merely shifts the problem to that of specifying L. However, as was mentioned in Part 1, we make use of laws in indirect measurement — the mass of a planet can be inferred from the acceleration it induces on a probe. Laws of nature therefore enable us to extend the range of measurement, and hence, indirectly, of our measurement operations. The matter can be expressed in this way: In general, we have direct access to subclass Q' of Q, and hence we can assign values to the members of Q'. On the basis of knowledge about values assigned to members of Q' and on the basis of knowledge about the correspondence between q and p, we can then make assignments to elements of Q that are not in Q', given that we can determine the requisite values of p.

But there will also be objects remote in space and time to which our measurement operations do not apply either directly or indirectly. For instance, we surmise that there are stars in the farthest reaches of the universe even though we have no information about them. Such objects have quantities like mass, volume, diameter, etc. Now if we had access to these objects — if only we had data about their size, composition, and so forth — then we could apply our measuring operations to them indirectly. For example, we could compute their masses from estimates of their size and composition, and in this way locate them in the order of mass. What underpins this counterfactual is the fact that objects do indeed have places in the order of mass and that the quantity stands in a lawful relation to the quantities volume and density, the latter being determinable from data about the composition of the objects in question. The truth conditions for the counterfactual therefore make reference to the order of mass. We are, however, quite at liberty to make reference to the quantity here as we are not trying to prove its existence. We have *posited* the existence of linear orders in nature; we are now stating conditions for the membership of the classes involved.

The second objection comes from Ellis himself. I have outlined a method by which we can define q. Ellis questions whether such a method ever reveals the real order associated with a quantity and hence whether we have any grounds for believing that there are orders of objects in nature.

What is the real order of a given quantity? And how do we know that there is any such order? Consider the first part of the question. The concept of a real order is, presumably, a kind of limit concept. It is conceived to be the order to which all physically obtainable orders (for a given quantity) approximate. But then if this ideal order is not physically obtainable, or otherwise knowable, it is also incomparable. Hence the degree to which a physically obtainable order may be said to approximate to the ideal one cannot be ascertained by the sorts of procedures we may use for deciding how closely physically obtainable orders approximate to each other. Hence, if the supposition of ideal quantitative orders is not to remain a purely metaphysical dogma, alternative criteria for degrees of approximation must be provided .... Perhaps this whole way of speaking about quantities should be rejected. (Ellis 1966, p. 49—50).

This passage expressed Ellis' reservations about the theory of quantities that I have attributed to him. He is correct, I think, when he states that we can never know for certain what the 'real' order of a quantity is. But it does not follow that this whole way of speaking about quantities should be rejected. Our measurement practices do in fact provide good grounds for believing that the are orders in nature.

The 'real' order of a quantity is not actually attainable for two sorts of reasons. First of all, our measurement techniques and operations do not always give the same order for a given class of objects. If we were to compare enough pairs of objects by means of both the equal arm balance and the spring balance, then for some x and y we would find an 'anomolous' result, namely either $D(x, y)$ and $SE(x, y)$, or $B(x, y)$ and $GE(x, y)$. In the second place, a class of ordering relationships will not always exhibit transitivity. For some x, y, z, for example, we will find $D(x, y)$, $B(y, z)$, but $B(x, z)$, or $B(x, y)$, $B(y, z)$ but $D(x, z)$, or perhaps even $D(x, y)$, $B(y, z)$ and $D(z, x)$. If it were then supposed that B and D revealed the actual order of mass, then this could not be a linear order. Ellis asked what justification we have for maintaining that ordering relationships such as these reveal an approximate order.

We know that measuring instruments are not always sensitive enough to discriminate between the masses of objects that are very close together in the order of mass. An equal arm balance, for instance, cannot be perfectly constructed. There will be friction about the knife that supports the arms of the balance and this will be sufficient to prevent deflection when two bodies of nearly the same mass are placed on the pans. Consider one of the 'anomalies', $D(x, y)$, $B(y, z)$, $B(x, z)$. If the balance is just sensitive enough to detect the difference between x and y, then it will not detect the difference between x and z if $x >_m z >_m y$ — hence the 'anomalous' result. This can be verified by using a more sensitive balance, one that can detect the differences between the masses of these bodies. A spring balance is also subject to physical effects that limit and impair its sensitivity. The elasticity of the spring may be reduced if the balance is used a good deal, particularly with large masses. This would also lead to failure to detect differences in mass and hence to a tendency to give the result $SE(x, y)$ as opposed to $GE(x, y)$. No scientific instrument is absolutely sensitive, and all instruments are subject to physical effects of one kind or another that limit their precision. The sort of results that we have been discussing are therefore not really anomalies, for they are to be expected.

What justifies the assumption that there are linear orders in nature is the fact that on this assumption we obtain just the results we would expect when comparing objects with one another using instruments that are not absolutely sensitive and techniques that are not infallible. For the most part, we find that relationships such as B, D and GE, SE are transitive and that they give the same order for the same classes of objects. When this does not happen we can often resolve the 'anomaly' by introducing a more sensitive instrument. When on occasion it turns out that we are unable to resolve the 'anomaly' with the available measurement technology, it is more reasonable to suppose that we do not have sufficiently sensitive measuring instruments than abandon the view that there are linear orders in nature. Indeed, the case discussed in the previous paragraph suggests a criterion for comparing approximate orders: for mass, assume that it is the balancing relationship B or the same extension relationship SG that is incorrect. As was mentioned, the insensitivity of the instruments will tend to manifest itself in failure to discriminate between objects close together in the order of mass. And as a general rule it may be suggested that attention be directed towards the interpretation of $=_q$, rather than the interpretation of $>_q$, when attempting to compare approximate orders obtained by a particular measurement technique in which these relations are interpreted. I do not, therefore, believe that supposition that there are linear orders in nature is a metaphysical dogma.

The third objection is that the existence of a linear order may be necessary for the existence of a quantity, but it is not sufficient. For example, the social security number of adults living in the United States gives a linear ordering, although this is clearly not a quantity. As another counterexample, objects can be linearly ordered with respect to their distance from Pittsburgh, but we would hesitate to call "distance from Pittsburgh" a quantity. There are two ways to strengthen the theory so as to deal with these counterexamples. The first is to impose the requirement that all genuine quantities are *extensive* structures. Mass is extensive in that the *composite* $x \circ y$ formed by concatenating x and y — for instance by placing one on top of the other — is the sum of their individual masses. This would certainly rule out social security number as a quantity, because there is, presumably, no way to concatenate adult Americans to form composites whose social security numbers is the sum of those of their components. Likewise, "distance from Pittsburgh" does not seem to be extensive. It is by no means altogether clear how we should form composites in this case, but consider the following. Suppose the objects that have the quantity are townships, and that these become concatenated when neighbouring municipalities grow and merge into one larger municipality, like Dallas-Fort Worth, or Albury-Wodonga in Australia. This concatenation does not give additivity. Dallas and Fort Worth are just about the same distance from Pittsburgh as Dallas-Fort Worth.

There are two drawbacks to the proposal just made. The first is that it requires us to give a general characterisation of the process of forming composites: When does a method of conjoining objects count as concatenation and when does it not? To give such a characterisation we need to know the formal properties of concatenation and we need to know what constitutes an acceptable interpretation of the formalism. Ellis, among others, has given a formal account of concatenation (Ellis

1966, p. 75). Less attention has been paid to the more difficult issue of specifying what constitutes an acceptable interpretation of the operation[5]. But it would not be profitable to go into the issue here in view of the second of the drawbacks to the proposal at hand, which is that there are genuine quantities that are not extensive. Temperature, pressure, density, etc., are all *bona fide* quantities, but none are extensive. To strengthen the Ellis theory by requiring all quantities to be extensive structures would therefore be to render it too restrictive.

This brings us to the second proposal. I have claimed that laws of nature are correspondences between quantities. It may, then, be suggested that an ordered class of objects is a genuine quantity only if it stands in some law of nature. On this supposition, there are no isolated quantities: all quantities are embedded in networks of correspondences. In this way we can eliminate the orders associated with social security number and distance from Pittsburgh, as there are no laws of nature in which these appear. But all the quantities familiar in science, mass, temperature, density, current, etc., appear in laws. This proposal does, therefore, give us the required strengthening without making the theory too restrictive. And this not just a matter of good fortune. It will be recalled that measurement practices are tied to laws in that interactions between objects and instruments are lawful interactions, and that laws are used in indirect measurement. Objects in an ordered class not embedded in a network of laws would not be measurable in the sense in which the objects investigated in science are measurable.

The fourth objection I want to consider is made much of by the realist. Theories that seek to identify properties with classes are often summarily dismissed because they are assumed to conflate different properties. Since everything that has a heart has a liver (except in some Lewis-Worlds), the class of things that have hearts is the same as the class of things that have livers, but having a heart is not the same as having a liver. Quantities, however, are *ordered* classes, and I am not aware of any examples of *orders* of objects associated with two or more quantities being the same. It is true that it was once believed that gravitational mass was different from inertial mass, although the order of these quantities is the same. But general relativity has the consequence that these two types of mass are in fact the same quantity, as is illustrated by Einstein's lift, which is subject first to an acceleration and then to a changing gravitational field. The effects of these two processes are identical.

If it is true that no two quantities actually have the same order, it does not follow that they could not have the same order. Consider a world consisting only of copper tubing of the uniform diameter. Then the orders associated with mass, length, resistance, conductivity, specific heat, volume, and much else besides, would be the same. Our theory has the consequence that in this world all these quantities would the identical. If this world were cited as part of an objection to our theory, my response would be that I do not see what it has to with our topic, namely quantities as these exist in our world, subject to our techniques of measurement. There would, for instance, be no natural science in the imaginary world, because presumably intelligent beings could not be made of copper tubing. Thus there could be no actual discriminations of one quantity from another. If intelligent life were suddenly to appear in the world, it is not clear what discriminations they would make. If the life forms were able to perform experiments on the copper tubing, they would soon detect differences in the orders associated with the quantities. For instance, if the beings took three pieces of tubing of the same length and welded two of them together, they would discover that equality of length does not entail equality of mass. Further experiments would reveal further differences.

I have already argued that imaginary worlds with strange populations have little or no relevance for our theory of quantities. Unless it can be shown that two distinct quantities invariably have the same order in their domains throughout our universe, then I do not believe we should be too concerned with the present objection. It should, however, be acknowledged that under certain conditions the same order will be found for certain quantities in given subclasses of their domains. This a consequence of laws of nature.

For example, consider the class V of mole samples of ideal gases that occupy vessels of exactely one litre. The order of the members of V in temperature will be the same as their order in pressure. But V is not the domain of a quantity, it is only a subclass of a domain.

The objection just considered may be termed "extensionalist" in that it states that there are different quantities with the same extension. Another, more serious, extentionalist objection to our theory is that the same quantity can have different extensions at different times. If the order associated with a quantity changes, then on our theory, as it has been developed thus far, so does the quantity. The possibility of this happening is by no means remote: indeed, it happens all the time. Ellis has a similar objection to Suppes' account of scales, although he does not apply it to the theory of quantities. Suppes takes a scale to be a mapping from a set of objects to the real numbers which preserves relations, such as order, defined on the set. So if s represents a scale from M to the real numbers, and if $s(x)$ and $s(y)$ are assigned to x and y, then if $x >_m y$, it is required that $s(x) > s(y)$. Ellis points out that if the membership of the set, e.g. M, changes, then so does the scale (Ellis 1966, p. 47). But if s is the gram scale for mass, then s is not converted into some other scale whenever the membership of M changes. Suppose, for example, that the object b in M is broken into three pieces b_1, b_2, b_3 at time t. It does not follow that the gram scale is thereby transformed into some other scale. Nor is it the case that the quantity mass is converted into some other quantity when such an event takes place. To deal with this objection we have to give up the idea that quantities are just orders of objects.

   To overcome this problem I suggest that a quantity be construed as a *sequence* of ordered classes. I shall outline how this might be done, although it will not be possible here to explore all the ramifications of the proposal. Instead of regarding a quantity q as an ordered class $q = \langle Q, \geqslant_q \rangle$, the idea is to treat it as a sequence of classes $q = \langle \langle Q_1, \geqslant_{q1} \rangle, \langle Q_2, \geqslant_{q2} \rangle, \ldots, \langle Q_n, \geqslant_{qn} \rangle, \ldots \rangle$. We want to ensure that there are interconnections between the terms of this sequence — otherwise there would be no grounds for maintaining that they are elements of one and the same quantity — which are such that the orders $\geqslant_{qi}$ can change without q thereby changing. There would appear to be four possible ways in which these terms might be interconnected. I shall state these, and then go on to consider whether it is plausible to claim that they are necessary conditions for quantities, using mass, once again, as the example.

(1) The domains Q_1, Q_2, $\ldots$, of the classes may have common members, in which case q will be said to be an *overlapping* sequence.
(2) The orders $\geqslant_{q1}$, $\geqslant_{q2}$, $\ldots$, may be made manifest by the same linear ordering relationships.
(3) If $x, y \in Q_i, Q_{i+1}$, then it may be the case that if $x \geqslant_{qi} y$ then $x \geqslant_{qi+1} y$.
(4) Suppose $x, y \in Q_{i+1}$ are the fragments of $z \in Q_i$, where z has undergone some process of fragmentation at time t_i. It may

be the case that for any object $u \in Q_i$, Q_{i+1}, if $z \geqslant_{qi} u$, then $x \circ y \geqslant_{qi+1} u$, where $\circ$ stands for the concatenation operation. (For extensive quantities, the ordered classes that make up the sequence have the form $\langle Q, \geqslant_q, \circ \rangle$).

Suppose we take mass to be the sequence $m = \langle\langle M_1, \geqslant_{m1}, \circ_1 \rangle$, $\langle M_2, \geqslant_{m2}, \circ_2 \rangle, \ldots, \langle M_n, \geqslant_{mn}, \circ_n \rangle, \ldots \rangle$. If an object $b \in M_1$ is broken into three pieces b_1, b_2, b_3 and these belong to M_2, then the order $\geqslant_{m2}$ will be different from the order $\geqslant_{m1}$. By including both orders within m, we can allow for the occurrence of events such as the fragmentation of b without being forced to admit the existence of a new quantity every time an event of this sort takes place. Let say, then, that a *development* occurs whenever there is a change in the membership of M_i which is such that there is a change in the order associated with this domain. And whenever a development takes place, there is introduced the order associated with the next ordered class in the sequence, in this case the ordered class with domain M_{i+1}. The index $1, 2, \ldots n, \ldots$ signifies the times at which developments take place. So, by interpreting m as a sequence of ordered classes, we can allow for the fact that orders of objects change over time. Furthermore, I think that it is plausible to maintain (1)—(4) as necessary conditions for a quantity like mass, although there are some quantities — those that are not extensive — for which (4) does not hold.

  If condition (1) did not hold for m, then there would be a development for which M_i and M_{i+1} have no common members. In which case the event at time t_i must have affected *every* body in the universe. It is not easy to say what we would make of an event of this kind. However, I do not think we could confidently assert that our quantity of mass was unaltered. With regard to condition (2), we have seen that linear ordering relationships are constituted by certain lawful interactions. Thus, if successive orders were not made manifest by the same linear ordering relationships, not only would we lack grounds for believing that we had the same quantity, but the laws in which the quantity figured before the development in question would presumably have changed. Given the close connection between laws and quantities, then any circumstance, how ever this might be imagined, that results in a change in the laws of nature will also change the pre-existing quantities. Condition (3) is Ellis' criterion for the identity of quantities applied to objects not involved in the development at t_i. This criterion has already been discussed and accepted in Part 2. Finally, I think we can accept condition (4) provided that no mass is liberated at t_i in the form of energy. This condition establishes a connection between the members that overlap M_i and M_{i+1} and the elements of M_{i+1} created as a consequence of the development at t_i.

The fifth objection has been somewhat more difficult to resolve than the previous four. Nevertheless, I believe that the solution just outlined goes some way to answering this objection, although it may be necessary it introduce some further modifications and changes into conditions (1)—(4) in order to provide a completely satisfactory answer[6].

CONCLUSION

In Part 3 I have attempted to defend Ellis' theory against five objections, and in so doing I have modified the theory in two respects. It has not, however, been necessary to introduce some notion of property into the theory. This does not, of course, mean that there can be no realist account that is in adequate in the sense that it does not conflate arbitrary conventions with facts about the world. But if the theory developed here is satisfactory, then it is not *necessary* to appeal to properties in our theory of quantities.

Griffith University, Queensland

NOTES

* A previous version of this essay was read at the Center for Philosophy of Science. I am grateful for the comments of those who attended, particularly those of Nick Rescher and John Norton. I would like to take this opportunity to thank the Center and its Director for the hospitality I received there in the spring of 1986.

[1] Byerley and Lazara accuse Ellis of 'neo-operationism'. In contrast, McGechie, in a review of *Basic Concepts* urges Ellis to make his realist stance more explicit, see McGechie 1967, p. 356. In fact, Ellis is neither an operationist or a realist.

[2] There are two major problems with operationism. The first is to say precisely what an operation of measurement is and hence what distinguishes one operation from another. The second is that there will be as many quantities as there are distinct classes of operations. In whatever way the first problem is solved, it appears that the second will lead to an unmanageable proliferation of quantities, see Hempel 1952.

[3] I shall assume throughout this essay that it is physical objects that possess quantities, how ever we understand "possess". By a physical object I mean a thing that exist for some definite time, possibly very short, and which has more or less definite spatial

location. Material bodies, fields, electrons, etc. are all physical objects. An alternative view, which I shall not consider here, is that it is events that possess quantities.

[4] This does *not* accord with the standard use of "direct" and "indirect" measurement. For instance, indirect measurement normally refers to measurements of a quantity which presupposes that some other quantity has been measured. Temperature is an example of a quantity that is only indirectly measurable in this sense.

[5] Hempel has claimed that we can find a concatenation for any quantity.

[6] I am less confident about condition (1) than about the other three.

BIBLIOGRAPHY

Armstrong, D. M. (1978), *A Theory of Universals*. Vol. 2. Cambridge: Cambridge University Press.

Armstrong, D. M. (1983), *What is a Law of Nature*? Cambridge: Cambridge University Press.

Byerley, H. C. and V. A. Lazara (1972), 'Realist Foundations of Measurement', *Philosophy of Science* **39**, pp. 5—26.

Ellis, B. (1966), *Basic Concepts of Measurement*. Cambridge: Cambridge University Press.

Hempel, C. G. (1952), *Fundamentals of Concept Formation in Empirical Science*. Chicago: Chicago University Press.

McGechie, J. E. (1967), 'Critical Notice of *Basic Concepts of Measurement*', *Australasian Journal of Philosophy*, pp. 353—370.

D. M. ARMSTRONG

COMMENTS ON SWOYER AND FORGE

COMMENTS ON FORGE'S THEORY OF QUANTITY

John Forge rejects the view that a quantity is a certain kind of property, one admitting of degrees. This view he calls *realism* about quantities. However, his own view, which he attributes a little uncertainly to Brian Ellis, might be described as a moderate realism about quantities.

On the property view, different degrees of the same quantity are different properties, properties of particulars. But from these different properties a one-dimensional order flows, an order which constitutes the properties as different degrees of the one quantity. Forge, however, dispenses with the properties. The order is something that exists objectively and independently, ordering certain particulars. The existence of the quantity is the existence of the order.

The one quantitative order may be, and regularly is, discovered and converged upon by different sorts of measuring operation. (For instance, scale-pans and spring-balances for mass.) But the operations in no way constitute the order. That the assigning of values in measurement does not constitute the quantity-order is shown, according to Forge, by the fact assignments can only be made on the assumption that various laws of nature hold. (This is particularly clear in cases where we extend the range of measurement by indirect methods.) But laws of nature, Forge says, are themselves correlations or correspondences holding between the orders that are the quantities.

Forge concedes that, although quantities are orders, not all orders are quantities. The order imposed upon certain persons by U.S. Social Security numbers is not a quantity. The orders that are quantities are constituted quantities by being embedded in a systematic network of correlations and correspondences with other orders. These correlations are what we call the laws of nature.

Quantities for Forge are thus theoretical entities, but entities realistically conceived, in the spirit of scientific realism. For myself, I like Forge's realism but think that he has not gone far enough in the realist direction.

One difficulty that can be raised for Forge's account is that it seems

311

John Forge (Ed.), Measurement, Realism and Objectivity, 311–317.

possible that a certain object should be the only thing in the universe, yet still have some determinate magnitude of some quantity. This suggests that the magnitude that the object has does not depend upon the place of the object in an order that involves other particulars.

Forge grants that it makes *sense* to say that such an isolated object would retain e.g. its particular mass. I am not sure whether he is allowing only that this is a conceivable supposition or is instead making the stronger concession that the case involves no contradiction. But in any case he thinks that the thought-experiment does not weigh much against his view: 'the theory is supposed to be about quantities in *our* universe subject to our measurement practices'.

I believe, however, that the adverse result for the Forge—Ellis theory of this apparently trivial thought-experiment is a sign of deeper trouble. We do not think that it is an entirely random matter that, for instance, the earth is greater in mass than the moon. Surely it has something to do with the constitution and structure of the two objects. Similarly for many other magnitudes in which the two objects can be compared. But if all these magnitudes and also the magnitudes associated with proper parts of the two objects dissolve into position in various orders, orders which involve particulars all over the universe, what can be said about the earth and the moon which will serve as a foundation for their particular places in the mass-order and other orders? Putting it in the language of properties, what properties will be left to the earth and the moon which will sustain and explain their places in the orders? I concede that the question where to stop demanding explanation is always one of great delicacy in such fundamental matters. But the Forge—Ellis theory appears to have stopped giving explanations too soon. Magnitude-properties would take us the extra explanatory step.

Suppose, however, that we concede to Forge that explanation of the phenomenon of quantity should stop with objective orders. How are we to interpret this order? One way to go would be to recognize *relations* (*greater in mass than*, etc.) in one's ontology, relations which hold between the ordered particulars and which constitute their order. One could in turn conceive of these relations as universals, the way that I would prefer to go, or else, following such philosophers as G. F. Stout and D. C. Williams, as particulars. (If the relations are particulars they might nevertheless be collected into equivalence-classes by a relation such as exact resemblance, equivalence-classes that would then be substitutes for universals.)

Forge, however, looks for a more reductive account than this. He favours a set-theoretical account where quantities are treated as *ordered classes* of the related particulars. This is made clear by the fact that he takes it to be a *prima facie* objection to his account that the very same ordered class might be associated with two or more co-extensive quantities. This would not be an objection to a view which admitted relations, for co-extensive quantities could then be treated as co-extensive networks of relations. Forge, however, meets the problem by saying that he is not aware that our world contains any co-extensive quantities. There were once good candidates: inertial mass and gravitational mass. But general relativity has the consequence that there is just one quantity here.

So for Forge quantities are nothing more than ordered classes of particulars. Presumably he would take the same view of all relations.

Such a treatment of relations is common among nominalistically minded philosophers, but if intended as ontology, as Forge appears to intend it, it seems to need further development. Is *order* to be taken as a single ontological primitive? *Prima facie*, it is a complex affair, to be analyzed in terms of *relations* of a certain sort among the ordered things. In any case, it is all to easy to find *some* sort of order in any class of things, so it is not clear that there is a genuine ontological distinction between an ordered and an unordered class. In the case of quantities, in particular, I think that Forge will need to appeal to *natural* orderings, and, indeed, his linking of quantities with laws of nature seems to be a step in that direction. But then it becomes unclear that natural orderings of particulars have any advantages over relations between particulars. (And they in turn, I would add, any advantage over properties from which these relations flow superveniently.)

Forge may wish to reduce ordered classes to unordered classes of classes by means of Wiener–Kuratowski device. The ordered pair $\langle a, b \rangle$, for instance, might be represented as $\{\{a\}, \{a, b\}\}$ and in theory this can be done for indefinitely large classes. But certainly this could only be contemplated if it were combined with a theory of natural classes. For the unordered class of classes of the sort indicated above exist in the case of *any two entities whatsoever.* Corresponding 'ordered' classes would then emerge with perfect promiscuity, unless it were held that only some classes of the form $\{\{x\}, \{x, y\}\}$ genuinely order their first-order members. But would the resulting natural classes of classes be any better than natural orderings?

In contrast to Forge, Swoyer wishes to give an account of quantities as families of properties. The word 'properties' is sometimes used to cover relations as well as properties in contrast to relations. But Swoyer, when faced with a choice between developing a theory of quantities as families of non-relational properties (henceforward just 'properties') or else as an order dependent upon mere relations between particulars, comes down on the side of properties. Properties (and relations) he conceives of as universals, and laws of nature as relations between universals. Laws of nature, empirical investigation has shown us, characteristically link distinct quantities rather than qualities. For Swoyer such laws involve systematic functional dependencies, counterfactually sensitive dependencies, holding between the families of properties that constitute the quantities. The result is what may be called a realist (but not platonist) scientific realism in contrast with Forge's nominalist scientific realism.

I applaud, because I share, these views of Swoyer's. They seem to me to be both natural and plausible. I also applaud the impetus that he has given by his paper to the further development of a realist theory of quantity. However, because of my firm agreement with the position just outlined, and because of the technical sophistication of his paper which builds upon earlier work on measurement (a sophistication that outruns mine), my comments will be desultory only.

Swoyer's idea is that to the family of determinate magnitudes associated with a certain quantity, such as mass, a certain order pertains. This order can be modelled mathematically, say by some of the relations between the real numbers. In measurement we try to catch certain magnitudes of certain particulars, and in so doing order the particulars with respect to a certain quantity. For extensive quantities two general relations are involved, an ordering relation, $>$, and a three-termed, concatenation relation, $\circ$, which holds when the result of concatenating the first two terms is equal to the third.

In a footnote Swoyer takes some favourable notice of a suggestion of mine which in effect identifies $>$ with the whole-part relation. He is dubious how far this identification will go, and so am I. But I hope that at least for the fundamental extensive quantities the suggestion may have merit. Every thing of two kilogram mass contains a thing (many things, in fact) of 1 kilogram as a proper part. This is surely relevant to the fact that *being two kilograms in mass* $>$ *being one kilogram in mass.* But not only do the particulars stand in a whole-part relation, it seems that the properties do also. *Being two kilograms in mass* is, among other things and in common with many other mass-properties, *being made up of disjoint parts having mass, one part being a kilo in mass.* This seems to be, in Swoyer's words, 'some sort of part-whole relation' between the two mass properties. It is from this relation between the universals, I think, that the corresponding relations between the particulars flow.

This last point links up with something that Swoyer discusses in a rather rich, if speculative and tentative, final section. Whatever the exact nature of the relations that hold together those families of properties which constitute a quantity, it is plausible that, like the part-whole relation, they are *internal*, flowing from the nature of the terms, in this case properties. That is why it is not uneconomical to have both the relations and the properties in the analysis of quantities. The relations flow internally from the properties and are explained by them. Every possible world that contains the properties contains the relations of the properties. Since the relations are thus supervenient upon the properties, we need not conceive of them as an ontological addition to the properties. I would draw the further conclusion from this (*contra* Swoyer, I think) that these relations are not *nomic*.

Now consider the relationships of co-variation that have been discovered to hold between different families of properties. They are laws of nature. In earlier work Swoyer has defended the view that the laws of nature are necessary rather than contingent. He still has sympathy with that view, he says. But now he makes an interesting concession:

... it would be natural to draw a line between the essential and the contingent higher-order relations among extensive properties in terms of those that hold within families (like > and ○), on the one hand, and those that can cut across families (perhaps like ∗), on the other. It is difficult to see how 1 meter might have been greater than 3 meters, but it may be easier to imagine that multiplicative relations across families could have been different. (p. 278)

Here, I take it, Swoyer is contemplating two sorts of law of nature, one sort necessary, the other contingent. By contrast, the moral that I should like to draw is that only the contingent, cross-family, relations are laws of nature.

This also links up with an idea that Swoyer canvasses at the end of his last section. Earlier in his paper he had treated a family of extensive properties as a *copy* of a mathematical structure. Now he suggests that it may be an actual *instance* of such a structure.

I think that this is a very promising idea. Consider the ratio that holds between a unit mass property, such as *being one kilogram in mass*, and the other mass-properties. Assuming that mass is a continuous quantity, one could find every sort of ratio: whole numbers, rationals and irrationals. The very same ratios are found in the case of other extensive quantities. Now why should we not identify these ratios, which are relations and are universals, with the real numbers? This is a traditional and plausible idea, recently revived by Peter Forrest and John Bigelow in so far unpublished work. (Though Bigelow takes quantities to be relational entities, and so the ratios to be relations between relations.)

But if within-the-family relations of the properties are mathematical relations, then, I would judge, they are not nomic relations.

One thing that might make us pause about this idea is something that we do not find in the structure of the real numbers: there may be 'missing values' in the families of properties. Swoyer has an independently interesting discussion of this situation. He points out that although the 'missing values' are, in default of uninstantiated properties, mere *possibilia*, yet they can be identified by 'rigid designators', designators which pick them out across all possible worlds. Suppose that the total mass in the universe never exceeds a certain quantity. If we reject uninstantiated properties, then we have only the *mere* possibility of there being a greater mass-property. But we can still designate it unambiguously by means of the internal relations that it has to actual, because instantiated, mass properties. This contrasts, Swoyer points out, with the description of merely possible individuals where it seems that we cannot pin down anything as essential to the 'missing individual'. Whatever we describe the individual as, it will lack the corresponding property, or that property will not individuate it, in some possible world.

I myself would explain the absence of missing values in the mathematical structures in the following way. They simply delineate possibilities rather than actualities. No wonder that, unlike the families of properties, the values are always present! But this explanation depends upon a deflationary doctrine of mathematical 'existence', to be defended on another occasion.

I will mention one final point among those discussed by Swoyer. He points out usefully that a property analysis of quantities will fit in happily with a causal/referential semantics, at least for those quantities of which we have some perceptual experience. If objects act on our sense-organs, acting as they often seem to in virtue of some magnitude of some quantity which they possess, then we have at least the beginnings of a plausible account of how we can *refer* to these properties. This causal link, while fixing the reference of the property, need not, and generally will not, fix its nature or real essence. This enables us to see how we can go on talking about the *same* property even while developing and altering our theories about the property, and the family that it is a member of. The causal link gives us a semi-archimedean point upon which our theories can stand unmoved. The reference of more purely theoretical quantities can then be fixed, however tenatively, by their postulated relations, causal and nomic say, to the more directly fixed properties encoutered in observation.

University of Sydney, N.S. W.

BRIAN ELLIS

COMMENTS ON FORGE AND SWOYER

The two authors whose papers I have been asked to comment on present contrasting theories of the nature of quantities and of quantitative relationships. One (John Forge) defends a version of the theory I presented in *Basic Concepts of Measurement* [1]; the other (Chris Swoyer) presents a realist theory of quantities of the sort I argued against. However, my position has since changed quite radically. For I no longer accept the theory of quantities Forge seeks to defend, and I find myself in very close agreement with Swoyer on most of the issues he discusses.

I do not now think it will do to regard quantities simply as objective linear orders. I was right, I am sure, to reject operationism, for all of the reasons I gave in my book. But my rejection of a realist theory of quantities was much too hasty. I now think that ordering relations, having the ontological status of universals, are needed to account for the existence of some objective linear orders. I also think that there are quantitative properties on which some quantitative relations depend. Therefore, I no longer wish to reject, as I once did, all talk of quantities having magnitudes which exist independently of our measuring operations [2].

I

The need to recognise the existence of quantitative universals is clear, I think, where the same linear order may be generated in any of a number of different ways. For on any other assumption, the existence of different ways of ordering things in respect of the same quantity seems entirely fortuitous.

Forge argues, correctly I think, that there are objective linear orders which are not those of genuine quantities. The orders of wind-chill corrected temperature, and the product of a person's height and age, may serve as examples [3]. In *Basic Concepts of Measurement*, I took the view that these are genuine quantities, but not theoretically interesting ones. However, I never faced up to the question of what makes a linear

319

John Forge (Ed.), Measurement, Realism and Objectivity, 319–325.

order theoretically interesting. What I would have said then is that it is
one which is related by laws to other objective linear orders. But I did
not consider why some quantitative orders should be related by laws,
while others are not.

To explain this, I now want to distinguish between real quantities,
phenomenal quantities, and quantitative constructs. For any real quantity q, there is a relation $>_q$ defining the order of q which is a universal.
To be more precise, the real quantity q exists iff there is an asymmetrical and transistive relation $>_q$ of being greater in respect of q
which imposes at least a partial ordering on things of a certain kind K.
The things of this kind K are then said to *possess* q. The ordering is
complete iff, for any two distinct things A and B of kind K, either $A >_q$
B or $B >_q A$. It is partial iff there are things A and B of kind K which
are equipollent in all q-relationships, so that neither '$A >_q B$' nor its
converse, '$A <_q B$', holds. Such things are said to be equal in q ($=_q$).
For real quantities, the relations $>_q$ and $=_q$ must be universals, and so
hold independently of the results of measuring operations and the
responses of measuring instruments (including people). That is, if A and
B are any two things of the kind K, then one or other of $>_q$, $<_q$ and $=_q$
must hold between A and B, even if we cannot say which.

For phenomenal quantities and quantitative constructs, there are no
such ontologically objective relations. In the case of phenomenal quantities, the linear orders are determined by the results of measuring
operations, or by the responses of complex systems or organisms, but
not by any single relational universal; and quantitative constructs are
just more or less arbitrary functions, having no ontological significance,
of the measures of other quantities.

Forge likewise wants to distinguish between the linear orders of real
quantities and other sorts of linear orders. He suggests that what
distinguishes real quantities is what I once thought made them interesting, viz. their involvement in laws of nature. But this clearly will not do.
For there are many quantitative concepts occurring in the statements of
laws of nature which do not refer to real quantities, and many more
which have a doubtful ontological status. On the other hand, there are
many real quantities, concerning the states of, or relationships between,
particular things (such as distance from Pittsburgh) which are not
involved explicitly in laws.

As an example of the first kind, consider the space-time coordinates
of the hypothetical events referred to in the Lorentz transformation

equations. Since there are no inertial systems, and the events referred to are supposed to be occurring in such systems, they cannot be the measures of real quantities. Consider also the case of electrical potential. James Clerk Maxwell once argued that potential, as opposed to potential difference, is not a genuine magnitude, but what he called 'a mere scientific concept'. He argued that all genuine magnitudes have properties other than those which enter into their definitions, and that electrical potential fails this test. He contrasted the concept of 'potential at a point in a field' with the formally analogous concept of 'temperature at a point in a conductor'. The latter, he said, is not a mere scientific concept, but a genuine physical magnitude, since temperature has effects independently of temperature differences. Now, I think he would have said much the same of direction, and of spatial and temporal location, as he said about electrical potential. For the physically relevant quantities here — the ones which have effects — are the *differences* in these respects, i.e. the angles, distances and time intervals.

Maxwell's insight is important. Real quantities have effects, depending on their magnitudes; they belong to the causal net. Swoyer is quite right about this. It may also be what Forge had in mind by involvement in laws of nature; for he may have thought that all laws are casual laws, or that involvement in a law implies causal efficacy. If so, he was wrong about this; not all laws are causal laws, and the fact that a quantitative expression occurs in the statement of a law does not simply that it refers to anything that is causally effective.

Spatial and temporal coordinates are cases in point. There is clearly a difference between the apparently fictitious coordinate representations of positions in space and time, on the one hand, and the undoubtedly real spatio-temporal intervals between events, on the other. My view is that while the relative positions of things in space and time are real quantities, their absolute positions are not. For the only spatio-temporal relationships which are causally relevant are those between real things or events, not the positions in space and time they may be depicted as having.

Some quantities, like time-interval, distance and angle, are fundamentally relational, and do not depend on the properties of the things they relate. For the time and place of an event are not properties that the event would have independently of other events; and the direction of a line is not a property of the line itself, but depends on how it is related to other lines. The ordering relations for these quantities must therefore be first order universals. However, the quantitative ordering relations for other quantities often seem to depend on the properties of the things they relate. Charge and mass, for example, seem to do so. My earlier view was that *all* quantities are fundamentally relational; and I claimed that it made no sense to speak of the charge or mass of an object except in relation to other objects. Thus, I held that to say that an object has a certain charge or mass is just to say that it bears certain relationships to other charges or masses.

However, I no longer think that this theory will do. For it does not explain how there can be things belonging to natural kinds which all have the same mass or charge, or why there should be fixed relationships between the charges or masses of things of different natural kinds. By contrast, there are no natural kinds which all exist at the same place, or all occur at the same time, or all point in the same direction; and there are no fixed relationships between the places, times or directions of any natural kinds. To explain these differences, I now want to say, as Swoyer does, that the ordering relations for mass and charge are relations between certain quantitative properties of the things they are said to relate. If these quantitative properties are first order universals, then the quantitative relations which depend on them must be relations between first order universals, i.e. they must be second order universals. On this point I also agree with Swoyer.

II

When I wrote *Basic Concepts of Measurement* I did so as an unrepentant conventionalist — writing very much in the tradition of Mach, Poincaré and Reichenbach. I still think there is something valuable in conventionalist analyses, as I have argued in a recent paper[4]. For such analyses often reveal the existence of false, or unnecessarily restrictive, theoretical presuppositions, and their revelation may help us to construct better theories. For example, I think the theory of measurement, and in particular the theory of dimensions, is in better shape now than it would have been without conventionalist analyses of the sort I undertook.

However, the original aim of conventionalism, to distinguish clearly the empirical from the conventional elements of scientific theories, is untenable. For no such clear distinction can be made. Moreover, the empiricist theory of meaning which motivated the conventionalist programme leads to anti-realism about theoretical entities, and a disposition to reduce properties and relations to their observational bases. Nevertheless, there is a good case for pursuing a modified programme of analysis and rational reconstruction — one which is compatible with a sophisticated scientific realism, and which does not depend on there being a clear empirical-conventional distinction[5]. Given this modified programme, it can be seen that many of the results in measurement theory obtained earlier by conventionalists are basically sound.

Swoyer's general position on scientific realism is almost identical with my present positions[6]. For, like him, I stress the importance of causal effectiveness as a criterion for what is real, and reject the rather naive view that if scientists quantify over Xs in currently accepted scientific theories, then it is most rational to believe that Xs exist. Moreover, Swoyer's argument for a discriminating realism about quantitative properties and relations is almost the same as mine. So I do not have much cause to disagree with him on any of the important ontological issues.

Swoyer's realist theory of quantities is also much the same as mine. For I accept that the results of measuring operations cannot be adequately explained unless it is assumed that there is an underlying structure of quantitative properties and relations. The world certainly behaves as if this structure existed. Therefore, by the main argument for scientific realism, we should believe it does exist.

The least satisfactory aspect of Swoyer's paper is his treatment of the relationships between scales and quantities. Swoyer proposes to use S. S. Stevens' criterion for invariance of scale form to discover the properties of the underlying concrete relational structures, arguing that such invariance requires an underlying reality to explain it. I think this is doubly unsatisfactory. First, it confuses two different concepts of invariance. What is physically real is plausibly what is independent of the perspective or physical state of the observer. The fact that a spatio-temporal interval is the same whatever the standpoint of the observer is a good argument for the reality of spatio-temporal intervals. It would be a good argument, even if measurements of space and time were always made on the same scales, because we know *a priori* that the same results would have been obtained if the measurements had been made on any similar scales for these quantities (i.e. scales related to the given scales by similarity transformations). But the fact that ratios remain invariant under similarity transformations is not a good argument for realism about ratios, unless one has some independent reason to think that more general transformations are not permissible. Even then, the whole weight of the argument would have to rest on the permissibility criteria, which Swoyer does not seriously discuss.

Secondly, sameness of scale form does not imply identity of quantitative structure. The Absolute temperature scale, for example, is plausibly a ratio scale by Stevens' criterion. For any other absolute scale (i.e. scale having the same zero, but differing in its measure of temperature interval) could 'serve all of the purposes' of the standard Absolute scale. For in place of 'T', the measure of the Absolute temperature, we should only have to write 'kT' (where k is a constant having the dimension $1/[T]$) in all expressions of quantitative relationships involving temperature. But the quantitative structure of thermometric relationships is surely different from that of say mass, which is also measured on a ratio scale. For there is no concatenation operation of heat states comparable to that of assembling masses.

The reason why the Absolute scale of temperature is a ratio scale is that the mathematical forms of the thermodynamic laws on which it is based are invariant only under similarity transformations[7]. But these laws are not isomorphic with those which make mass a ratio scale. Therefore, if one is interested in the underlying structure of quantitative relationships, Stevens' criterion for invariance of scale form is a fairly crude instrument. As a realist about quantities and quantitative relationships, Swoyer should be seeking to classify scales of measurement according to the kinds of structures they represent, and how they seek to represent them, not by the restrictions on scale transformations these structures would seem to impose. That is, he should be using C. H. Coombs' system of classification, rather than Stevens'[8].

La Trobe University, Victoria, Australia

NOTES

[1] Cambridge: Cambridge University Press, 1966.
[2] *Ibid.*, p. 24.
[3] These are not Forge's examples, but I use them because I am not happy with those he provides. Distance from Pittsburgh *is* a quantity, in my view, because there is a set of distance relations between Pittsburgh and other places which holds independently of our measuring operations.
[4] 'Conventionalism in Measurement Theory', to appear in a volume of essays on

measurement (title not yet known) being edited by C. Wade Savage and Paul Erlich for *Minnesota Studies in the Philosophy of Science.*

[5] The new programme is outlined in 'Conventionalism in Measurement Theory'.

[6] See my paper 'The Ontology of Scientific Realism', to appear in *Mind, Morality and Metaphysics: Essays in Honour of J. J. C. Smart,* edited by Philip Pettit, Richard Sylvan and Jean Norman. Oxford: Basil Blackwell, 1987.

[7] See *Basic Concepts of Measurement,* p. 66.

[8] For a discussion of Coombs' way of classifying scales see *Basic Concepts of Measurement,* pp. 63—67.

INDEX OF NAMES

INDEX OF SUBJECTS

AUSTRALASIAN STUDIES IN HISTORY AND PHILOSOPHY OF SCIENCE

General Editor:

R. W. HOME, *University of Melbourne*